Police Integrity in South

Policing in South Africa has gained notoriety through its extensive history of oppressive law enforcement. In 1994, as the country's apartheid system was replaced with a democratic order, the new government faced the significant challenge of transforming the South African police force into a democratic police agency—the South African Police Service (SAPS)—that would provide unbiased policing to all the country's people. More than two decades since the initiation of the reforms, it appears that the SAPS has rapidly developed a reputation as a police agency beset by challenges to its integrity.

This book offers a unique perspective by providing in-depth analyses of police integrity in South Africa. It is a case study that systematically and empirically explores the contours of police integrity in a young democracy. Using the organizational theory of police integrity, the book analyzes the complex set of historical, legal, political, social, and economic circumstances shaping police integrity. A discussion of the theoretical framework is accompanied by the results of a nationwide survey of nearly 900 SAPS officers, probing their familiarity with official rules, their expectations of discipline within the SAPS, and their willingness to report misconduct. The book also examines the influence of the respondents' race, gender, and supervisory status on police integrity.

Written in a clear and direct style, this book will appeal to scholars and students of criminology, policing, sociology, political science, as well as police administrators interested in expanding their knowledge about police integrity and enhancing it in their organizations.

Sanja Kutnjak Ivković is Professor at the School of Criminal Justice, Michigan State University, U.S.A. For more than two decades, her research has focused on comparative and international criminology, criminal justice, and law. She received the 2017 Mueller Award for Distinguished Contributions to International Criminal Justice.

Adri Sauerman is legal counsel in the mineral, energy, trading, and commerce industries as well as a former Fulbright and current independent scholar from South Africa. As legal counsel, he navigates the legal and governance intricacies of several African and European commercial systems. His research and publications focus on international criminology, criminal justice, and law.

Andrew Faull is Senior Researcher in the Justice and Violence Prevention Programme at the Institute for Security Studies, and a Research Associate at the Centre of Criminology, University of Cape Town, South Africa. He has over 15 years of experience researching policing and related subjects in South Africa.

Michael E. Meyer is Emeritus Professor of Criminal Justice at the University of North Dakota, U.S.A. He was recipient of a Fulbright Scholarship to South Africa and two National Endowment for the Humanities Fellowships. His research interests focus on policing in transitional societies.

Gareth Newham is Head of Justice and Violence Prevention Programme at the Institute for Security Studies (ISS), an independent organization in South Africa that provides authoritative policy research and technical support to enhance human security across Africa (www.issafrica.org). He leads a team of people who work to better understand, respond to, and prevent violence, and strive to strengthen the criminal justice system in South Africa.

"This is a compelling 'must-read' for anyone interested in understanding the complexity of South Africa. In its own words, 'the book offers a unique, in-depth and systematic empirical exploration of the contours of police integrity in South Africa.' As such it should be made obligatory reading material for all persons in the police engaged in planning and research. This book underscores the need for and importance of understanding the historical development of the police in South Africa and the challenges it faces in the future to reaching its ultimate goal as a service for all people."

Director/Brigadier Stef Grobler (ret. 1963–2003),
former head of the South African Police Service's
Anti-Corruption Unit (1996–2002), South Africa

"*Police Integrity in South Africa* pulls no punches in establishing the state of South African policing. Police misconduct is widespread, endemic, deeply embedded, routinely accepted, endorsed, and encouraged at all levels—including by governments. All South Africans are victimised. Solutions will be hard to find and will be even harder to realise. Extraordinary leadership is required."

Clifford Shearing, Professor and Leader of the Global Risk
Governance Programme, Public Law Department,
Law Faculty, University of Cape Town, South Africa

"When renowned scholars get together to co-author a book, it is usually an endeavor worthy of attention. In the case of *Police Integrity in South Africa*, this effort produced an invaluable addition to the study of policing from a comparative perspective in countries experiencing political turmoil and transitions. In addition, it is a must-read for scholars of policing, as it touches upon the most profound concepts embedded in this profession, namely the interplay between politics, the public, and effective policing."

Maria (Maki) Haberfeld, Professor of Police Science,
John Jay College of Criminal Justice, U.S.A.

"This is a major book on the transformation of policing and society over time. It excels at placing the analysis of policing in a political framework, at unveiling long term trends in the transformation of societies and policing, and, last but not least, at understanding policing as a means of political domination (cf. Max Weber) and a tool for political abuse. It also appropriately situates policing in the context of extreme levels of interpersonal violence, which in part derives from the regime's establishment of apartheid. It is not only a contribution to the study of policing in South Africa, but in divided countries in general. The focus on integrity is not only useful in order to rebalance the study of trust and legitimacy of police: accounting for those requires that the performance of policing is studied, and such performance is first and foremost destroyed by the lack of integrity in a large sense, and the use of overt violence in the framework of policing."

Sebastian Roché, Professor of Policing,
University of Grenoble-Alpes, France

"Police integrity has been the focal concern for many researchers, police administrators, and policy makers, particularly in transitional societies where old and new cultures and policies clash, creating an environment conducive to misconduct and corruption. Although South Africa's transformation from apartheid to democracy is viewed as one of the most important political events of the 1990s, the country has unfortunately suffered a steady decline in public confidence in the police. Kutnjak Ivković and colleagues' book on police integrity in South Africa makes a significant and valuable contribution to the literature by tracking the historical development of policing and, most importantly, identifying factors related to the problem of police integrity in the transitional democracy. This book is not only theoretically rich but also exceptionally well-written.

The central theme of the book is that variation in police integrity is influenced by a broad range of historical, social, political, and economic factors. Drawing upon survey data collected from the police forces during two time periods, the authors provide a comprehensive examination of the organizational theory of police integrity along the four key dimensions of the framework, including the existence and enforcement of policies and rules, the approaches to misconduct control, the taming of the code of silence, and the influence of the larger environment. The book offers the reader a new way of thinking about police misconduct and corruption by concentrating on theorizing and measuring police integrity.

Overall, the book represents a compelling and thoughtful examination of the theoretical arguments and empirical findings regarding police integrity in a relatively new democracy. Given the authors' extensive teaching and research experience in the subject area, *Police Integrity in South Africa* should be an indispensable reference for those who want to understand the connections between external environmental factors and internal organizational integrity. This excellent book will be of particular interest to government officials and police administrators interested in the development of strong police forces and better police–community relations."

Ivan Y. Sun, Professor at the Department of Sociology and Criminal Justice, University of Delaware, U.S.A.

"This book presents an in-depth study, describing the complex nature of police integrity in South Africa, which depends on recent social changes and police reforms in the country. The authors' contributions add significant value to studies on police integrity and police deviance in young democracies in a timely and critical manner. I recommend it to students, researchers, and policymakers interested in police integrity and the development of democratic policing globally."

Gorazd Meško, Professor and Head of the Institute of Criminal Justice and Security, University of Maribor, Slovenia

"This book provides a detailed, thought-provoking look at police integrity in South Africa. The authors provide an excellent historical overview of the South African police. They also describe how social, cultural, and political changes in South Africa over time influenced changes within police organizational structure, as well as how the South African police are perceived by the citizens they serve. An entire chapter is devoted to a discussion of the internal control mechanisms related to police misconduct. The authors include important findings and policy implications from a recent study they conducted on police integrity in South Africa. This book is a must-read for students, scholars, and policing professionals interested in police integrity from a comparative perspective."

Carol A. Archbold, Walter F. and Verna Gehrts Endowed Professor, Department of Criminal Justice and Political Science, North Dakota State University, U.S.A.

Routledge Frontiers of Criminal Justice

Police Corruption in Comparative Perspective
Building Trust in the Police in India and China
George Chak Man Lee

Violence Against Children in the Criminal Justice System
Global Perspectives on Prevention
Edited by Wendy O'Brien and Cédric Foussard

The Development of Transnational Policing
Past, Present and Future
Edited by John L.M. McDaniel, Karlie E. Stonard and David J. Cox

'Crossover' Children in the Youth Justice and Child Protection Systems
Susan Baidawi and Rosemary Sheehan

Critical Reflections on Evidence-Based Policing
Edited by Nigel Fielding, Karen Bullock and Simon Holdaway

Policing and Mental Health
Theory, Policy and Practice
Edited by John McDaniel, Kate Moss and Ken Pease

The Making of a Police Officer
Comparative Perspectives on Police Education and Recruitment
Edited by Tore Bjørgo and Marie-Louise Damen

Police Integrity in South Africa
Sanja Kutnjak Ivković, Adri Sauerman, Andrew Faull, Michael E. Meyer and Gareth Newham

For more information about this series, please visit: www.routledge.com/Routledge-Frontiers-of-Criminal-Justice/book-series/RFCJ

Police Integrity in South Africa

Sanja Kutnjak Ivković

Adri Sauerman

Andrew Faull

Michael E. Meyer and

Gareth Newham

Routledge
Taylor & Francis Group
LONDON AND NEW YORK

First published 2020
by Routledge
2 Park Square, Milton Park, Abingdon, Oxon OX14 4RN

and by Routledge
52 Vanderbilt Avenue, New York, NY 10017

Routledge is an imprint of the Taylor & Francis Group, an informa business

British Library Cataloguing-in-Publication Data
A catalogue record for this book is available from the British Library

Library of Congress Cataloging-in-Publication Data
Names: Kutnjak Ivković, Sanja, 1965– author. | Sauerman, Adri, author. | Faull, Andrew, author. | Meyer, Michael E., 1946– author. | Newham, Gareth, author.
Title: Police integrity in South Africa / Sanja Kutnjak Ivković, Adri Sauerman, Andrew Faull, Michael E. Meyer and Gareth Newham.
Description: New York City : Routledge, 2020. | Includes bibliographical references and index.
Identifiers: LCCN 2020001161 | ISBN 9781138639652 (hardback) | ISBN 9781315637068 (ebook)
Subjects: LCSH: South African Police Service. | Police ethics—South Africa. | Police—South Africa.
Classification: LCC HV7924 .K866 2020 | DDC 363.20968—dc23
LC record available at https://lccn.loc.gov/2020001161

ISBN: 978-1-138-63965-2 (hbk)
ISBN: 978-0-367-49383-7 (pbk)
ISBN: 978-1-315-63706-8 (ebk)

Typeset in Bembo
by Apex CoVantage, LLC

To all the South African police officers who shared their stories and views with us over many years, we truly appreciate your indulgence. And to those officers who continue to serve with integrity under incredibly difficult conditions, thank you.

The authors

To my loving family—thank you for your patience, understanding, and unconditional love.

Sanja Kutnjak Ivković

For Eileen, Wollie, Suzaan, and Walter, with whom I am perpetually window shopping, in a desert town, on a Sunday night. Smiling.

Adri Sauerman

For Kori, Rei, and Elona, and the dedicated, honest members of the South African Police Service.

Andrew Faull

To my wife, Teresa, who has shared a phenomenal 47-year journey. We have had a great ride. And to my South African colleagues and friends, too many to name, you opened your doors to me, both figuratively and literally, over the past 20 years. The warmth of your friendship is very much felt on the winter plains of North Dakota.

Michael E. Meyer

To my colleagues at the Institute for Security Studies and those members of the South African Police Service who remain committed and honest under difficult circumstances.

Gareth Newham

Contents

Figures

Tables

Acknowledgments

I'm indebted to Henk van Vuuren for his tireless belief and assistance in our research and to Hermann Giliomee for guiding me through the intricacies of South African history. They, with the other special people who listened, advised, and helped, made finding the words so much easier.

Adri Sauerman

We are immensely grateful to the Hanns Seidel Foundation and to the Bavarian State Chancellery, whose generous financial support made our participation in this project possible.

Andrew Faull and Gareth Newham

Chapter 1

Studying police integrity in South Africa

A democratic police for South Africa

The year 1994 was critical in the recent history of South Africa and a turning point in the lives of the country's citizens. The decades-old apartheid regime had gradually crumbled and was replaced in April of that year by a new, representative government established after the country's first democratic elections. Enthusiasm for Nelson Mandela, then the leader of the African National Congress (ANC) and the first black South African president, and his ANC-led coalition government abounded across the country. In a 1995 survey of 2,000 black residents, more than 80% of respondents agreed that Mandela was leading the country "very well" or "fairly well" (Duke, 1995). He was popular among the majority of white citizens too; a survey of 920 white residents revealed that more than 60% approved of Mandela's leadership (Duke, 1995).

This overwhelming support for the government extended to the police as well; about three-quarters of the South African respondents participating in the World Values Survey from 1994 to 1998 stated that they had a substantial degree of confidence in the police (Boateng and Lee, 2018). Considered against more recent opinion polls, this was the *highest* level of confidence in the police during *any* post-apartheid period (Boateng and Lee, 2018). During apartheid, the South African Police were used as an enforcer of the government's discriminatory laws and subsequently oppressed the majority race groups, while the minority white population could generally rely on paramilitary police protection and services. It is not surprising that blacks expressed much lower levels of confidence in the police during apartheid's final years (1981–1993) and much higher levels of confidence in the police in the immediate post-apartheid period (e.g., 1994–1998) than whites did (Boateng and Lee, 2018).

President Mandela and the first democratically elected government that took office in 1994 chartered a path to transform the South African Police into an organization respectful of all citizens' human rights, regardless of race (e.g., Gastrow and Shaw, 2001; Rauch, 2000; Jensen, 2014). At the dawn of what was soon called the "new South Africa," a police agency that would provide services to *all* residents was a powerful idea. A substantial change, both symbolic and operational, occurred with the merging of apartheid state's 11 police agencies, and the rebranding of the South African Police (SAP) into the South African Police *Service* (Makda et al., 2012; Van Graan and Ukpere, 2012). In addition to a dramatic change in the way policing has been carried out across the country, a number of scholars believe that the establishment of the Truth and Reconciliation Commission (TRC) in 1995—entrusted to investigate human rights violations committed by all sides during the apartheid period—further reaffirmed the rationale for strong support of the South African Police Service (e.g., Alexander et al., 2005; Vora and Vora, 2004; Wiebelhaus-Brahm, 2010).

However, the surge in black residents' confidence in the police immediately following apartheid's demise did not last long (Boateng and Lee, 2018). Compared to the levels of support during the first years of democracy, the most recent World Values Survey (2005 to 2014) revealed that only about 52% of South Africans surveyed—barely half—expressed confidence in the police (Boateng and Lee, 2018). Similar results were found in a 2007 survey of public confidence in South African public institutions (Roberts, 2008), where only 39% of the respondents "trusted" or "strongly trusted" the police. The results of the 2015 Afrobarometer survey (Institute for Justice and Reconciliation, 2017) further reveal the relatively low levels of support for police in recent times: 54% of the respondents stated that they do *not* trust police at all or trust them "just a little." The results of the 2018 Afrobarometer (Plus 94 Research, 2018)—the most recent Afrobarometer available—are even more discouraging: Two thirds (66%) of respondents expressed no trust in police at all or stated that they had "just a little" trust.

Empirical studies suggest that the decline in confidence in the police was not sudden. Instead, confidence in the SAPS withered away over time, from the mid-1990s to the mid-2010s. The percentage of respondents with confidence in the police started to decrease from about three-quarters in the immediate post-apartheid period (74% in 1994–1998), to about two-thirds in the early 2000s (60% in 1999–2004), to barely one-half in the late 2000s and the early 2010s (52%; 2005–2014; Boateng and Lee, 2018). Similarly, data from Statistics South Africa show a decline in satisfaction with police service, from a 64.2% approval in 2011 to 54.2% in 2017–2018 (Stats SA, 2011, p. 23, 2018, p. 86). Conversely viewed, the percentage of respondents who stated that they do *not* have confidence in police has increased from about one-quarter (27%) in the immediate post-apartheid period (1994–1998), to over one-third (40%) in the middle period (1999–2004), to close to half (48%) in the most recent period (2005–2014).

Clearly, enthusiasm for the new democratic police has waned. The pressing question is what prompted this decline in confidence. Because a strand of extant research has indicated that citizens' perceptions of the police agency's zeal to deal with crime are related to the citizens' evaluations of the police (e.g., Boateng, 2019), one explanation may lie in South Africa's very high violent crime rates (e.g., UNODC, 2019). To be sure, the extent to which the citizenry relates evaluations of the police with crime rates is far from clear in the South African context. For example, Faull (2010, pp. 39–40) found in the discussions with civilian focus-groups that participants were not as interested in the police who were effective in "fighting crime" as they were in the police who were "polite and respectful." However, using survey analysis, Bradford and colleagues later found that fighting crime and respectful policing were considered equally important to South Africans (Bradford et al., 2014). More recently, Bello and Steyn (2019) found negative perceptions of the police among a sample of university students, based both on students' perceptions and experiences of corruption and abuse, as well as perceived ineffectiveness in relation to crime.

Moreover, the feasibility of the relevance of the SAPS is challenged by the crime rates themselves. Specifically, although South Africa's murder rate is still very high compared to most other countries (United Nations Office on Drugs and Crime, 2019), the post-apartheid period was not characterized by a sharp increase in murder. In fact, murder had been increasing from the mid-1970s until 1994, and declined substantially from 1994 to 2011 (Kriegler and Shaw, 2016a; UNODC, 2019, p. 25). Specifically, the murder rate in 2019 was 36.4 per 100,000 residents, about one-half of the murder rate of 78 per 100,000 recorded in 1993 (Dixon, 2018; SAPS, 2019). Kriegler and Shaw (2016b, p. 1) reasoned in 2016 that "the past two decades have shown unprecedented *improvement* in the physical safety of the vast majority of people in the country [emphasis added]." Despite this improvement in safety since 1994, surveys conducted since the late 1990s show roughly half of South Africans consistently believing that crime has increased (Pharoah, 2008; Stats SA, 2018). According to 2017 opinion polls, a common perception held that the ANC government with Jacob Zuma as president had been doing a poor job in addressing South Africa's crime problems: 77% of Afrobarometer respondents believed Zuma's government was controlling crime "fairly badly" or "very badly" (Institute for Justice and Reconciliation, 2017).

While the country's endemic crime may damage public perceptions of police, an equally plausible reason for the decline in confidence may be related to perceptions of widespread misconduct in government in general and the police in particular. Studies indicate that trust in, and the legitimacy of, the police are generated through fair, community-oriented, and inclusive policing (Trinkner et al., 2016). Hence, perceptions of police officers' misconduct could represent a strong impediment to public trust in, and confidence of, the police. The 2017 Afrobarometer findings (Institute for Justice and Reconciliation, 2017, p. 8) indicate a negative relation between perceptions of corruption and trust. A great majority of the respondents (83%) included in the 2015 opinion poll (Institute for Justice and Reconciliation, 2017) believed that corruption had increased in the previous year, though this number declined to 65% in 2018 (Plus 94 Research, 2018), possibly as a result of Zuma's maligned presidency having ended. Furthermore, about one-half of the citizens included in the 2017 survey responded that "all" or "most" police officers (48%) were corrupt. In 2018, this increased slightly to 49% (Plus 94 Research, 2018). The corresponding fraction of respondents assessing judges and tax collectors as corrupt in 2017 was substantially smaller, about 25% (Institute for Justice and Reconciliation, 2017), increasing to 32% in 2018 (Plus 94 Research, 2018).

Perceptions that police misconduct is widespread in South Africa and the firsthand and vicarious experiences with police misconduct further erode the SAPS's legitimacy. A 2001 study found that the primary reason respondents thought policing services had deteriorated was that "the police are corrupt" (Louw and Pelser, 2001), while of the 25% of Afrobarometer respondents who requested help from police in the year preceding the 2018 survey, 12% said that they paid a bribe, gave a gift, or did a favor to get their assistance. Of the 27% who interacted with the police at a checkpoint, during a traffic stop, investigation, or similar encounter, 26% said that they gave a bribe, gift, or favor to avoid a problem (Plus 94 Research, 2018). Similarly, the National Victims of Crime Survey 2017–2018 revealed that, of the 8% of victims who experienced corruption, 12% experienced it in "policing" and 61% in "traffic" contexts (Statistics South Africa, 2018, p. 63).[1] Empirical research further demonstrates that perceptions of police corruption and lowered confidence in the police may be related. One of the top two reasons for low levels of trust identified by Grobler (2013, p. 92) was the fact that the respondents experienced police misconduct.

The problem is further aggravated by the belief among the respondents that reports of corruption would not be investigated by the government (Plus 94 Research, 2016). In the end, there is a widely shared perception that governmental officials, including the police, are "above the law" and that they do not face the punishment they deserve. The 2015 Afrobarometer demonstrated that two-thirds of respondents agreed that officials who commit crimes either "often" or "always" go unpunished (compared to a belief that about one-third of ordinary citizens go unpunished; Institute for Justice and Reconciliation, 2017).

If police misconduct in South Africa is as systematic and widespread as some studies suggest (e.g., Grobler, 2013), the level of trust the police enjoy is undoubtedly adversely affected. One victim of a police assault explained that the attack "completely changed my idea of cops—I'm now scared of them" (Grobler, 2013, p. 45). Another citizen, whom police forced to run back and forth across a busy road when he refused to pay them a bribe, reported that the experience left him feeling "like beating them" whenever he saw the police (Faull, 2010, p. 37).

Whether the South African public really is fearful of its police, as 35% seemed to be according to one study (Grobler, 2013, p. 27), they are undoubtedly less likely to go to the police to report crimes. This trend is clear in the victim survey data, which show that in 2017–2018, just 47% of assault victims, 51% of burglary victims, and 59% of house robbery victims reported these crimes to the police (Stats SA, 2018). Coupled with the fact that in the 2000s police administrators in some stations were accused of manipulating crime statistics (e.g., Grobler, 2013, p. 28), while ethnographic research has also documented the perverse incentives officers face to alter their data (Matshedisho, 2009; Faull, 2016, 2017), it is not surprising that the official crime rates might be artificially low.

Furthermore, potentially close relationships between police officers, organized crime, and corrupt politicians would impede the SAPS's ability to successfully investigate crimes or service communities, and would most certainly result in impunity for criminals (e.g., Grobler, 2013; Pauw, 2018; Myburgh, 2019). In 2008, journalist Wonder Hlongwa argued that "the involvement of police officers in organised crime—including violent crimes and sophisticated syndicates—is hampering the country's fight against criminals." Justice Malala, a political commentator, put it more bluntly in 2011, "Our cops can't be trusted: Too many police [are] linked to too many gangsters and killers" (2011). Alarmingly, both statements were made at a time Zuma's capture of the SAPS was just getting started (e.g., Pauw, 2018; Basson and du Toit, 2018).

Police misconduct in South Africa

In 2017, former South African president Jacob Zuma argued that "[n]o one has proof that all the cops are corrupt. We must rather say someone [sic] of the cops are part of the crime" (Mngadi, 2017, p. 1). Already in 2008, Joseph Ngobeni, a SAPS Assistant Commissioner, reasoned (cited in Bruce, 2008):

> Surely, a wrong by an individual or a few individuals cannot realistically be interpreted as a general phenomenon or practice in the entire 173,161 employees of the SAPS. The perception should therefore not be created in the media that the SAPS is the most corrupt or brutal police service in the world, simply because this is not true.

Indeed, the 2019 SAPS Annual Report documented that 1,888 employees—less than one percent of all SAPS members—were subject to disciplinary hearings for offenses in the previous accounting year, including murder, rape, robbery, extortion, and kidnapping. Forty-seven percent of these complaints (n = 887) were withdrawn during the hearing or resulted in a "not guilty" finding, leading to only 53% (n = 1001) of accused officers being sanctioned (SAPS, 2019, p. 324). These figures may be interpreted as implying that police misconduct is not a serious problem for the SAPS, in line with the reasoning of Zuma and Ngobeni.

However, opinion polls, scholarly research, newspaper accounts, and NGO reports challenge this conclusion. An editorial in *The Times* summarized the prevailing sentiment in 2012: "If the police are no better than thugs, who will save us?" The report continues, "It seems torture and outright violation of human rights is becoming the order of the day for some of our police officers and experts warn that the line between criminals and our law enforcement officers is 'blurred'" (The Times, 2012, p. 1). In 2011, Amnesty International also raised serious concerns about police torture, death in police custody, extrajudicial killings, and rape, stating that "[i]n South Africa, numerous cases of torture and ill-treatment by police were reported, many of which were investigated by the Independent Complaints Directorate. Among the methods reported were beatings, electric shocks, suffocation, and death threats" (Amnesty International, 2011, p. 4). Conway-Smith (2012) wrote in the editorial titled "South Africa Troubled by Corrupt Cops,"

> There is growing frustration over this country's corrupt and incompetent police at seemingly every level of the system, from traffic cops soliciting a bribe for a "cold drink" to the controversial appointments of some of the country's highest-ranking police officials.

In 2018, a range of books by investigative journalists were published, revealing the dark bond between South Africa's predatory elite and corrupt police leaders (e.g., Pauw, 2018; Basson and du Toit, 2018; Wiener, 2018). In their aftermath, one prominent news editor suggested that the only way to save the country would be to disband the SAPS's elite corruption fighting unit—The Hawks—and start afresh (Daily Maverick, 2019). Clearly, public trust in the police is weak.

Opinion polls about police misconduct

Various opinion polls suggest that citizens have long perceived the post-apartheid police as corrupt. A questionnaire inserted into a November 1995 issue of *Sowetan*, one of the most widely read South African newspapers, revealed that two-thirds of respondents (67%) assumed that police officers accepted bribes (Lodge, 1998, p. 171). Several administrations of the Afrobarometer survey (2006, 2008, 2015, 2018) consistently show that about half of the respondents believe that "all" or "most" police (46% to 49%) are corrupt (Faull and Newham, 2011; Institute for Justice and Reconciliation, 2017; Plus 94 Research, 2018). Furthermore, the 2018 South African Citizens' Bribery Survey showed that one-quarter of residents surveyed said they personally knew somebody who had paid a bribe, most often to deal with a traffic offense (Ethics Institute, 2018). In addition, about one-quarter of Afrobarometer respondents who had contact with the police in the year preceding its 2018 survey said they had paid a bribe to the police *at least once* to avoid problems (Plus 94 Research, 2018).

Studies exploring police officer views lead to similar conclusions. A majority of 200 detectives interviewed in the late 1990s expressed the belief that *most* police were corrupt (Newham, 2000a). One-third of respondents surveyed at the Hillbrow Police Station in 2000 stated that they personally knew of a police officer who engaged in corruption (Newham, 2000b). Furthermore, most of the surveyed police officers in Johannesburg (Newham, 2005) and the larger Gauteng province (Faull, 2009) also perceived that corruption was a serious issue facing the SAPS and/or their station. In addition, a 2005 survey of 375 SAPS supervisors revealed that "[a]t least one of the four supervisors (25%) would allow police bribery, theft from a crime scene, and theft of money from a found wallet to continue without reporting it and/or expected that others would not report it either" (Kutnjak Ivković and Sauerman, 2012).

Media accounts of police misconduct

In 1998, Syed and Bruce (1998) analyzed newspaper stories, identified instances of police corruption, and classified them into 22 different types of corruption. The conclusion of their analysis illustrated the nature and severity of the problem at the time (Syed and Bruce, 1998, p. 14):

> the fact that, of the twenty-one practices that had international examples (which were drawn from book surveys, media sources, and analytical literature published over the last two decades), there were reports of sixteen in the South African press over the fifteen-month period from February 1996 to April 1997 alone, does seem to indicate that police corruption is quite widespread in South Africa. Although one needs to be cautious, it seems reasonable to assert that the publication of reports on so many different types of corrupt activities in the South African press in little over a year may indicate that police corruption in South Africa is fairly extensive, as well as varied in nature.

More recently, Grobler scoured decades of news clippings and research reports, concluding that "[(t)he SAPS] has a significant problem with corruption and criminality" (Grobler, 2013, p. 246). Our own review suggests much the same. We searched four internet news sites, the most frequently read sources of daily news in South Africa, covering crime-related stories (News24, Sowetan Live, Times Live, and IOL News) over a period of about one year. We used the keywords "police/cop corruption," "police/cop brutality," "police/cop bribery," and "police/cop dismissal/dismissed." Clearly, not all reports of police misconduct are published in newspapers; only the offenses that come to journalists' attention, and which their editors deem potentially interesting to readers for certain reasons (e.g., number of police officers involved, rank of the police officers involved, severity of the rule violation, size of the gain, *modus operandi*), find their way into the newspapers. However, we are not using the newspapers to ascertain the *extent* of police misconduct in the country (which would suffer from a selection bias) but rather to illustrate its *diversity*.

Our search revealed both allegations of police misconduct and court decisions pertaining to cases covering a range of violations. In particular, as is visible in the newspaper stories summarized below, police corruption and abuse of force frequently appeared in our search.

Police corruption

- **Protection of illegal activities** (citizens engaged in illegal activities bribe police officers to avoid a sanction; Barker and Roebuck, 1973). In addition to the instances of the bribes given to the police officers by drug dealers in Johannesburg witnessed by journalist Paul McNally (News24, 2016), a story described how a specialist investigator was found guilty in a disciplinary hearing, received a conditional dismissal, and was charged criminally for not disclosing her relationship with a high-ranking gang member while her unit sought to arrest the gang member for murder (Sain, 2017).
- **Shakedown** (a police officer learns about a criminal violation/suspect and demands a bribe for not making an arrest; Barker and Roebuck, 1973). A police officer investigating an Uber rape, assault, and robbery case was allegedly bribed by the two suspects to drop the investigation (Madibogo, 2016). Similarly, two Johannesburg Metro Police Department officers were caught on camera soliciting a bribe from a motorist (Cowan, 2017) and another Tshwane Metro Police Department officer solicited a bribe from an undercover national traffic anti-corruption officer (Ndlazi, 2017). A Polokwane municipality traffic officer demanded thousands of rands from a female driver after they crashed into each other (Mahopo, 2016). The officer first threatened the driver with arrest and suspension of her driver's license if she did not pay an immediate bribe of R1,000. The following day, the officer went to the driver's place of work and demanded an additional R4,700 (310 USD) (Mahopo, 2016). In another case, a police officer demanded R50,000 (3,300 USD) from a dentist claiming they were investigating the dentist for submitting false claims to a medical aid/insurance provider (Chambers, 2016).
- **Fix** (quashing the prosecution or "taking up" of traffic tickets; Barker and Roebuck, 1973). Based on a preliminary investigation, a man arrested for possession of "dagga" (marijuana), seems to have bribed two police officers and a prosecutor to release him "without putting the matter on the court roll" (Wicks, 2017, p. 1).
- **Corruption of authority** (a police officer received unauthorized material gain because of his position as a police officer, without violating the law *per se*; Barker and Roebuck, 1973). A Western Cape commissioner and three brigadiers received gifts, including clothes and accommodation for themselves and their families, worth R2.5 million (167,000 USD) from a Cape Town businessman "interested in ensuring efficient service from the SAPS" (Hyman, 2017, p. 1).
- **Opportunistic theft** (a police officer steals from arrestees, victims, crime scenes, and unprotected property; Barker and Roebuck, 1973). A police

officer was convicted of stealing drugs entered as evidence in the drug-related cases he was investigating (TMG Digital, 2016a).

Falsification of official documents/tampering with the crime scene

- Following an ambush by armed robbers which resulted in a police officer accidentally shooting a colleague, it seems that the officer tampered with the scene and covered up evidence (IOL News, 2017; Bornman, 2017). In another case, a major-general was suing the SAPS for R15 million (1 million USD) for "malicious investigation and instigation of prosecution" (Dolley, 2017). According to the plaintiff, he was "unlawfully investigated and arrested" (Dolley, 2017). When the plaintiff was convicted on a number of charges, he was dismissed from the SAPS, a decision he successfully appealed. Following his acquittal, he was reinstated in the SAPS, but at a lower rank, prompting him to litigate against the SAPS on the grounds that "his dignity and political rights suffered" (Dolley, 2017).

Failure to act

- Police officers present at a scene where one police officer had shot another officer, potentially tampered with the evidence to cover up the shooting (Bornman, 2017).

Threats to use excessive force

- A man charged with theft of a computer claimed the police left intimidating messages on his cell phone and with his children. They had allegedly threatened to kill him (Jordann, 2017).

Excessive force

- **Empty-hand control** (a police officer uses bodily force; soft technique includes grabs, holds, and locks, while the hard technique includes punches and kicks; National Institute of Justice, 2009). An Unemployment People's Movement organizer received R120,000 (8,000 USD) in damages because the police slapped and kicked him in view of his six-year old son when he went to a police station to report that the police were harassing his mother (Ndamase, 2016). In another case, illegal occupants of a piece of land were allegedly beaten by Cape Town police officers (Wolf, 2017). Police officers allegedly kicked in doors at a student residence after protests at the University of KwaZulu-Natal (Milambo, 2016).
- **Less-lethal methods** (a police officer uses less-lethal technologies; blunt impact includes a baton or projectile; chemical weapons include chemical sprays, such as a pepper spray, and energy-conducting devices, such as tasers; National Institute of Justice, 2009). A protesting University of the

Witwatersrand's student was allegedly shot in the back with rubber bullets up to 13 times at close range. At the same time, the police also used tear gas and stun grenades to deal with protestors (TMG Digital, 2016b). In another case, a citizen suspected of murder claims he was kicked and punched several times while his hands were handcuffed behind his back (Masilela, 2016).

- **Lethal force** (a police officer uses lethal weapons, such as firearms; National Institute of Justice, 2009). Two police officers were investigated in connection with the death of a man following an argument on New Year's Day. According to the statement given by his sister (Maponya, 2017):

> [t]he police van drove at high speed with the cop holding my brother's hands. He was dragged for a very long distance and sustained injuries to his head. Both his legs were also broken and he was bleeding from the mouth and ear.

A Tshwane Metro Police Department officer was applying for bail after allegedly shooting a man trying to restart his own car by connecting wires. The officer allegedly believed the man was reaching for a weapon (Tihabye, 2017).

Violation of procedural rules

- Three people tried for a double murder and rape claimed that police officers "severely assaulted them and pressured them to give statements implicating them in the crime" (Masilela, 2016).

Police sexual violence

- A female student reported that she had been raped by a SAPS officer during a protest at the University of KwaZulu-Natal (Milambo, 2016).

Although we do not provide these stories to draw inferences about the prevalence of police misconduct in South Africa, they nonetheless paint a picture of the nature of police misconduct. Cases of police violations of official rules are undoubtedly serious. When Barker and Roebuck (1973) designed the police corruption typology, they classified cases based on five factors (acts and actors involved; norms violated; support from peer group; organizational degree of deviant practices; and the police department's reaction). Barker and Roebuck (1973) maintained that shakedowns, opportunistic thefts, and protection of illegal activities—all examples of police corruption uncovered during our one-year news search—are serious and, if detected by the police agency, would result in a dismissal, the most serious internal disciplinary option in each police agency. Even the case of corruption of authority detected in our survey of newspaper accounts, typically evaluated as the least serious form of police corruption among the eight forms of police corruption, has certain characteristics—a large gain (R2.5 million), a high degree of organization, and involvement of a high-ranking police officer—that make it much more serious than a typical case classified into this form of police corruption. Cases of the use of excessive force (e.g., allegations of beating up a person who came to the police station to report harassment of a member of his family; beating up a suspect to confess) and police sexual violence (e.g., allegedly raping a student during a protest) show a complete disregard for human rights, even a malicious character (e.g., killing a suspect by dragging him behind a police pick-up vehicle).

Cases of police misconduct by police administrators

The report of a major-general suing the SAPS for a sizable amount because he was "unlawfully investigated and arrested" (Dolley, 2017) in an effort to dissuade him from pursuing an investigation into police corruption highlights the organizational nature of police misconduct, and clearly illustrates transgressions among supervisors and senior managers alike. Particularly poignant is the story of South Africa's National Police Commissioners. The SAPS National Commissioner is appointed on a five-year, renewable contract. Yet, between 1995 and 2019, only one has finished a full term, albeit not renewed in the position. Astonishingly, as a result of a litany of integrity violations, nine different people have occupied the post: Six as permanent appointments and three in acting capacities.

The first National Commissioner of the SAPS, General George Fivaz, was appointed in 1995 by Nelson Mandela in his capacity as president of the country. Fivaz was tasked with shepherding apartheid's 11 police agencies into the democratic era, merging them to form the SAPS, and introducing community and democratic policing to South Africa. Considering the challenges at hand, his term was relatively successful, and he retired after five years, in 2000.

Fivaz was followed by Jackie Selebi, who was appointed by the then president, Thabo Mbeki. Selebi, a former teacher, was a long-time member of the ANC who spent much of his life in exile, returning to South Africa only in 1990. When he took over from Fivaz, he had no prior police experience, but was instead a loyal ally of Mbeki's faction in an increasingly divided ruling party. As we explain in Chapter 3, Selebi's tenure coincided with the initial erosion of the criminal justice system as ANC members wielded it as a weapon in their battles with one another (Steinberg, 2017; Pauw, 2018). Although initially protected by President Mbeki, Selebi was placed on special leave in 2008 and eventually prosecuted and found guilty of corruption in 2010.

With Selebi suspended, Tim Williams was appointed to lead the SAPS. During his one-year term, Nathi Mthethwa, the Minister of Police, instructed Williams to appoint murder-accused and Zuma-ally, Richard Mdlulu, as the head of the SAPS Crime Intelligence division. Williams thought this irregular because Mdluli had not been subjected to the standard promotion processes. Nevertheless, Mthethwa manipulated the selection process to secure the appointment. This had grave consequences in the subsequent years as Mdluli abused his position to prop up a corrupt president and his allies (Pauw, 2018).

Williams resigned in 2009, upon which South Africa's new president, Jacob Zuma, appointed his own political ally—Bheki Cele—to head the organization. Cele led the SAPS during its return toward military ranks and became known for his rhetorical encouragement of police use of force. He was generally popular among the police as a result of this, despite having no prior policing experience. Cele was suspended by President Jacob Zuma in October 2011 and dismissed in June 2012 after a finding of "maladministration" and "improper, unlawful" behavior against him by the Public Protector. A separate board of inquiry found he had acted dishonestly and had an undeclared conflict of interest in relation to police leasing contracts (Moloi Board of Inquiry, 2015). Following his dismissal, Cele appealed both the Board's findings and President Zuma's decision to fire him to the High Court. And in April 2019, the North Gauteng High Court ruled that the discharging of Bheki Cele as National Police Commissioner be declared invalid and to have no force and effect, although the Public Protector's findings stood. While his case was on appeal, President Zuma resigned and was replaced as state president by Cyril Ramaphosa who then appointed Cele as Minister of Police in 2018.

Nhlanhla Mkhwanazi was appointed in an acting capacity following Cele's removal. Respected by his peers, Mkhwanazi took steps in 2012 to suspend Zuma-ally Richard Mdluli, which may explain why Mkhwanazi's appointment remained interim.

In June 2012, Riyah Phiyega replaced Mkhwanazi at the helm of the SAPS. Phiyega was neither a politician nor a police officer, but was the first woman to lead to the SAPS. Just three months into her term, SAPS officers killed 34 striking mine workers at Marikana. While she was not responsible for this abuse of force, she led the SAPS's efforts to cover it up. The Farlam Commission of Inquiry into the massacre found that Phiyega and associates conspired to produce a tailored narrative of events preceding the massacre in an effort to absolve the SAPS and its officers of wrongdoing (Farlam, 2015). This is an example of the police code of silence—an informal social norm common to police agencies that prohibits the reporting of misconduct (see Chapter 6)—operating at the highest levels within the organization, championed by a brand-new National Commissioner. During her tenure, Phiyega was involved in a number of scandals and was eventually suspended on the recommendation of the Farlam Commission pending a separate Board of Inquiry into her fitness to hold office. This board, headed by retired judge Neels Klaasen, found her unfit to hold the position of the SAPS National Commissioner and recommended that she be dismissed. Nevertheless, President Zuma allowed her to wait out the remaining two years of her contract on suspension, as if rewarding her for her service.

While Phiyega remained on suspension, Zuma appointed Lieutenant-General Khomotso Phahlane to act in her place. An experienced career officer,

Phahlane's appointment and early "Back to Basics" efforts to improve SAPS performance were praised. However, like his predecessor, Phahlane further bloated top management by appointing roughly 70 new senior members. Even more worryingly, it emerged that Phahlane was being investigated by the Independent Police Investigative Directorate (IPID) for fraud and corruption for allegedly taking kickbacks to secure police tenders for a company during his tenure as head of the police's forensics division (Eyewitness News, 2017a). Additional charges of defeating the ends of justice were later dropped by the Director of Public Prosecutions (SABC News, 2017). Other senior officers accused with Phahlane included Gauteng Commissioner Deliwe De Lange and the national divisional commissioner in charge of supply chain management, Lieutenant-General Ramahlapi Mokwena. In June 2017, Police Minister Fikile Mbalula suspended Phahlane, who was then immediately replaced with Lieutenant General Lesetja Mothiba as interim Commissioner by President Jacob Zuma (The Citizen, 2017). At the time of this book going to press, Phahlane's criminal trial was pending.

In November 2017, President Zuma appointed General Khehla Sitole as the new permanent National Commissioner. Sitole's tenure was tainted early in 2018, when the news broke that he was being investigated by the IPID for allegedly helping to launder R45 million (3 million USD) in an effort to influence the outcome of the ANC's tightly contested National Conference in December 2017 (Eyewitness News, 2017b). Sitole hampered the IPID investigation by classifying documents central to the case, arguing that they related to "national security" matters. At the time of writing, a legal battle was underway after the IPID turned to the courts to force Sitole to declassify the documents and enable the investigation to continue unimpeded (All Africa, 2019).

It is clear from this brief review that corruption, abuse of power, and political interference began affecting police top management less than a decade after the SAPS was formed. Given the visibility of these scandals, their effect was felt far beyond the office of the National Commissioner. Some junior officers likely felt ashamed to be associated with the organization as a result of transgressions by the senior ranks, while others may have seen the abuses as a way to justify their own. Many were likely left uncertain about which of their colleagues to trust or whether their seniors even deserved respect. The implications are important for the success of any organizational reforms which the SAPS pursues. It is hard to clean a complete flight of stairs without first cleaning the top part. In the SAPS, it remains unclear just how spotless the highest steps truly are.

Difficulties in studying police misconduct

These narratives clearly show police misconduct to be a serious challenge for South Africa. As rank and file engage in various violations and mostly escape any punishment for such behavior, police tarnish their image, trust in the police decreases, citizens' willingness to report crime diminishes, the ability to obtain information from citizens deteriorates, and offenders increasingly avoid sanction. Because of its serious consequences, police misconduct should be explored systematically and in-depth using scholarly methods. Yet the topic of police misconduct in South Africa remains under-researched (e.g., Grobler, 2013, p. 258). This is hardly surprising, because the empirical study of police misconduct remains riddled with challenges (e.g., Klockars et al., 2000; Kutnjak Ivković, 2003, 2005; Syed and Bruce, 1998).

The code of silence often renders police officers unwilling to talk about misconduct committed by their peers, as they are frequently prepared to protect their colleagues with silence even when they do not approve of the rule-violating behavior. In the course of its investigation of police corruption in the New York City Police Department (NYPD) in the 1990s, the Mollen Commission (1994, E6, p. 6) discovered that police officers who wanted to report misconduct were ill-treated (e.g., being ostracized and harassed, becoming targets of complaints and physical threats), and that honest police would "silently hope that such officers [we]re removed from the job," but would do nothing to make it happen. In this way, the NYPD's code of silence flourished.

The pressure to adhere to the code of silence prevents misconduct-related information from being shared not only with police supervisors but also with scholars (e.g., Kutnjak Ivković, 2003). Three U.S. studies illustrate this point. Fabrizio (1990) surveyed experienced police officers from across the United States who were attending the Federal Bureau of Investigation (FBI) National Academy. Even when guaranteed confidentiality, none of the respondents answered a question asking them to provide "specific examples of graft or corruption in [their] department, without revealing names or places, and what [they] think could have been done to prevent it" (Fabrizio, 1990, p. 39, 65), although they readily answered other questions. Similarly, in a study funded by the U.S. Department of Justice that explored the extent and nature of police misconduct (Knowles, 1996; Martin, 1994), scholars faced significant hurdles in collecting data. The Fraternal Order of Police in Pennsylvania opposed the study and openly discouraged their members from participating (Martin, 1994, pp. 1–2). As a result, Pennsylvania was not included in the project. The Fraternal Order of Police in Chicago took the same stance, and the Chicago Police Department, representing about one-half of all sworn police officers in the state, was dropped from the Illinois component of the study (Martin, 1994, p. 4).

The exploration of police misconduct is further complicated when police management is reluctant to allow scholars access to their agencies to collect data. Because of their adherence to the "rotten-apple theory," police managers may present misconduct as the failures of individual officers, allowing them to disregard organizational and systemic drivers of abuse. Research findings indicating that misconduct exists within a police agency may raise difficult questions about the competency of its management and may even result in a public scandal (Kutnjak Ivković, 2003). Consequently, police managers often have no motivation to allow scholars to study police misconduct in their agencies. Instead, as documented by several independent commissions (e.g., Knapp Commission, 1972; Pennsylvania Crime Commission, 1974; Mollen Commission, 1994), some police managers may sweep misconduct under the rug, deny its existence for as long as possible, and, if misconduct is discovered, downplay its extent and importance as much as possible while, at best, swiftly punishing a few officers publicly accused of such behavior (the "bad apples").

Organizational resistance to accountability and research may also emerge after scholars have successfully gained access to police organizations. Liza Grobler, a South African criminologist who studied police misconduct for her doctoral dissertation, tells a poignant story underscoring this phenomenon. The publication of an excerpt from her dissertation, in which she illustrated the depth and breadth of police misconduct in South Africa, was immediately followed, in Grobler's words, "by a personal attack on me and an obligatory request by SAPS management to focus only on the good cops. My phone was tapped" (Grobler, 2013, p. xii).

Theory of police integrity

To avoid the potential problems associated with a direct study of police misconduct, Klockars and Kutnjak Ivković (e.g., 2004) have developed an approach that instead studies its opposite—police integrity. Such a strategy facilitates the study of police officers' views and opinions, thereby avoiding the resistance from police officers and managers that a direct study of police misconduct may arouse (e.g., Fabrizio, 1990; Grobler, 2013; Knowles, 1996; Martin, 1994). The police integrity approach provides a systematic means to test theoretical arguments without such obstacles.

Klockars et al. (2004, p. 1) define police integrity as "the normative inclination among police to resist temptations to abuse the rights and privileges of their occupation." From this perspective, police integrity is viewed as a normative inclination, that is, a belief, rather than an actual behavior (Klockars et al., 2006, p. 10). Police integrity could be a feature of individual police officers, police agencies, and police subunits.

Theories of police integrity or its opposite—police misconduct—can be broadly classified into several categories. According to Delattre (2011), hypotheses explaining police corruption have fallen into three main positions: The society-at-large hypothesis, the structural or affiliation hypothesis, and the rotten-apple hypothesis. Pollock (2007) also classifies potential theoretical arguments into three groups of explanations of police misconduct/police integrity: Individual explanation (e.g., the rotten-apple theory), institutional (organizational) explanations, and systematic (societal) explanations. Klockars and Kutnjak Ivković's (2004) theory of police integrity, the theoretical basis for this book, is both organizational and societal in nature.

Organizational theory of police integrity

Because of its organizational nature, Klockars and Kutnjak Ivković's (2004) theory does not frame police misconduct as a problem of individual officers and their failures (i.e., "rotten apples," Knapp Commission, 1972; Punch, 2003), but instead as a problem of a police organization and "system failure" (i.e., "rotten orchards," Punch, 2003). Punch (2003, p. 194) explained that

> deviance in the apparatus of formal social control may become systematic, whereupon it is no longer possible to speak of "rotten apples" which conveys a model of individual rule-bending by weak or bad people predisposed to deviance. Here the emphasis has been on how systems can become the contexts for widespread and "institutionalized" rule-bending and illegality. It is less a matter of bad people and more an issue of bad systems. When that happens it is more appropriate, I would maintain, to speak of "rotten orchards" as a metaphor of wider "system failure."

As Klockars and Kutnjak Ivković (2004, p. 1.4) have argued, this approach creates opportunities for "a substantially different understanding of the problem, suggest[s] alternative control mechanisms, and allow[s] for the development of a novel methodological approach to the study." Rather than seeking to detect the characteristics of individual police officers who lack integrity, this theory explores organizational and societal features that erode integrity both in police organizations and among their employees. The basic propositions of the organizational theory of police integrity are outlined in four theoretical dimensions.

The first dimension of the theory targets the police agency's organizational rules and how they are made by the police administration and communicated to and understood by officers (Klockars and Kutnjak Ivković, 2004, p. 1.4). The rules regulating police misconduct typically include constitutional norms, various legal codes, and the police agency's internal rules. Police agencies of high integrity are expected to develop rules that prohibit police misconduct and outline behaviors that constitute violations of official rules. Then they are expected to teach and communicate these rules to their officers and enforce them when violations occur.

The second dimension of the police integrity theory explores the overall control efforts that the police agency is making to address police misconduct (e.g., Klockars and Kutnjak Ivković, 2004; Klockars et al., 2000, 2004, 2006). While the first dimension of the theory emphasizes the need to develop and teach official rules, the second dimension focuses on the mechanisms the police agency has put in place to enforce them. Such systems are primarily reactive—they are initiated once police managers learn that rules have been violated (e.g., Kutnjak Ivković, 2005). They typically include an internal investigation and, if the case merits it, a criminal investigation.

The third dimension of the theory of police integrity explores the code of silence and the police agency's efforts to curtail it (e.g., Klockars and Kutnjak Ivković, 2004; Klockars et al., 2000, 2004, 2006). The code of silence, "the blue curtain" (Goldstein, 1970), "the blue code" (Skolnick, 2002), and "the blue wall of silence" (Chin and Wells, 1998; Kleinig, 2001) are all terms describing the same concept: An informal social norm, often ingrained in police culture, that prohibits police officers from reporting misconduct committed by other officers. Because of the quasi-military nature of the police, the code is present to some extent in every police agency (Klockars et al., 2000, 2004, 2006). Because the norms of police culture vary across agencies and countries, the contours of the code of silence follow suit (e.g., Klockars et al., 2004, 2006; Kutnjak Ivković and Haberfeld, 2015). Compared to the police agencies of low integrity, police agencies of high integrity are expected to have a much weaker code that does not protect serious forms of misconduct.

The fourth dimension of the theory of police integrity proposes that the level of integrity in an agency is influenced by the social, economic, legal, and political environment in which it operates (Klockars et al., 2000, 2004, 2006). In a nutshell, societies enact the rules governing the conduct of governmental employees, develop expectations regarding appropriate conduct, maintain cultures (in)tolerant of misconduct, establish external mechanisms of conduct control, and punish civil servants when their conduct violates the rules (e.g., Kutnjak Ivković, 2015a). If the society at large has relaxed attitudes toward government misconduct in general, it is more likely that such attitudes will similarly apply to the police.

Methodology for measuring police integrity

Shifting the topic from the study of police misconduct to the study of police integrity, as Klockars and Kutnjak Ivković (2004) have done, means the empirical study based on this methodological approach is much less likely to trigger resistance from police than direct inquiries about police misconduct would. Instead of directly asking police officers whether they or their colleagues have engaged in misconduct, the police integrity approach poses questions of fact and opinion related to hypothetical scenarios describing rule violations. The validity and reliability of this methodological approach have been tested by recent studies (e.g., Kutnjak Ivković et al., 2019; Maskaly et al., 2019a, 2019b; Van Droogenbroeck et al., 2019), which indicated a substantial degree of both validity and reliability.

In the mid-1990s, Klockars and Kutnjak Ivković (2004) designed the first questionnaire using this method—the police corruption questionnaire. Because it was envisioned as the opposite of police corruption, the questionnaire contained 11 hypothetical scenarios containing examples of police corruption (10 out of 11 scenarios) and the use of excessive force (1 out of 11 scenarios). The police corruption questionnaire has been successfully applied, with minor adjustments, in more than 20 countries (for an overview, see Kutnjak Ivković, 2015a).

The second police integrity questionnaire (Klockars and colleagues, 2006) was based on the same principles as the police corruption questionnaire: It contained a dozen hypothetical scenarios describing police misconduct, including examples of police corruption, use of excessive force, falsification of official records, and a failure to execute an arrest warrant (Klockars et al., 2006). The police integrity questionnaire has been used to study police integrity in more than a dozen countries, including Armenia (Khechumyan and Kutnjak Ivković, 2015), Australia (Porter et al., 2015), Croatia (e.g., Kutnjak Ivković, 2015b), Estonia (Vallmuur, 2015), Russia (Cheloukhine et al., 2015), Slovenia (Lobnikar and Mesko, 2015), South Korea (Kang and Kutnjak Ivković, 2015), Thailand (Phetthong and Kutnjak Ivković, 2015), and the United States (Kutnjak Ivković et al., 2015). Yet the police integrity questionnaire has not previously been used to survey police in South Africa.

Measuring police integrity in South Africa

This book offers a unique, in-depth, and systematic empirical exploration of the contours of police integrity in South Africa. Relying on the theory of police integrity and related methodology (Klockars and Kutnjak Ivković, 2004; Klockars et al., 2000, 2004, 2006), we analyze the complex set of historical, legal, political, social, and economic circumstances that affect levels of police integrity in South Africa. The discussion of our theoretical framework is accompanied by empirical results from a unique nationwide survey of SAPS officers.

South African sample

We surveyed commissioned and non-commissioned SAPS officers from all nine provinces. We relied on the face-to-face or in-person survey in which a member of the research team distributed questionnaires to police officers and invited them to participate in the study. The first administration of the police officer survey was conducted between 2011 and 2013. Ultimately, we collected 757 questionnaires. The second administration of the police officer survey occurred between 2017 and 2019,[2] yielding an additional 111 respondents. Our combined sample includes 868 SAPS officers. The response rate is 87.5%, an above average response rate for face-to-face police surveys (e.g., Nix et al., 2017). Answers provided by the officers who participated in our study are the basis of the empirical analyses presented in this book.

Although our sample is a convenience sample, it is reflective of the demographics of the SAPS (we relied on the SAPS's 2017–2018 Annual Report to calculate the demographic profile of the organization). First, about one-quarter of both our sample and the SAPS population is female (Table 1.1), while most

Table 1.1 Sample and population

Sample	*Number of police officers*	*Percent*	*SAPS (March 31, 2018)*	*Number of police officers*	*Percent*
Gender			*Gender*		
Male	616	71.5%	Male	108,160	71.7%
Female	246	28.5%	Female	42,666	28.3%
Race			*Race*		
White	142	16.4%	White	14,803	9.8%
Black	505	58.2%	Black	116,396	77.2%
Other	221	25.5%	Other[1]	19,627	13.0%
Supervisory position			*Supervisory position*[2]		
Yes	360	41.6%	Yes	24,936	16.5%
No	505	58.4%	No	125,890	83.5%

Source: SAPS Annual Report 2017–2018, p. 30

[1]Includes Indian and Colored.
[2]Includes executive authorities, top management, senior management, and commissioned officers.

respondents and SAPS officers are male. Second, most of our sample and most SAPS officers are black (Table 1.1). While the proportion of black officers in the SAPS is higher than it is in our sample, they nevertheless constitute the majority in both the sample and the population. Third, the majority of police officers in our sample and the majority of police officers in the SAPS population do not occupy supervisory posts (Table 1.1). Although our sample has a higher proportion of supervisors than the SAPS population does, this may be a consequence of our supervisor calculation. Our respondents self-identified as supervisors on the questionnaire, while our calculation of the proportion of supervisors is based on the SAPS' officially recorded data for executive authorities, top management, senior management, and commissioned officers. As such, sergeants or warrant officers who supervised constables may have identified themselves as supervisors, even though they are not commissioned officers.

The overwhelming majority of our respondents are employed in either patrol/community policing (38.4%) or investigative/detective units (36.4%; Table 1.2). The size of the unit in which they work varies greatly across the respondents: About one-third (37.7%) are employed in very large or large police stations or units, one-third (35.8%) in small or very small police stations, and about one-quarter (26.5%; Table 1.2) in medium-sized police stations.

In terms of their experience, very few respondents have been police officers for fewer than five years (15.7%; Table 1.2). In fact, more than 84% have been police officers for five years or more (Table 1.2); about one-half (50%) have been police officers for 15 years or longer. Although SAPS officers do not spend their whole careers in the same police station, they generally do not move frequently either; only about one-quarter (27.5%; Table 1.2) have been police officers in their current station for fewer than five years.

Table 1.2 Sample characteristics

	Number of police officers	*Percent*		*Number of police officers*	*Percent*
Assignment			*Length of service*		
Patrol and community policing	334	38.4%	Below 1 year	8	0.9%
Investigation	316	36.4%	1–2 years	36	4.1%
Communications	48	5.5%	3–5 years	93	10.7%
Traffic	34	3.9%	6–10 years	211	24.3%
Administrative	60	6.9%	11–15 years	86	9.9%
Other	76	8.8%	16–20 years	124	14.3%
Size of police agency			Over 20 years	310	35.7%
Very large (over 500)	209	24.1%	*Length of service in this station*		
Large (201–500)	118	13.6%	Below 1 year	43	5.0%
Medium (76–200)	230	26.5%	1–2 years	75	8.6%
Small (25–75)	167	19.3%	3–5 years	121	13.9%
Very small (fewer than 25)	143	16.5%	6–10 years	225	25.9%
			11–15 years	124	14.3%
			16–20 years	108	12.4%
			Over 20 years	172	19.8%

Questionnaire

The respondents were asked to fill out the police integrity questionnaire (Klockars et al., 2006). The original police integrity questionnaire contains 11 scenarios describing different forms of police misconduct (Table 1.3), including police corruption, use of excessive force, planting evidence, and falsifying official documents. In the South African version of the questionnaire, we changed the wording of two scenarios to adapt them to local conditions (e.g., replaced the description of "a felony warrant" with "an issued warrant of arrest" in Scenario 2; changed the reason a police officer stops a driver from "for speeding" to "for reckless driving" in Scenario 7). We also added three scenarios from Kutnjak Ivković's (2009) study of the Croatian police. Two of these scenarios were designed with the Croatian police to capture misconduct by omission (a failure to note on a report that a crime could be classified as a hate crime; a failure to intervene when the police officer sees juveniles writing graffiti on the wall). The third additional scenario, taken from the police corruption questionnaire (Klockars and Kutnjak Ivković, 2004; Klockars et al., 2000, 2004), describes a bribe accepted during a traffic stop, a situation of particular relevance in South Africa where traffic-related bribes are the most common type of bribe solicited by government officials (Statistics South Africa, 2018). However, the SAPS does not generally enforce traffic rules, so most roadside bribery involves metropolitan and traffic police rather than SAPS officers. To adjust it to the local condition, we changed the rationale (i.e., instead of "a police officer stops a motorist for

Table 1.3 Scenario descriptions and abbreviations

	Scenario Abbreviation	*Scenario Description*
Scenario 1	Free meals, gifts from merchants	A police officer is widely liked in the community. Local merchants and restaurant owners regularly show their appreciation for his attention by giving him gifts of food, cigarettes, and other items of small value.
Scenario 2	Failure to arrest friend with warrant	A police officer is aware that there is an issued warrant of arrest for a longtime friend of his. Although he sees his friend frequently over a period of more than a week and warns his friend of its existence, he does not arrest him.
Scenario 3	Theft of knife from crime scene	A police officer discovers a burglary of a hardware store. The display cases are smashed, and many items have obviously been taken. While searching the store, he takes an expensive pocketknife and slips it into his pocket. He reports that the knife has been stolen during the burglary.
Scenario 4	Unjustifiable use of deadly force	An officer, who was severely beaten by a person resisting arrest, has just returned to duty. On patrol, the officer approaches a person standing in a dimly lit alley. Suddenly, the person throws a gym bag at the officer and begins to run away. The officer fatally shoots the person, striking him in the back. It was later determined that the person was unarmed.
Scenario 5	Supervisor offers holiday for errands	A police officer is scheduled to work during coming holidays. The supervisor offers to give him these days off, if he agrees to run some personal errands for the supervisor. Evaluate the SUPERVISOR'S behavior.
Scenario 6	Officer strikes prisoner who hurt partner	In responding with her male partner to a fight in a bar, a young female officer receives a black eye from one of the male combatants. The man is arrested, handcuffed, and, as he is led into the cells, the male member of the team punches him very hard in the kidney area saying, "hurts, doesn't it."
Scenario 7	Verbal abuse	A police officer stops a motorist for reckless driving. As the officer approaches the vehicle, the driver yells, "What the hell are you stopping me for?" The officer replies, "Because today is 'Arrest an Asshole Day.'"
Scenario 8	Cover-up of police DUI and accident	At 2:00 A.M. a police officer, who is on duty, is driving his patrol car on a deserted road. He sees a vehicle that has been driven off the road and is stuck in a ditch. He approaches the vehicle and observes that the driver is not hurt but is obviously intoxicated. He also finds that the driver is a police officer. Instead of reporting this accident and offense, he transports the driver to his home.

(*Continued*)

Table 1.3 (Continued)

	Scenario Abbreviation	*Scenario Description*
Scenario 9	Auto body shop 5% kickback	A police officer has a private arrangement with a local auto body shop to refer the owners of cars damaged in accidents to the shop. In exchange for each referral, he receives a payment of 5% of the repair bill from the shop owner.
Scenario 10	False report on drug dealer	A police officer arrests two drug dealers involved in a street fight. One has a large quantity of heroin on his person. In order to charge them both with serious offenses, the officer falsely reports that the heroin was found on both men.
Scenario 11	Sgt. fails to stop beating	A police sergeant, without intervening, watches officers under his supervision repeatedly strike and kick a man arrested for child abuse. The man has previous child abuse arrests. Evaluate the SERGEANT'S behavior.
Scenario 12	Failing to report hate crime	In responding to a stabbing, a police officer arrests a suspect under circumstances that suggest that the attack was racially motivated. The police officer fails to write on the official report that the attack was racially motivated.
Scenario 13	Bribe from reckless motorist	A police officer stops a motorist for not obeying a red traffic light. The officer agrees to accept a personal gift for half of the amount of the fine in exchange for not issuing a citation.
Scenario 14	Failing to react to graffiti	While on patrol in his patrol car, a police officer notices a group of juveniles writing graffiti on a building. The police officer decides not to intervene and instead drives on.

speeding," we wrote, "a police officer stops a motorist for not obeying a red traffic light").

Upon reading each hypothetical scenario, the respondents were asked seven identical questions. Three questions related to the first theoretical dimension. Respondents were asked to assess how seriously they, and how seriously most police officers in their agency, would evaluate the behavior described in the questionnaire, as well as whether the described behavior violates official rules. Two questions focus on the second theoretical dimension, asking the respondents to state what they think the appropriate discipline should be, and what discipline they expect would result from the behavior described in the questionnaire. Finally, the last two questions explore the third dimension of the theory by inquiring about the respondents' willingness to report misconduct and their estimates of the willingness of their colleagues to do the same. The original questionnaire developed by Klockars and colleagues (2006) contained no questions measuring the fourth dimension of the theory.

Structure of the book

This book offers a unique perspective through an in-depth analyses of police integrity in South Africa. It sets out to explore systematically and empirically the contours of police integrity in the country. Utilizing the organizational theory of police integrity (Klockars and Kutnjak Ivković, 2004), we analyze the historical, legal, political, social, and economic circumstances shaping the levels of police integrity in South Africa. We study the relation between police integrity and race, gender, and rank. The discussion of our theoretical framework is accompanied by the results of our nationwide survey of about 900 SAPS officers.

The first part of the book—Chapters 2 and 3—provide a historical and contemporary analysis of social, political, economic, and legal factors underlying policing in South Africa. They are based on the fourth dimension of the police integrity theory that explores the influence of the society at large on police integrity. Chapter 2 explores the historical development of policing in South Africa up until 1994. It analyzes the social and political circumstances that led to the establishment of the South African state, including the emergence of the South African Police (SAP) in 1913. The chapter discusses the pivotal role of policing in the "total strategy" response of the later apartheid government to actions perceived as a "total onslaught" against its oppressive nationalist policies. Particular emphasis is also placed on the allegiances, influences, and intricate strategies that would shape post-apartheid South Africa and its new police service.

Chapter 3 reviews the two-and-a-half decades following the country's first democratic election in 1994, including efforts to transform the SAP into a service-oriented police agency, the South African Police Service (SAPS). We survey the relevant social, political, and economic conditions that necessitated the transition and reflect on those that facilitated its successes and failures. We explore the development of the SAPS over time and discuss the immediate challenges facing the organization and country, including persistent violence, inequality, unemployment, and political corruption.

The second part of the book—Chapters 4, 5, and 6—provides an in-depth exploration of three organizational dimensions of the theory of police integrity in the South African context. Chapter 4 explores the first dimension and presents the existing legal framework for the South African police. The discussion focuses not only on criminal substantive and procedural law but also on the specific laws and regulations underpinning the work of police officers. In the process, we provide a contextual analysis of existing legal rules and regulations and their enforcement. The chapter includes an empirical analyses of our survey data, focusing on SAPS officers' knowledge of official rules and the seriousness with which they evaluate various forms of misconduct.

Chapter 5 focuses on the second dimension of the police integrity theory and provides a detailed analysis of the SAPS's internal mechanisms of control for police misconduct. It discusses the effectiveness of institutions such as the Independent Complaints Directorate (ICD), Independent Police Investigative Directorate (IPID), the SAPS Anti-Corruption Unit (ACU), Organized Crime Units (OCUs), the "Scorpions," and the "Hawks." It also contains an analysis of officers' views regarding appropriate discipline for various forms of misconduct and the discipline they expect the agencies to mete out for such violations, as well as their estimates of discipline fairness.

Chapter 6 explores the third dimension of the theory of police integrity by delving into the code of silence (the prohibition against reporting misconduct by fellow police officers). It discusses the contours of the code of silence in the South African context and analyzes the societal and organizational conditions that strengthen the code. The chapter presents the results of our survey and analyzes the strength of the code of silence. It also discusses the SAPS management's efforts to curtail the code and the effectiveness of these efforts.

The third part of the book—Chapters 7, 8, and 9—discusses the contours of police integrity across specific demographic groups within the SAPS. Chapter 7 focuses on race in South Africa. Race has been central to the history of South Africa and the SAPS, and remains a sensitive and critical subject in the country. This chapter discusses its implications for contemporary policing. The empirical part of the chapter compares the contours of police integrity among the three racial groups in our sample (black, white, and other). The analysis explores similarities and differences in the views regarding misconduct seriousness, appropriate and expected discipline, and willingness to report misconduct expressed by officers of different races.

Chapter 8 explores gender, South African policing, and police integrity. Although women have been employed as police officers since 1972, statistics indicate that they are yet to be fully integrated within the SAPS. This chapter explores the construction of gender in South Africa and how this shapes gender (in)equality among SAPS officers. It includes a discussion of extant research investigating the relation between gender and police integrity. The empirical part of the chapter analyzes gender differences in officers' views regarding misconduct seriousness, appropriate and expected discipline, and willingness to report.

Chapter 9 studies the contours of police integrity across the SAPS's rank structure. Like race, rank in the SAPS is a contentious, political, and emotive subject. We discuss recruitment and promotion practices within the SAPS that emphasize the attainment of particular race/gender quotas and the related allegations of institutional bias. We study potential differences among non-commissioned officers, commissioned officers, and police managers in their views on police integrity, namely, familiarity with official rules, views about appropriate and expected discipline, and expressed willingness to report.

In the last chapter—Chapter 10—we discuss the research findings in the contextual framework of social, political, legal, and economic factors, as well as prior empirical research on police integrity. With a focus on the future evolution of critical issues related to South African policing, we discuss the research and policy implications of our findings. The policy implications cover general themes, including police officer knowledge of official rules, perceptions of disciplinary fairness, and the contours of the code of silence. At the same time, we formulate tangible policy recommendations regarding specific groups within the South African police along the dimensions of gender, race, and rank.

Notes

1 In South Africa, the South African Police Service does not generally enforce traffic laws. As such, "traffic"-related bribes would usually be solicited by traffic and metro police. The latter are much smaller than the SAPS.

2 Although the data collection had a four-year gap, we argue that the level of police integrity did not change substantially across this time period. Comparisons of the respondents' answers did not yield large and substantive differences between the two subsamples.

References

Alexander, K., D. Batchelor, A. Durand, and T. Savage (2005). Truth commissions and transitional justice: Update on a select bibliography on the South African Truth and Reconciliation Commission debate. *Journal of Law and Religion*, 20(2): 525–565.

All Africa (2019). South Africa: State capture inquiry: Zondo to question top cop on classified crime intelligence documents. *All Africa*. Available at: https://allafrica.com/stories/201909240036.html

Amnesty International (2011). *Amnesty International Report 2011: The State of the World's Human Rights*. Available at: www.amnesty.org/download/Documents/32000/pol100012011en.pdf

Barker, T. and J. Roebuck (1973). *An Empirical Typology of Police Corruption*. Springfield, IL: Charles C. Thomas.

Basson, A. and P. du Toit (2018). *Enemy of the People: How Jacob Zuma Stole South Africa and How the People Fought Back*. Cape Town: Jonathan Ball.

Bello, P. and J. Steyn (2019). Students' perceptions of the South African Police Service in Durban, South Africa: A non-judgmental assessment. *International Journal of Social Sciences and Humanity Studies*, 11(2): 61–85.

Boateng, F.D. (2019). Fearfulness in the community empirical assessments of influential factors. *Journal of Interpersonal Violence*, 34(3): 562–584.

Boateng, F.D. and J.Y. Lee (2018). Apartheid and post-apartheid analysis of public confidence in the police: A longitudinal analysis across time. *Policing: An International Journal of Police Strategies and Management*, 41(6): 766–781.

Bornman, J. (2017). SA's top cop "shocked and disturbed" by video showing apparent SAPS shooting cover-up. *Sowetan Live*. Available at: www.sowetanlive.co.za/news/2017/01/17/sas-top-cop-shocked-and-disturbed-by-video-showing-apparent-saps-shooting-cover-up

Bradford, B., A. Huo, J. Jackson, and B. Roberts (2014). What price fairness when security is at stake? Antecedents of police legitimacy in South Africa. *Regulation and Governance*, 8: 246–268.

Bruce, D. (2008). Benign neglect? The politics and practise of controlling police corruption in South Africa. Paper presented at the Workshop Crime, Poverty and Police Corruption Annual Bank Conference on Development Economics Cape Town, 9–11 June. Available at: http://siteresources.worldbank.org/INTABCDESOUAFR2008/Resources/DavidBruce.pdf

Chambers, D. (2016). Cop arrested after extracting R50,000 blackmail cash from dentist. *Sowetan Live*. Available at: www.sowetanlive.co.za/news/2016/10/29/cop-arrested-after-extracting-r50000-blackmail-cash-from-dentist

Cheloukhine, S., S. Kutnjak Ivković, Q. Haq, and M.R. Haberfeld (2015). Police integrity in Russia. In S. Kutnjak Ivković and M. Haberfeld (eds.), *Police Integrity across the World*. New York: Springer, pp. 153–182.

Chin, G.J. and S.C. Wells (1998). The "blue wall of silence" as evidence of bias and motive to lie: A new approach to police perjury. *University of Pittsburgh Law Review*, 59(2): 233–300.

The Citizen (2017). Phahlane booted as top cop over allegations against him, says Mbalula. *The Citizen*. Available at: https://citizen.co.za/news/south-africa/1531233/phahlane-booted-top-cop-allegations-says-mbalula/

Conway-Smith, E. (2012). South Africa troubled by corrupt cops. *MinnPost*. Available at: www.minnpost.com/global-post/2012/05/south-africa-troubled-corrupt-cops/

Cowan, K. (2017). JMPD officers suspended after "bribe" video posted online. *Sowetan Live*. Available at: www.sowetanlive.co.za/news/2017/03/20/watch-jmpd-officers-suspended-after-bribe-video-posted-online

Daily Maverick (2019). Editorial: To rebuild South Africa's investigative capacity, disband Hawks. *Daily Maverick*, 24 June. Available at: www.dailymaverick.co.za/article/2019-06-24-to-rebuild-south-africas-investigative-capacity-disband-hawks/

Delattre, E. (2011). *Character and Cops: Ethics in Policing*, 6th ed. Washington, DC: AEI Press.

Dixon, B. (2018). South African crime statistics need to be put into context. *Guardian, Letters*, 13 September. Available at: www.theguardian.com/world/2018/sep/13/south-african-statistics-need-to-be-put-into-context

Dolley, C. (2017). The Mafioso, the ex-Mandela cop and the "malicious" police: Case heads to court. *News24*. Available at: www.news24.com/SouthAfrica/News/the-mafioso-the-ex-mandela-cop-and-the-malicious-police-case-heads-to-court-20170308

Duke, L. (1995). Mandela bridging white-black divide. *Washington Post*, 21 July, p. A23. Available at: www.washingtonpost.com/wp-srv/inatl/longterm/s_africa/stories/bridge0695.htm?noredirect=on

Ethics Institute (2018). *South African Citizens' Bribery Survey 2017*. Available at: www.tei.org.za/phocadownload/CBS%202017_29%20November%202017_Upload%20to%20web.pdf

Eyewitness News (2017a). Former top cop Phahlane to appear in court on fraud, corruption charges. *Eyewitness News*. Available at: https://ewn.co.za/2018/02/08/former-top-cop-phahlane-to-appear-in-court-on-fraud-corruption-charges

Eyewitness News (2017b). Surveillance footage confirms top cop Sitole's meeting with fraud suspect. *Eyewitness News*. Available at: https://ewn.co.za/2018/03/14/surveillance-footage-confirms-top-cop-sitole-s-meeting-with-fraud-suspect

Fabrizio, L.E. (1990). *The FBI National Academy: A Study of the Change in Attitudes of Those Who Attend*. Chicago: University of Illinois Chicago's Office of International Criminal Justice.

Farlam, I.G. (2015). *Marikana Commission of Inquiry: Report on Matters of Public, National and International Concern Arising Out of the Tragic Incidents at the Lonmin Mine in Marikana, in the North West province*. Available at: www.sahrc.org.za/home/21/files/marikana-report-1.pdf

Faull, A. (2009). Need or greed? Corruption and integrity management in a Gauteng police station. *South African Crime Quarterly*, 28(June): 11–19.

Faull, A. (2010). "When I see them I feel like beating them": Corruption and the South African Police Service. *South African Crime Quarterly*, 34(December): 33–40.

Faull, A. (2016). Measured governance? Policing and performance management in South Africa. *Public Administration and Development*, (36): 157–168.

Faull, A. (2017). *Police Work and Identity: A South African Ethnography*. Abingdon: Routledge.

Faull, A. and G. Newham (2011). Protector or Predator? Tackling Police Corruption in South Africa Institute for Security Studies Research Monograph. Available at: https://issafrica.org/research/monographs/protector-or-predator-tackling-police-corruption-in-south-africa

Gastrow, P. and M. Shaw (2001). In search of safety: Police transformation and public responses in South Africa. *Daedalus*, 130(1): 259–275.

Goldstein, H. (1970). *Police Corruption: A Perspective on its Nature and Control*. Washington, DC: Police Foundation.

Grobler, L. (2013). *Crossing the Line: When Cops Become Criminals*. Auckland Park, South Africa: Jacana Media.

Hyman, A. (2017). Corruption trial of senior Western Cape police officers delayed by defence lawyers. *Times Live*. Available at: www.timeslive.co.za/local/2017/04/18/Corruption-trial-of-senior-Western-Cape-police-officers-delayed-by-defence-lawyers

Institute for Justice and Reconciliation (2017). *Afrobarometer*. Available at: www.ijr.org.za/home/wp-content/uploads/2012/07/SAF_R6_South-Africa-Round-Results-and-Analysis_20Mar17.pdf

IOL News (2017). Police slam "insensitive" video of fatal shooting. *IOL News*. Available at: www.iol.co.za/capetimes/news/police-slam-insensitive-video-of-fatal-shooting-7435367

Jensen, S. (2014). Conflicting logics of exceptionality: New beginnings and the problem of police violence in post-apartheid South Africa. *Development and Change*, 45(3): 458–478.

Jordann, N. (2017). "Police threatened to kill me", chief justice computer theft suspect tells court. *Sowetan Live*. Available at: www.sowetanlive.co.za/news/2017/03/31/police-threatened-to-kill-me-chief-justice-computer-theft-suspect-tells-court

Kang, W. and S. Kutnjak Ivković (2015). Police integrity in South Korea. In S. Kutnjak Ivković and M. Haberfeld (eds.), *Police Integrity across the World*. New York: Springer.

Khechumyan, A. and S. Kutnjak Ivković (2015). Police integrity in Armenia. In S. Kutnjak Ivković and M. Haberfeld (eds.), *Police Integrity across the World*. New York: Springer, pp. 37–66.

Kleinig, J. (2001). The blue wall of silence. *The Journal of Applied Philosophy*, 15(1): 1–23.
Klockars, C.B. and S. Kutnjak Ivković (2004). Measuring police integrity. In A.R. Piquero, J.R. Greene, and M.J. Hickman (eds.), *Police Integrity and Ethics*. Belmont, CA: Wadsworth Publishing, pp. 1.3–1.20.
Klockars, C.B., S. Kutnjak Ivković, and M.R. Haberfeld (eds.) (2004). *The Contours of Police Integrity*. Newbury Park, CA: Sage Publication.
Klockars, C.B., S. Kutnjak Ivković, and M.R. Haberfeld (2006). *Enhancing Police Integrity*. New York: Springer.
Klockars, C.B., S. Kutnjak Ivković, W.E. Harver, and M.R. Haberfeld (2000). *The Measurement of Police Integrity*. Research in Brief. U.S. Department of Justice, Office of Justice Programs, National Institute of Justice. Washington, DC: Government Printing Office.
[Knapp Commission]. Commission to Investigate Allegations of Police Corruption and the City's Anti-Corruption Procedures (1972). *Report on Police Corruption*. New York: G. Braziller.
Knowles, J.J. (1996). *The Ohio Police Behavior Study*. Columbia, OH: Office of Criminal Justice Services.
Kriegler, A. and M. Shaw (2016a). *A Citizen's Guide to Crime Trends in South Africa*. Johannesburg, South Africa: Jonathan Ball Publishers.
Kriegler, A. and M. Shaw (2016b). Facts show South Africa has not become more violence since democracy. *The Conversation*. Available at: https://theconversation.com/facts-show-south-africa-has-not-become-more-violent-since-democracy-62444
Kutnjak Ivković, S. (2003). To serve and collect: Measuring police corruption. *Journal of Criminal Law and Criminology*, 93(2–3): 593–649.
Kutnjak Ivković, S. (2005). *Fallen Blue Knights: Controlling Police Corruption*. New York: Oxford University Press.
Kutnjak Ivković, S. (2009). The Croatian police, police integrity, and transition toward democratic policing. *Policing: An International Journal of Police Strategies and Management*, 32(3): 459–488.
Kutnjak Ivković, S. (2015a). Studying police integrity. In S. Kutnjak Ivković and M. Haberfeld (eds.), *Police Integrity across the World*. New York: Springer.
Kutnjak Ivković, S. (2015b). Police integrity in Croatia. In S. Kutnjak Ivković and M. Haberfeld (eds.), *Police Integrity across the World*. New York: Springer, pp. 97–124.
Kutnjak Ivković, S., I. Cajner Mraović, and D. Sudar (2019). The speed of progress: Comparing citizen perceptions of police corruption in Croatia over time. In S. Kutnjak Ivković and M. Haberfeld (eds.), *Exploring Police Integrity: Novel Approaches to Police Integrity Theory and Methodology*. New York: Springer.
Kutnjak Ivković, S. and M.R. Haberfeld (eds.) (2015). *Police Integrity across the World*. New York: Springer.
Kutnjak Ivković, S., M.R. Haberfeld, and R. Peacock (2015). Police integrity in the United States. In S. Kutnjak Ivković and M. Haberfeld (eds.), *Police Integrity across the World*. New York: Springer.
Kutnjak Ivković, S. and A. Sauerman (2012). The code of silence: Revisiting South African police integrity. *South African Crime Quarterly*, 40(June): 15–24.
Lobnikar, B. and G. Meško (2015). Police integrity in Slovenia. In S. Kutnjak Ivković and M. Haberfeld (eds.), *Police Integrity across the World*. New York: Springer, pp. 183–212.
Lodge, T. (1998). Political corruption in South Africa. *African Affairs*, 97: 157–187.
Louw, A. and E. Pelser (2001). *Community Policing and Police Service Improvement Study*. Unpublished Report for the Department of Safety and Security.

Madibogo, J. (2016). Cop probing Uber rape case was arrested for taking a bribe from suspects. *Sowetan Live.* Available at: www.sowetanlive.co.za/news/2016/12/06/cop-probing-uber-rape-case-was-arrested-for-taking-a-bribe-from-suspects

Mahopo, Z. (2016). Public urged to trap bribe cops. *Sowetan Live.* Available at: www.sowetanlive.co.za/news/2016/04/13/public-urged-to-trap-bribe-cops

Makda, T., M.S. Bayat, and W.I. Ukpere (2012). Risk management within the events industry in Cape Town, South Africa. *African Journal of Business Management*, 6(20): 6243–6253.

Malala, J. (2011). Our cops can't be trusted. *Times Live*, 27 March. Available at: http://whatishappeninginsouthafrica.blogspot.com/2011/03/our-cops-cant-be-trusted.html

Maponya, F. (2017). Cops used van o drag and kill my brother. *Sowetan Live.* Available at: www.sowetanlive.co.za/news/2017/01/03/cops-used-van-to-drag-and-kill-my-brother

Martin, C. (1994). *Illinois Municipal Officers' Perceptions of Police Ethics.* Chicago, IL: Illinois Criminal Justice Information Authority, Statistical Analysis Center.

Masilela, B. (2016). Rhodes park accused repeatedly denied assault by cops. *IOL News.* Available at: www.iol.co.za/news/crime-courts/rhodes-park-accused-repeatedly-denied-assault-by-cops-2093547

Maskály, J., C.M. Donner, and T. Chen (2019a). Improving the measurement of police integrity: An application of LTM to the Klockars et al. (1997) scales. In S. Kutnjak Ivković and M. Haberfeld (eds.), *Exploring Police Integrity: Novel Approaches to Police Integrity Theory and Methodology.* New York: Springer.

Maskály, J., S. Kutnjak Ivković, M. Haberfeld, C. Donner, T. Chen, and M. Meyers (2019b). Assessing the Validity of Police Integrity Scale in a Comparative Context. *Policing & Society.* Forthcoming.

Matshedisho, B. (2009). Nothing succeeds like success: The manipulation of crime statistics. In D. Vigneswaran and J. Hornberger (eds.), *Beyond "Good Cop"/"Bad Cop": Understanding Informality and Police Corruption in South Africa.* Johannesburg: Forced Migration Studies Programme and Open Society Foundation.

Milambo, S. (2016). UKZN students claim police brutality and rape. *IOL News.* Available at: www.iol.co.za/news/crime-courts/ukzn-students-claim-police-brutality-and-rape-2065322

Mngadi, M. (2017). Zuma urges residents to report corrupt cops. *News24.* Available at: www.news24.com/SouthAfrica/News/zuma-urges-residents-to-report-corrupt-cops-20170314

[Mollen Commission]. New York City Commission to Investigate Allegations of Police Corruption and the Anti-Corruption Procedures of the Police Department (1994). *Commission Report.* New York: Author.

Moloi Board of Inquiry (2015). Africa Criminal Justice Reform, 2012. President Zuma "releases" Cele as police chief, appoints Phiyega. Available at: https://acjr.org.za/resource-centre/president-zuma-releases-south-african-police-chief

Myburgh, P. (2019). *Gangster State: Unraveling Ace Magashule's Web of Capture.* Johannesburg: Penguin Books.

National Institute of Justice (2009). *Use of Force Continuum.* Available at: www.nij.gov/topics/law-enforcement/officer-safety/use-of-force/pages/continuum.aspx

Ndamase, M. (2016). Activist wins R120, 000 in cop brutality case. *Sowetan Live.* Available at: www.sowetanlive.co.za/news/2016/10/20/activist-wins-r120000-in-cop-brutality-case

Ndlazii, S. (2017). Metro cop who demanded R100 bribe to know fate in May. *IOL News.* Available at: www.iol.co.za/news/crime-courts/metro-cop-who-demanded-r100-bribe-to-know-fate-in-may-8413187

Newham, G. (2000a). Towards understanding and combating police corruption. *Crime and Conflict*, 19: 21–25. Available at: http://citeseerx.ist.psu.edu/viewdoc/download?doi=10.1.1.574.8598&rep=rep1&type=pdf

Newham, G. (2000b). *Management of Police Integrity at the Hillbrow Police Station*. Unpublished Masters Dissertation. Johannesburg, South Africa: Graduate School of Public and Development Management, University of the Witwatersrand.

Newham, G. (2005). *The SAPS Johannesburg Area Police Transformation Survey Results*. Johannesburg, South Africa: Centre for the Study of Violence and Reconciliation.

News24. (2016). Crooked cops and drug dens in West Johannesburg. *News24*. Available at: www.news24.com/Video/SouthAfrica/News/watch-crooked-cops-and-drug-dens-in-west-johannesburg-20160914

Nix, J., J.T. Pickett, H. Baek, and G.P. Alpert (2017). Police research, officer surveys, and response rates. *Policing and Society*. DOI: 10.1080/10439463.2017.1394300

Pauw, J. (2018). *The President's Keeper: Those Keeping Zuma in Power and Out of Prison*. Cape Town: Tafelberg.

Pennsylvania Crime Commission (1974). *Report on Police Corruption and the Quality of Law Enforcement in Philadelphia*. Saint Davids, PA: Author.

Pharoah, R. (2008). National Victims of Crime Survey: Overview of Key Findings. Institute for Security Studies Papers 2008 (175). Available at: https://issafrica.org/south-african-crime-statistics/01-nov-2008-iss-paper-175-national-victims-of-crime-survey-robyn-pharoah

Phetthong, N. and S. Kutnjak Ivković (2015). Police integrity in Thailand. In S. Kutnjak Ivković and M. Haberfeld (eds.), *Police Integrity across the World*. New York: Springer.

Plus 94 Research (2016). *Afrobarometer: South Africa Results*. Available at: www.ijr.org.za/home/wp-content/uploads/2012/07/SAF_R6_South-Africa-Round-Results-and-Analysis_20Mar17.pdf

Plus 94 Research (2018). *Afrobarometer: South Africa Results*. Available at: www.afrobarometer.org/publications/south-africa-summary-results-2018

Pollock, J. (2007). *Ethical Dilemmas and Decisions in Criminal Justice*, 5th ed. Belmont, CA: Thompson Higher Education.

Porter, L.E., T. Prenzler, and K. Hine (2015). Police integrity in Australia. In S. Kutnjak Ivković and M. Haberfeld (eds.), *Police Integrity across the World*. New York: Springer, pp. 67–96.

Punch, M. (2003). Rotten Orchards: "Pestilence", police misconduct and system failure. *Policing & Society*, 13(2): 171–196.

Rauch, J. (2000). Police reform and South Africa's transition. *Centre for the Study of Violence and Reconciliation*. Available at: www.csvr.org.za/docs/policing/policereformandsouth.pdf

Roberts, B. (2008). Between trust and skepticism: Public confidence in Institutions. *HSRC Review*, 6(1).

SABC News (2017). NPA drops charges against Phahlane. *SABC News*. Available at: www.sabcnews.com/sabcnews/npa-decides-against-pursuing-charges-against-phahlane/

Sain, R. (2017). Cop accused of "protecting" gangster lover fined. *IOLNews*. Available at: www.iol.co.za/news/crime-courts/cop-accused-of-protecting-gangster-lover-fined-8314269

SAPS (2019). SAPS Annual Report 2018/2019. Available at: https://www.saps.gov.za/about/stratframework/annual_report/2018_2019/saps_annualreport2018_2019v2.pdf

SAPS (2019). *Crime Statistics: Crime Situation in the Republic of South Africa, Twelve (12) Months (April to March 2018/19)*. Presentation to Portfolio Committee of Police, 11 August.

Skolnick, J. (2002). Corruption and the blue code of silence. *Police Practice and Research*, 3(1): 7–19.

Statistics South Africa (2011). *Victims of Crime Survey 2010/11*. Pretoria: Statistics South Africa.
Statistics South Africa (2018). *Victim of Crime Survey, 2017/18*. Statistics South Africa.
Steinberg, J. (2017). Policing during and after apartheid. In J. Beek, M. Gopfert, O. Owen, and J. Steinberg (eds.), *Police in Africa: The Street Level View*. London: Hurst & Company.
Syed, T. and D. Bruce (1998). Inside and outside the boundaries of police corruption. *African Security Review*, 7(2): 3–21. Available at: www.iss.co.za/Pubs/ASR/7No2/SayedAndBruce.html
Tihabye, G. (2017). We want cop to see the pain he caused. *IOL News*. Available at: www.iol.co.za/news/crime-courts/we-want-cop-to-see-the-pain-he-caused-7363316
The Times (2012). Editorial: If the police are no better than thugs, who will save us? *The Times*. Available at: http://citizenalertzablogspotcom-tango.blogspot.com/2012/09/if-police-are-no-better-than-thugs-who.html
TMG Digital. (2016a). Harsh sentence handed down to corrupt police officers. *Times Live*. Available at: www.timeslive.co.za/local/2016/05/17/Harsh-sentence-handed-down-to-corrupt-police-officer1
TMG Digital. (2016b). Wits students to meet on Friday, as Ipid probes claims police brutality. *Sowetan Live*. Available at: www.sowetanlive.co.za/news/2016/10/21/wits-students-to-meet-on-friday-as-ipid-probes-claims-police-brutality1
Trinkner, R., T.R. Tyler, and P.A. Golf (2016). Justice from within: The relations between a procedurally just organizational climate and police organizational efficiency, endorsement of democratic policing, and officer well-being. *Psychology, Public Policy, and Law*, 22(2): 158–172.
United Nations Office on Drugs and Crime (2019). *Global Study on Homicide 2019: Homicide Trends, Patterns and Criminal Justice Responses 2019*. Vienna, Austria: United Nations Office on Drugs and Crime. Available at: www.unodc.org/unodc/en/data-and-analysis/global-study-on-homicide.html
Vallmuur, B. (2015). Police integrity in Estonia. In S. Kutnjak Ivković and M. Haberfeld (eds.), *Police Integrity across the World*. New York: Springer, pp. 125–152.
Van Droogenbroeck, F., B. Spruyt, S. Kutnjak Ivković, and Mr. R. Haberfeld (2019). The effects of ethics training on police integrity. In S. Kutnjak Ivković and M.R. Haberfeld (eds.), *Exploring Police Integrity: Novel Approaches to Police Integrity Theory and Methodology*. New York: Springer.
Van Graan, J. and W.I. Ukpere (2012). The role of impact evaluation on service delivery within the public-sector organizations. *African Journal of Business Management*, 6(39): 10,445–10,457.
Vora, J.A. and E. Vora (2004). The effectiveness of South Africa's Truth and Reconciliation Commission: Perceptions of Xhosa, Afrikaner, and English South Africans. *Journal of Black Studies*, 34(3): 301–322.
Wiener, M. (2018). *Ministry of Crime: An Underworld Explored*. Johannesburg, South Africa: Pan Macmillan South Africa.
Wicks, J. (2017). State prosecutor, two cops arrested for accepting bribe. *Sowetan Live*. Available at: www.sowetanlive.co.za/news/2017/03/03/state-prosecutor-two-cops-arrested-for-accepting-bribe
Wiebelhaus-Brahm, E. (2010). *Truth Commissions and Transitional Societies: The Impact on Human Rights and Democracy*. New York: Routledge.
Wolf, R. (2017). Squatters claim police brutality. *IOL News*. Available at: www.iol.co.za/capetimes/squatters-claim-police-brutality-816611

Chapter 2

The origin and functions of the police in South Africa

From pre-colonialism to democracy

Introduction

In the previous chapter (Chapter 1), we argue that the theory of police integrity (Klockars and Kutnjak Ivković, 2004; Klockars et al., 2000, 2006) falls within the structural/institutional tradition and is based on Goldstein's (1975) organizational approach, with the basic propositions of the organizational theory of police integrity outlined in four theoretical dimensions. In Chapters 4, 5, and 6 we discuss findings regarding the integrity of the South African Police Service (SAPS) in the contexts of the first three dimensions of the theory of police integrity: Familiarity with organizational rules (Chapter 4), enforcement of those rules (Chapter 5), and the contours of the code of silence (Chapter 6). Here, and in the next chapter, we explore aspects of the fourth dimension of the theory of police integrity, namely, the social, economic, legal, and political environment in which a police organization operates. While this dimension is multifaceted and complex, placing straight-line and causal connections is beyond the scope of this work. We believe that contextualization of this environment both for greater understanding of the specific case of police integrity in South Africa and for the theory of police integrity in general is critical.

In accounting for the origins and functions of the police in South Africa from the pre-colonial period to the democratic elections in 1994, we distinguish between the concepts of police and policing. According to English-Ingles.com (2019), the etymology of both concepts originates from the Greek *politeuo*, meaning "live as a citizen." *Politeuo* evolved into *politeia*, meaning "life of a citizen," and, when adopted in Latin, became *politia* meaning "government" or "state," changing the meaning from the ancient Greek. In its broadest application, then, *policing* is "the power of a government to exercise control over persons and property within its jurisdiction in the interest of the general security, health, safety, morals, and welfare" (Merriam-Webster Online Dictionary, 2019) without reference to the character of the governing power. In the more restrictive meaning, *policing* refers to the inherent power of the state.

Police, in contrast, refer to institutions or individuals granted authority by a governing body to use coercive force in its territory. Definitions/descriptions adopted from two influential police scholars provide the foundation for our discussion. Bittner (1970; 1991, p. 48) described the role of the police as "best understood as a mechanism for the distribution of non-negotiable coercive force." Klockars (1985, p. 12) expanded on Bittner's writings, defining police as "institutions or individuals given the general right to use coercive force by the state within the state's domestic territory." Notable in both these definitions/descriptions is the centrality of "coercive force" in understanding the concept of the police in an institutional context. This right to use coercive force and the potential to abuse it are likely the most significant characteristics for understanding policing in the South Africa.

Defining these terms is important for two reasons. First, they identify forms of police and policing that do not necessarily fit within contemporary usage of the term police (e.g., private security; regulatory agencies). Second, the broader application of the concepts allows for an evaluation of earlier forms of policing in South Africa that bear influence on our discussion of contemporary policing (the SAPS) in the country and, more specifically, on police integrity in the SAPS.

The legitimate exercising of this authority to use coercion is at the center of general police legitimacy in the context of the social, economic, legal, and political environments in which a police organization operates. As Comaroff (2013) noted, "[W]hat precisely constitutes legitimate coercion in any real-life situation rarely goes uncontested" (p. xiii), and as Bittner (1991) stated, even the legitimate exercise of coercion by police organizations is antithetical to the very values that created this authority. In discussing the exercise of this right, Comaroff (2013) further notes:

> There are many parts of the world in which cops are regarded, ambivalently, as both a positive presence, a "prerequisite of social order," and a popular nightmare, a dark, often excessively violent force that undermines that very order, blurring the line between those who administer the law and those who evade, erode, and exploit it.
>
> (p. xii)

It is against this backdrop that we turn to our discussion of the origin and functions of the police in South Africa—from pre-colonialism to democracy.

Pre-colonial South Africa

The region now called South Africa has been home to modern humans for more than ten millennia. About 10,000 years ago, San (Bushmen) hunter-gatherers migrated to the area via contemporary Botswana, where they intermittently eked out an existence or enjoyed the country's abundant wildlife. They were later joined by Khoikhoi[1] pastoralists who had for a long time occupied land across Southern Africa, and around 300 AD, Bantu-speaking African farmers and herders joined the fray as they moved out of central and east Africa (Vansina, 1995; Nattrass, 2017). Around 1100 AD, some of the latter settled at Mapungubwe, just south of the present Zimbabwean border, to form what is believed to have been the region's first major state (Huffman, 2009). Archaeological findings suggest that Mapungubwe supported artisans, miners, and farmers, and is thought to have been connected to trade routes as far afield as South Asia (Steyn, 2007; Nattrass, 2017, Loc 656).

By the fifteenth century, Bantu farmers were likely the dominant group in the region, occupying as much of the land as their wet-climate crops, cattle, and iron tools allowed (Diamond 1997, Loc 6420), and, by the eighteenth century, they had forged numerous polities that were patriarchal, chief-led societies tied to fixed homesteads but without individual ownership (Nattrass, 2017, Loc 656).

Because of their mobile existence, the Khoikhoi's political and social structures were weak and their economy volatile (Elphick and Malherbe, 1989, p. 3). With wealth hierarchies in the region thought to have been characterized by the principle that the man with the most cattle could exercise control over those who provided tribute and labor (Hall, 2009, p. 118), the Khoikhoi's affluence was constantly at the mercy of external factors like theft, disease, and drought. In describing the consequences when these influences would render a Khoikhoi group impoverished, Elphick and Malherbe write (1989, p. 6):

> Individuals and groups who had suddenly lost their livestock had two options other than "going to war." They could hire themselves out as herders for wealthy Khoikhoi, who would pay them in animals suitable for rebuilding their herds. Undoubtedly this was their preferred response. Or, if the disaster had affected their neighbours as well, they could fall back on hunting and gathering. Even in times of prosperity the Khoikhoi continued to hunt and gather, and in times of distress they could rely on these as their sole means of subsistence.

These "ecological cycles," with pastoralists at one end and hunter-gatherer on the other, were entrenched in the existence of the Khoikhoi living along the south-west coast when Europeans set foot in the region (Elphick and Malherbe, 1989, p. 7). The newcomers' culture of commercial livestock farming exacerbated the frequency of these Khoikhoi cycles, and the resulting conflict between farmers and hunter-gatherers would necessitate law enforcement. Through circumstance, coincidence, and consequence, the era of prejudiced policing in South Africa had begun.

Early societal influences—the European arrivals (1488–1830s)

In 1488, Portuguese explorer Bartolomeu Dias landed in Mossel Bay after inadvertently sailing past the southernmost point of Africa, blown astray by an unusually strong wind. According to Portuguese accounts, on arrival, the crew offered conciliatory gifts to members of the local Khoisan group. Instead of accepting those gifts, the Khoisan fled from the strangers and later, while the Portuguese ship was replenishing its water supply, threw stones at the European sailors, who responded with crossbow fire (Couzens, 2013, p. 10). One Khoisan was left dead from this first documented European skirmish in South Africa, an ominous harbinger of the violent conflicts that would permeate the country's future.

Roughly 254 nautical miles southwest of this tragedy, the area now known as Cape Town would rapidly become a favored layover for the fatigued crews of mostly Dutch and British ships traveling to Asia. The Khoisan gradually grew accustomed to trading beef and mutton with the malnourished sailors in return for tobacco, copper, and iron, and, by 1600, "Table Bay" was a regular, albeit informal and disorderly, trade interlude for ships from several seafaring nations (Elphick and Malherbe, 1989, p. 8). When the Netherlands East India Company (*Verenigde Oost-Indiesche Compagnie* or VOC) eventually established a permanent refreshment station in Table Bay in 1652, "its goals were simply to regularise the benefits which sailors had long derived from the Cape stopover" (Elphick and Malherbe, 1989, p. 10).

A powerful trading company with ostensible business acumen, the VOC had little appetite for expensive conflicts in foreign countries, and its directors insisted on a policy of benign coexistence with the Khoisan. Alas, seamless cohabitation did not last long. Reports of VOC metal and livestock thefts increased. The Dutch suspected the Khoisan and reacted in 1655 by forming a paramilitary organization—the Dutch Watch (Coutsoukis, 2004)—which utilized watch houses and mounted patrols to lend to the colony "a new sense of security" (Elphick and Malherbe, 1989, p. 13). Portentously, as this rudimentary protection group would evolve to shoulder a public law and order function, the roots of South African policing were laid. For much of the subsequent 350 years, the protection and defense of European migrants and their descendants to the general detriment of indigenous populations would remain a characteristic of the South African law and order community. More immediately, the continuous presence of armed soldiers at the Cape trading post would strengthen the VOC's resolve not only to interfere directly with Khoisan activities and adjudicate disputes but also to use force to instill the company's will, centered on its need for steady and inexpensive trade supplies (Elphick and Malherbe, 1989, p. 14).

As was the case in the other territories where the VOC operated, the Cape birthed a culturally intricate society in which the individual's prosperity was largely determined by their social class. Already by 1658, VOC servants, free burghers (former VOC employees turned farmers), slaves, and "natives" formed distinct groups with different rights and obligations pertaining to domicile, land ownership, military service, and freedom of movement. Although they did not enslave local people, in 1657 the VOC began importing slaves to the

Cape from Madagascar, Indonesia, Sri Lanka, and elsewhere in Africa (Nattrass, 2017, Loc 827). Elphick and colleagues reflect (2007, p. 47):

> From the start the Cape was a slave society and not one in which slave labour was simply one of several forms of labour. A slave society is one in which all institutions—the labour market, the economy, the legal system, the family and the church—are permeated by slavery. The master-slave model served as the model for all other relationships.

A small number of slaves served as assistants to the VOC official tasked with prosecuting criminal cases in the Cape. They also carried out some basic policing functions. Clothed in distinctive uniforms with noblemen waistcoats, these prosecutor' assistants were in fact permitted to bear arms, and would "roam around the Cape day and night, enforcing rough justice on all—also Europeans who were guilty of the smallest infringement of Cape law" (Elphick et al., 2007, p. 54). Peculiarly, race may have determined less than class in the enforcement of law in these early years. To illustrate, during a heightened wildfire alert period in the Cape, the prosecutor's assistants were under strict instruction to "soundly" beat any person caught smoking a pipe in the street, "without distinction" (Elphick et al., 2007, p. 54). Increasingly, their duties were broadened to include other judicial functions and, by the turn of the century, accounts show them carrying out brutal court sentences. Indeed, judging by incidents recounted by Elphick and colleagues (2007), justice during this time seems measured in increments of the macabre. For instance, a 1724 court record shows a slave found guilty of murder and arson, with the police executing the sentence to "cut off his right hand (the murder weapon), after which he was to be 'half-strangled' and killed on a slow fire" (the sentence for arson) (Elphick et al., 2007, p. 54).

During the period between 1707 and 1717, the VOC implemented three agricultural decisions that would entrench white supremacy in the colony. First, all agriculture would be in the hands of free burghers, who were all European; second, slave labor would be preferred over white workers; and third, "acceptance of the loan farm system and unsystematic colonisation" would be established (Elphick et al., 2007, p. 60). When the burghers clashed with nearby Khoisan on whose land they intended to settle, the VOC encouraged the former to arm themselves and form a militia. Raids and conflict between the two groups would become common occurrences under the VOC's rule (Nattrass, 2017, Loc 819). Such was the nature and extent of the enforcement of the VOC's determinations that several decades later, as the British succeeded the Dutch as rulers of the colony in 1795, there had been no serious challenge to the assumption that landholding was a prerogative of Europeans and that coerced labor was the natural fate of everyone else (Elphick and Giliomee, 1989, p. 560).

Under British rule, the next century proved formative in shaping identities still found in contemporary South Africa. It was also the period in which the majority of those inhabiting what is now South Africa were reduced to a semi-proletariat amid rapid, white and black social expansion and the establishment of new economies (Landau, 2009, p. 92). Britain was the world's leading industrial power, and its focus was on broadening its influence in the modern world. To achieve this goal, it believed that African societies should serve the British economy, and racial hierarchy and white domination were subsequently strengthened under British control (Legassick and Ross, 2009, p. 314). Through an increase in slave labor, the production of wheat and wine in the Cape, for example, would gradually triple during the period between 1790 and 1820 (Elphick et al., 2007, p. 80).

By 1760, the region north and east of the colony was likely inhabited by hundreds of decentralized chiefdoms. By the 1830s, through conflict and conquest, three groups had risen to dominance—the Zulu, Gaza, and Ndebele (Wright, 2009, p. 249)—with kingdoms comprised of the many smaller chiefdoms absorbed into these groups (Wright, 2009, p. 229). Under British rule, 4,000 of its citizens were settled almost a thousand kilometers east of Cape Town, forming a buffer on the eastern frontier against the Xhosa and nearby groups upon whose land the settlers had encroached (Trapido, 2011, p. 69). From 1779 to 1836, nine separate wars were waged against the Xhosa on this border, involving both civilian commandos and British troops (Ross, 2009, p. 209).

In pursuing its imperial goals, societal unification was not a British priority, and, by 1814, "the transitional governments had simply reaffirmed the essentials of Cape social structure as it had existed prior to 1795" (Peires, 1989, p. 472), with inhabitants classed into groups with different rights and freedoms. Even the Dutch Reformed Church was considered as established and left to continue in its role.

The British gradually implemented their own laws over the Dutch law of the VOC. They also introduced measures to separate the powers of state, including an independent judiciary and a system of case law (Nattrass, 2017, Loc 1033). During the 1830s, they introduced the principle of equality before the law in an attempt to build a "post-slavery society in which the key point of coercion lay in the law rather than in the personal authority of the master" (Legassick and Ross, 2009, p. 315). By then, however, policing and justice functions in the colony staunchly maintained class distinctions within the Cape, preventing any radical policy change. Not surprisingly, as the Cape society appeared by the 1830s to shy away from the European tendency in rural areas to identify themselves by race and stigmatize other races, official records of the executive branch at the time point to an entrenched policy of segregation. Reflecting on a study of court sentences during 1844, for example, Elphick and Giliomee (1989, p. 557) observed that "while Europeans comprised 45% of those convicted in the police court, they received only 17% of corporal punishments. Moreover, black and European prisoners were separated in gaol." John Montagu, the colonial secretary at the Cape for the period from 1843 to 1852, expressed concern that "colour had ceased to be the badge of civil disabilities," but that it "still forms a bar to social intercourse and intimate relationships far more formidable than any rising either from diversity of origin, language or religion" (Elphick and Giliomee, 1989, p. 558). When the governor of the Cape (1850 to 1853) called for the "extermination" of the Xhosa and other indigenous people in order to fight wars on the colony's eastern border (Legassick and Ross, 2009, p. 315), the British in 1854 granted the Cape its own constitution and parliament, with all males owning land to the value of £25 or earning a salary of £50 per year, eligible to vote, regardless of race (Nattrass, 2017, Loc 1054). Although more liberal than any other such constitution at the time, the qualifying criteria excluded most indigenous people. Dominated by the interests of the wealthy, the new parliament generally sided with white interests, while also producing early divisions between groups that progressively viewed themselves as "coloreds" and "Africans" (Legassick and Ross, 2009, p. 318).

Policing in the colony would remain a paramilitary function, and, as new districts were established and border wars stretched through much of the century, rural police assemblages gradually assumed military functions on the colony frontier. In fact, Brogden (1989) suggested that the degree of military influence on early policing in South Africa was unique in the colonial experience. Nevertheless, in urban centers, a system of policing modeled on Sir Robert Peel's London Metropolitan Police was slowly introduced (Nasson, 1991, p. 238). By 1825, the Dutch Watch was replaced with the Cape Constabulary, which would become the Cape Town Police Force in 1840. To the northeast, where a second British settlement was growing, the Durban Police Force was established in 1846, morphing into the Natal Mounted Police by 1861 (Coutsoukis, 2004).

Division, secession, and border expansion (1830s–1898)

During the mid-1830s, struggling Afrikaner farmers (Boere), by then of a mixed European descent, left the Cape to "plant new societies in the interior of southern Africa," in an attempt to "restore the traditional social order of the Cape" (Elphick and Giliomee, 1989, p. 560). Rapidly forming a "Voortrekker," or pioneer identity on the outskirts of the colony, they sought permission to establish two independent republics northeast of the Cape. The British consented and, following an inland migration and defeat of the Ndebele, the Boer republics of the Oranje Vrystaat (Orange Free State) and the Zuid-Afrikaansche Republiek (ZAR) were established in 1837 and 1844, respectively, with the short-lived republic of Natalia proclaimed in 1839 in the aftermath of a bloody conflict with the Zulu (Etherington et al., 2009, p. 336). Laws governing Boer territories included the banning of "non-European" residential settlements close to white towns. Indigenous people living in the territory were either forced off their land or were required to work it in return for tenancy (Etherington et al., 2009, p. 339). The secession of the Voortrekkers thus not only expanded modern agriculture in the country, but also "ensured the maintenance for several more generations of the European-dominated social order which had solidified at the Cape in the eighteenth century" (Elphick and Giliomee, 1989, p. 561).

In 1843, Britain annexed the Voortrekker Natalia and established the Natal colony nearly 2,000 kilometers east of Cape Town. The already strained relationship between the Voortrekkers and the British worsened with an 1848 British push for sovereignty over settled land north of the Orange River, leading to direct conflicts. As a result, Britain recognized the Transvaal Republic as an independent Afrikaner area in 1852 and granted independence to the Orange Free State in 1854. However, Britain would continue to view these states as impediments toward its economic growth and would wait for an opportunity to add the Afrikaner republics to the empire (Manson et al., 2007, p. 164).

In 1867, in one of these areas to which the British cape colony had no claim, a child picked up a diamond, and within three years more than 10,000 diggers were busy in the region, desperately burrowing for the evasive stones. As a result of the influx of people and demand for labor, a pass system was introduced in 1872, requiring that black men seeking work on the mines be registered with a labor bureau, which authorized them to work only in certain areas. Those not registered could be "deported" back to their rural homes. Hut taxes were also imposed in rural areas to force their residents into the cash economy, and so compel men to sell their labor, first digging for diamonds, and later in search of gold (Smalberger, 1976; Etherington et al., 2009; Marks, 2011a). The pass system would ultimately have a long, damaging effect on the South African society, with pass enforcement a notorious feature of policing during apartheid.

Ostensibly the driving force behind a land claim registered by a Griqua chief, the British appointed one of their own lieutenant-governors to mediate the dispute, resulting in an award to the chief and the almost immediate annexation of the land by the British. Having served their purpose of initializing the claim, the Griquas would see no benefit from their land appropriation, and, when they attempted an insurrection to reclaim the land from the British, their defeat saw "hundreds of prisoners . . . marched off . . . and forced to work as labourers on the mines that had been stolen from them" (Manson et al., 2007, p. 159).

The discovery of diamonds shifted the strategic value of the country for Britain—instead of the coastline, mineral riches had become its primary concern. With this focus shift, Britain set about abolishing the Afrikaner republics and Zulu kingdoms. Both these groups, for decidedly different reasons, would have nothing to do with Britain's ideas, only making the empire more resolute to control the mineral resources. With a British official boldly describing the

Zulu kingdom: "[H]ad . . . [the Zulu's] . . . thirty thousand warriors been in time changed to labourers working for wages, Zululand would have been a peaceful prosperous country instead of what it now is, a source of perpetual danger to itself and its neighbours" (Manson et al., 2007, p. 164), it became evident that the Zulu-controlled areas would be the first to go. The conflicting agendas of the Afrikaners, British, and Zulus beckoned bloodshed that would see the British victorious in the 1879 Anglo-Zulu War but defeated in the First Anglo-Boer War (Manson et al., 2007, p. 164; Etherington et al., 2009, p. 384). Even though the latter conflict was short-lived (from December 1880 to March 1881), Britain encountered such fierce resistance and subsequent losses that it withdrew its forces and grudgingly granted the Republics the right to self-rule. At this time, the ZAR established the South African Republic Police (Zuid-Afrikaansche Republiek Politie or ZARP), which, in step with the rural and frontier nature of the republic, would embrace strong paramilitary characteristics, carrying out both law enforcement and civil order functions.

Half a decade later, in 1886, the discovery of a major gold deposit in the ZAR would see Britain reevaluate its armed conflict policy in terms of the independent republics. When the ZAR revealed its vision of using the new mineral wealth to modernize the republic and become financially free to pursue its own destiny (Callinicos et al., 2007, p. 206), Britain sensed the threat to its exclusive power in the region. With the ZAR further strengthening its political and commercial bonds with Germany—one of Britain's global power rivals—a domino-effect was set in motion that would culminate in the Second Anglo-Boer War. This war was also named, due to the participation in the conflict of several South African cultures, the "Great South African War" of 1899 to 1902. Briefly assessed in this chapter, this conflict would remain a "powerful event in shaping the history of twentieth-century South Africa" because it resulted in "the consolidation of a single powerful South African state, politically dominated by white people and economically dependent on the controlled supply of cheap black labour" (Manson et al., 2007, p. 164; Callinicos et al., 2007, p. 206). In fact, the consequences of this war would be so tumultuous and pervasive that Farwell (2009, Preface) reflects:

> South Africa is a land in whose history bizarre, improbable events have occurred with astonishing frequency, but no event has left a more lasting imprint on the minds and hearts of her people than the Second Anglo-Boer War, and no one can hope to understand the country or those who today rule it without an understanding of this conflict.

The Great South African war (1899–1902)

As Britain encroached on the republics in 1896, Paul Kruger, the president of the ZAR, presciently warned that conquering the Boer republics would be a tough endeavor. Still, the war would be unexpectedly costly for Britain. In fact, it would be the biggest and most expensive conflict ever fought by the British during the entire century preceding 1914. From 1899 to 1902, Britain and the Boer republics would be at war to determine who would have "supreme authority over the country's minerals, productive lands and dispossessed black labour" (Callinicos et al., 2007, p. 217). Throughout the war, Boer commandos were devastatingly effective in their guerrilla attacks on British lines but would nevertheless surrender in May 1902 (Marks, 2011b, p. 159). Peculiarly, the full republican force never engaged Britain's troops, as roughly 40% of fighting-fit commandos were withheld to protect farmsteads from possible attacks by surrounding chiefdoms. In fact, this "atmosphere of fear and panic at spreading black hostilities against republican civilians and soldiers" would become one of the reasons given by the republic leadership for its surrender (Callinicos et al., 2007, p. 217). A more obvious reason, however, was the immoral war tactics employed by the British, which included summary public executions, farm burnings, and the use of concentration camps where the horrendous conditions of captivity caused the deaths of approximately 28,000 Boer women and children (Callinicos et al., 2007, pp. 220–222) and probably a similar number of black and colored people, imprisoned in similar, though separate, camps (van Heyningen, 2008; Nattrass 2017, Loc 2589). In the aftermath of the war, the casualties of this conflict would extend across South African race groups, with many black soldiers also fighting and perishing on both sides of the conflict (Callinicos et al., 2007, p. 223).

Internationally, the anti-colonial struggle of the Boer republics garnered considerable sympathy from Britain's global rivals—the Netherlands, Germany, France, and Russia—who reviled British imperialism. Nationally, Britain lost its appetite for further expansion because of the expensive price tag of the war. While subsequent immigration to South Africa slowed and British imperialism waned, Afrikaner nationalism waxed in a void of general global empathy (Manson et al., 2007, p. 185). The deep psychological wounds left by the war on the Afrikaner people and their postwar bitterness at the unacknowledged atrocities committed by the British would ultimately shape the political landscape of the country. In no small part, the war would contribute the "ideological tools" of recollecting and documenting the Afrikaners' suffering and agony which would find eventual form in the "future rebuilding of an Afrikaner nationalist identity and republican political culture" (Callinicos et al., 2007, p. 223). With British war propaganda stereotyping the Boers as "a lower form of humanity" (Callinicos et al., 2007, p. 217), relations between Afrikaans- and English-speaking whites deteriorated rapidly during and after the war to the extent that they not only viewed themselves as increasingly different from one another (Marks, 2011a, p. 148), but even considered themselves belonging to different races (Brits et al., 2007a, p. 230). Of course, they did not consider themselves of the same race as black Africans, and so the postwar environment also promoted black nationalism. While the British committed financial support to the Boer republics to aid their recovery, nothing of the sort was provided to the black majority, who were partly united by this omission (Nattrass, 2017, Loc 2935).

Nationalist politics, organized labor, and the SAP (1904–1920s)

Driven by the republican political culture and nationalist sentiments, the leading white politicians soon set about officially discussing the terms of unification in the new colonies, and, by 1909, without any input from black representatives (blacks outnumbered whites by five to one at the time), a draft constitution had been submitted to the parliaments of the Cape, Natal, Orange River, and Transvaal colonies (Marks, 2011b; Nattrass, 2017). Article 137 of this document would magnanimously specify the official language of the Union as both English and Dutch, the result of a former president's plea to the drafters to purge the country of "the devil of race hatred." With Boers and British at the time referred to as different races, this measure somewhat allayed their underlying loathing, while wholly creating a new exclusion of black citizens and their languages (Brits et al., 2007a, p. 230).

The Union of South Africa was officially declared in 1910, merging the four colonies into one, and marking political independence from Britain. The South African Native National Congress (SANNC) was established in 1912, calling for equal opportunities and participation in government for blacks, demands that were consistently denied (Nattrass, 2017, Loc 2974). By 1923, the organization would be renamed the African National Congress (ANC).

In 1913, the South African Police force (SAP) was established out of the structures already carrying out policing functions in the four former colonies, with many of those already employing both black and white officers (Shear, 1998). This police force would officially assume responsibility for domestic law and order in 1926, while still maintaining its close relationship with the military. As such, throughout its early days, SAP officers often called on the army for support during emergency situations and during World War II, a SAP brigade even served with the Second Infantry Division of the South African Army in North Africa (Coutsoukis, 2004). The infamous Natives Land Act was passed in 1913, effectively making it illegal for blacks to buy land other than within scheduled "reserves" and totaling less than eight percent of the Union's landmass (Freund, 2011, p. 237). While the Union adopted a policy of restricting blacks to reserves ruled by chiefs and traditional customs (Freund, 2011, p. 213) and the exclusion of non-whites from full citizenship (Dubow, 2011, p. 34), the SAP was tasked with the enforcement of various discriminatory laws. Nevertheless, according to Nasson, policing in Cape Town at least remained fairly liberal in the first few decades of the SAP's existence. In his account, when blacks started moving into white areas in the 1920s, the police generally let them be or responded with leniency (1991, p. 245). With the labor market favoring white employees, the SAP struggled from its inception to attract the number of whites it required. As such, black officers were recruited from rural areas where the state believed they would be less likely to have been politicized. Between 1913 and 1939, between 30% and 50% of officers were black (Shear, 1998, p. 287). For much of the twentieth century, black police were armed with spears and knobkerries (firearms were provided to white officers), were paid less than whites, and held no authority over white colleagues or civilians (Hornberger, 2011, p. 32).

The Union would soon find itself on shaky ground economically, and its economic situation progressively worsened after World War I. Between 1920 and 1932, the gross domestic product decreased, while industrial production showed little growth (Brits et al., 2007b, p. 245). With Union constituents becoming rapidly unemployed and destitute, and unbridled inflation increasing prices up to 50% between 1917 and 1920, the security and stability of the Union would face serious challenges from organized labor, in particular those active within the mining industry. With the older, more conformist mining workforce returning to their European countries to participate in World War I, Afrikaners were the dominant white workforce in the sector by 1918.

When the SAP was founded, a sizable portion of whites—particularly Afrikaners—remained poor, and the government steadily skewed policies in their favor to grow white wealth. For example, semiskilled, white gold miners earned five times more than their black peers did, but still only one-half of their skilled white counterparts' earnings (Nattrass, 2017, Loc 3130). When mine owners announced that they wanted to do away with the color bar and use black miners more effectively, white miners rebelled (Freund, 2011, p. 225). At the time, however, these white miners were lacking any clear rights to strike or form labor unions, limiting their ability to stand up to mine owners (Davies, 1976). Subsequently, in 1913, they went on strike in the Johannesburg area to demand the recognition of their trade union. As the protest intensified, it soon became clear that neither the SAP nor the military could maintain order in the region. A portent of future labor strikes within the mining sector, the protest ended in bloodshed with 100 strikers and onlookers shot and killed by the armed forces (Brits et al., 2007b, p. 245). The new government adapted quickly, and when a countrywide strike was subsequently threatened, most of the English-speaking trade union leaders were arrested, with some illegally deported. To strengthen its security forces, the government called on rural Afrikaners to perform Active Citizen Force duties. Still smarting from the British atrocities during the Anglo-Boer War, most conscripts "enthusiastically took up their weapons 'to shoot Englishmen' in Johannesburg" (Brits et al., 2007b, p. 246). By 1922, the government's liberal use of force to quell dissent in general, and trade union protests in particular, led to the bloodiest-ever suppression of a white miner strike in the Witwatersrand (the Rand). Following three days of increasing civil unrest, martial law was declared and a seven-day battle ensued involving the SAP, infantry, air strikes, artillery, and tanks. When the smoke cleared, 81 civilians, 29 policemen, and 43 soldiers had been killed, with more than 650 injured. In due course, 853 people were charged with various crimes, and of the 46 prosecuted for murder, four were hanged. With this violent settlement of the mining wage dispute in favor of mine owners, the resulting public blowback would cost the sitting president his position in the 1924 election (Brits et al., 2007b, p. 247).

In 1923, the Native Urban Areas Act was passed. Building on the existing pass system, the Act paved the way for racial segregation in urban areas. As with employment on the mines, blacks were permitted to enter urban areas only to work for whites, but could not permanently reside, nor own land, in such areas. Instead, they were housed in dormitory towns, also known as locations and townships, outside of the white centers. As relations between English and Afrikaans-speaking whites improved marginally—aided in part by laws intended to uplift whites in general, restrictions on blacks were increasingly formalized (Bonner, 2011).

Founded in 1921, the South African Communist Party (SACP) would be another political consequence of this period that would reverberate over nearly a century into present day South Africa. Less than a year after its launch, the SACP's banner reading, "Workers of the world fight and unite for a white South Africa" was displayed at the Rand strike, and consensus held that the party was headed by "a militant Council of Action consisting of Afrikaner nationalists and 'Marxist socialists.'" By 1924, however, the SACP had become disillusioned with politics and had reversed its racial makeup and rhetoric to such an extent that, within two years, more than 80% of its members were black. This brought the party into contact with the ANC, which it considered a movement with reactionary potential, despite the latter's poor fiscal state and its relative disorganization at the time (Brits et al., 2007b, p. 249).

While co-attending a 1928 communist-sponsored conference in Belgium, the secretary of the SACP's Cape Town branch and the ANC president tabled "a motion endorsing the 'right of self-determination through the complete overthrow of capitalism and imperialist domination—the principle of Africa for the Africans.'" After a visit to Moscow at the invitation of the Communist International, the SACP adopted a new policy, suggesting the realization of socialism in South Africa through a two-phased approach. First, "a nationalist democratic revolution in which communities would forge an alliance with 'reformist' blacks to overcome white domination" and following this—"a socialist revolution" (Brits et al., 2007b, p. 249). Although very little came of this policy in the years immediately following its inception, the Soviet Union had taken an interest in South Africa. As the country's strategic significance grew in Moscow's global struggle against capitalism, the Soviet Union would steadily augment its various forms of support to both the SACP and ANC in order to create a communist ally in South Africa (Jeffery, 2019, Loc 234).

World War II, a nationalist government and its apartheid policy, political opposition, repressive legislation, and social order policing (1930s–1950s)

Although the 1934 Statute of Westminster ensured South Africa's autonomy and, thus, the country's neutrality in the event that Britain joined a looming European war, the South African leadership was deeply divided on the issue of joining the Allied war effort. After all, there were two white communities competing for political dominance within the ruling minority, and the general consensus within the political sphere remained that previous attempts at unity had failed. During the parliamentary debate concerning joining the war effort, those against warned that if another attempt to unify the English and Afrikaner communities failed "as a result of a war with which we have nothing to do it will affect our national life for fifty or even a hundred years" (Grundlingh and Giliomee, 2007, p. 294). It was also stressed that, if South Africa has to fight in every war Britain entered, "talk as much you wish about freedom, but we are a country of slaves" (Grundlingh and Giliomee, 2007, p. 294). When the anti-war motion was narrowly defeated, "largely along the ethnic division in the white community," and the request for a general election on the matter was refused, Afrikaner nationalism was further fueled (Grundlingh and Giliomee, 2007, p. 294). Within political circles, it was long believed that the English community was stronger than the Afrikaner community and that "an authentic republican party would be the only means of reviving a national spirit [among Afrikaners]" (Elphick et al., 2007, p. 279).

During the war, the economy of South Africa grew steadily, creating employment opportunities for women and black workers alike. Rapid urbanization resulted, and, by the end of the war, 30% of the population was living in cities across the country (Freund, 2011, p. 218). The growing Afrikaner nationalist movement sought not only to reverse these societal changes, but already, halfway through the war, introduced both in the media and parliament the concept of "apartheid"—that would give "the various races the opportunity of uplifting themselves on the basis of what is their own" (Lodge et al., 2007a, p. 308). Coined in 1929 by the Dutch Reformed Church, the term apartheid now had policy potential that resonated philosophically with those opposing the war and, in real world terms, with farmers concerned about losing black labor to the growing cities (Brits et al., 2007b, p. 259; Lodge et al., 2007a, p. 310).

By the end of the war, 190,000 white, 76,000 black, and 45,000 colored soldiers returned home to their fundamentally unchanged societal hierarchies. Political views had either remained intact or had hardened, and a strong opportunity for the country to leverage the war effort's ideals of freedom and democracy was lost at a critical time of development in its history (Grundlingh and Giliomee, 2007, p. 294). In 1947, the Fagan Commission recommended that blacks be allowed to live more freely in towns and cities, which concerned many poorer whites. The pro-Afrikaans National Party (NP) seized the political opportunity to sow fear about a "black danger," and promised to protect whites if the party would be elected to rule the country (Nattrass, 2017, Loc 3444).

It was against this backdrop that the NP came to power in a whites-only election in 1948. The party would operate on two fronts. First, its ideology of apartheid would curb the urbanization of blacks and base the economy on migrant labor from the reserves. Second, the party hoped to expand its supporter base by promising South African sovereignty and "the cultural and economic advancement of the Afrikaner people" (Grundlingh and Giliomee, 2007, p. 307). Acknowledging the crucial role of supporting legislation to its operational strategies, the party rapidly promulgated the majority of apartheid laws between 1950 and 1953. In reaction to these laws, the ANC would, at this time, commit itself to overcome apartheid through mass action demonstrations (Feit, 1972; Mandela, 1994).

Concerning its first operational front, the NP set about engineering a state in which residence, migration, employment, education, health care, and marriage were delineated through four categories of race, determined by the Population Registration Act (1950) as white, black, Asian/Indian, and colored (Posel, 2011, p. 321). Using these labels, apartheid effectively separated every facet of daily South African life along color lines.

The Group Areas Act (1950) and the Prevention of Illegal Squatting Act (1950) prohibited residence in urban areas outside of those designated to a person's race group. With many blacks born or settled in cities allowed to remain (albeit at times forcibly relocated to new, designated areas), efforts were made to prevent further black migration into urban municipalities, chiefly by encouraging industries to employ workers from only the established residential communities. Under the auspices of "ethnic self-determination," the Bantu Authorities Act (1951) sought to create "Bantustans," or homelands, with each intended to be—in conjunction with legislation like the Bantu Education Act (1953)—ethnically homogenous and self-governed. Slightly bigger in collective size than the reserves of the 1913 Natives Land Act, 13% of the country's landmass was allocated to the homelands, while the rest of the country was designated to whites, coloreds, and Indians, who, in due course, would also be allocated specific residential areas (Posel, 2011, pp. 337, 342). Ethnic self-determination meant in reality that, when a homeland accepted its independence, its citizens would lose their South African citizenship, leaving blacks with "no claim to political rights outside their respective homelands" (Jeffery, 2019, Loc 1163).

Incongruous with the concepts of urbanization and subsequent industrial and economic growth, this system of population control—much of which was left to the SAP—would not only prove near impossible to maintain in the long run, but would also leave other inadvertent social disasters in its wake. For instance, the incessant stopping, searching, and coercing of black men by police could likely have led to feelings of belittlement and emasculation. Some of these men could have attempted to reclaim their masculinity through violent contests with other men in the mine compounds, township taverns, or even through violence against women and children as they returned to their rural homes. In such circumstances, state violence could have unwittingly promoted interpersonal violence.

Between 1950 and 1960, the urban black population grew by 50% in comparison to the overall black population's growth rate of 27% during the same period (Lee, 2011, p. 229). The ensuing housing shortage lead to many makeshift, informal, and unserviceable shacks being erected on urban fringes, mostly incompatible with municipal rules and regulations. An absence of opportunities for young men saw many forming gangs and engaging in petty crime—actions made simpler by the scarcity of police in these townships (Glaser, 1998).

Most migrants considered themselves temporary city visitors who planned to return to their homes in the countryside (Mager and Mulaudzi, 2011, p. 370). In what would become notorious, tangible symbols of the apartheid ideology, legislation was passed in 1952 requiring blacks 16 and older to carry at all times their pass books that proved their authorization to reside and work in urban areas.

During this time, the SAP was "the most important institution for upholding the apartheid social order, enforcing the influx control regulations along with more general political repression. Fighting 'communism and terrorism' were the overriding priorities and police ruthlessly put down protests and demonstrations" (Marks, 2005, p. 36). In enforcing other infamous legislation, the SAP would also become adept in clandestine investigations and techniques ranging in degrees of legality. For example, in their research of police enforcement of the Immorality Amendment Act (1950) and the Prohibition of Mixed Marriages Act (1949), which sought to prevent sex and marriage between the delineated race groups, Lodge and colleagues (2007a, p. 316) delivers this disturbing account:

> [T]he police used binoculars, tape recorders and cameras, burst into bedrooms, and instructed district surgeons to examine the genitals of suspects. Those caught included not only ordinary people but also church ministers, schoolteachers, and even a private secretary to a prime minister. By 1985 when the law was repealed, some 11,500 people had been convicted and many more had been charged. Some white offenders committed suicide or emigrated.

Although black SAP officers were tasked with enforcing apartheid's laws, Kynoch suggests they remained respected within black communities until the time of the 1976 protest actions (Kynoch, 2003). With apartheid laws restricting black aspirations, government positions in the police, nursing, and teaching professions were considered some of the best jobs available to black workers, and as such black police were viewed as symbols of progress.

On its second operational front, the NP progressed rapidly in removing the societal symbols of British dominance. In the aftermath of ending British citizenship and removing *God save the Queen* as one of the national anthems and the Union Jack as a national flag, Afrikaner nationalist supporters felt triumphant (Lodge et al., 2007a, p. 310).

When the ANC launched its first "defiance campaign" in 1952 in response to the newly promulgated legislation, the SAP arrested some 8,500 people "involved in resistance politics." Then, the government wrote more draconian laws. The Criminal Law Amendment Act (1953) forbade anyone from committing an offense "by way of protest, or in support of any campaign against the law," which also extended to the use of words intended to move anyone to commit an offense as a means of protest. In the same year, the Public Safety Act sanctioned the Governor General to declare a state of emergency whenever he perceived the public order to be under "serious threat." During a state of emergency, the SAP had the authority to summarily arrest and detain people, with the singular obligation to inform parliament within 30 days of the identity of the arrested. Following a 1955 revival of the ill-fated defiance campaign, protests against the pass laws and other dissents flared up, prompting the following year's enactment of the Riotous Assembly Act (1956). This allowed the Minister of Justice to act against anyone he suspected of "promoting feelings of hostility between whites and non-whites" (Karis and Carter, cited in Marks, 2005, p. 34), while gatherings of more than ten black people were allowed only with prior permission. To curb rolling protest actions, this law decentralized authoritative powers. Both local magistrates and the Minister could prohibit a public gathering whenever it was perceived as a "serious threat to public peace." The Minister naturally wielded more absolute and invasive authority and could even prohibit a specific person from attending a public gathering. Where official forbiddances failed, the SAP was given authority to disperse any prohibited gathering by force (Marks, 2005, pp. 34, 37, 38). In comparing the policing of protests during this time as an approach similar to that of warfare, Marks points out that "[p]rotests were regarded as having been organized by communist agitators who had to be silenced in the most 'effective' way possible" and that these practices and tactics would evolve through the 1960s to "set the stage for public order policing for the next two decades" (2005, p. 35).

Apartheid in bloom—the ANC withers, the SAP evolves (1960s)

Despite South Africa remaining the last major Western country with a racial policy rooted in white supremacy, the West neither abandoned nor turned on the NP government following the introduction of apartheid. Perceived as indispensable in the battle with the Soviet Union as a counterbalance to the spread of communism on the continent, South Africa was protected to such an extent from Western economic and diplomatic sanctions that it even benefited from the West's strong economic growth—stimulated by Europe's postwar reconstruction. Not only did it achieve one of the highest growth rates globally during the 1960s (Grundlingh and Giliomee, 2007, p. 307), but its economy grew tenfold in real terms between 1930 and 1970 (Nattrass and Seekings, 2011, p. 518).

With India gaining independence from Britain in 1947, a global push for African and Asian decolonization was set in motion, and the Soviet Union jumped at the opportunity to fill the vacuum created by the West's departure from these countries, both in ideology and leadership. To speed up this process in South Africa, the head of the Soviet Union's International Department would, for several decades, cultivate close relationships with the SACP and the ANC, while gradually dictating the policies, strategies, and actions of the two organizations (Jeffery, 2019, Loc 942). During 1954, the ANC began a process of collecting "demands" from across the country to include in its Freedom Charter. Drafted at the last minute by a prominent SACP member, Lionel Bernstein, the Charter was approved in 1955 by some 3,000 delegates gathered on the outskirts of what would become the southwestern townships (Soweto) of Johannesburg. It contained several basic human rights (Lodge et al., 2007a, p. 329) as indictments to the NP's apartheid iniquities. It would also call for the nationalization of mines, banks, and "monopoly industry" as well as the redistribution of land "among those who work it" (Lodge et al., 2007a, pp. 329–330; Jeffery, 2019, Loc 954).

The drafting of the Freedom Charter caused government alarm. In December 1956, the SAP's elite Security Branch swooped on premises around the country and arrested 140 people, including Nelson Mandela, under the Suppression of Communism Act. In 1958, 91 of the arrestees—black, white, colored, and Indian—were tried for treason, but prosecutors failed to prove their case, and in 1961 the accused were found not guilty (Mandela, 1994).

As the "treason trial" ground on, on March 21, 1960, a sizable crowd of protesters gathered at a police station in Sharpeville, a township on the Rand. The group, inspired by Robert Sobukwe, who left the ANC to form the Pan African Congress (PAC), was demonstrating against the restrictive pass laws. What would have been a "routine" protest was complicated by a recent event in Durban, [Kwa-Zulu] Natal province, where protesters had murdered nine policemen. At some stage during the demonstration, the police panicked and opened fire, killing 69 protesters, including women and children. The public outcry was immediate and sufficiently extensive that the international community criticized the severe methods of the apartheid state. The ANC called for a countrywide workplace boycott, resulting in a near standstill of industries, stock market losses, and the outflow of much-needed capital. Government action would be swift. By March 30, a state of emergency was declared, and, by April 8, the ANC and the PAC were banned. Within a few weeks, some 18,000 people were arrested and detained (Lodge et al., 2007a, p. 335).

Although it had followed a path of nonviolence, in November 1961 the ANC's younger leadership formed its military wing—Umkhonto weSizwe (MK). Contemplated by these leaders since 1953, the decision to implement an armed struggle would be, according to Jeffery (2019, Loc 227), "largely prompted by the SACP and encouraged by the Soviet Union." By December, MK detonated homemade bombs, commencing with its short-lived sabotage campaign. As MK's commander-in-chief, Nelson Mandela had helped plan this first phase of the operation, intended to be followed with guerrilla warfare. By late December, he left South Africa as the ANC's delegate to a conference in Addis Ababa, Ethiopia.[2] During his subsequent interactions with African officials, Mandela learned that the ANC's collaboration with white and Indian communists in the SACP was deeply troubling to leaders in the region (Mandela, 1994, Loc 4924). Lesson learned, Mandela would on his return to South Africa project an image of ANC independence and "emphasise its dominant role in any alliances it formed with other organisations."

Acting on information provided by America's Central Intelligence Agency (CIA), the SAP arrested Mandela on August 5, 1962, on charges of incitement and having left the country illegally (Lodge et al., 2007a, p. 337). In July 1963, with Mandela still imprisoned, other prominent ANC and SACP members—some involved with the earlier sabotage campaign—were arrested on a farm in Rivonia, north of Johannesburg. During the raid, police seized hundreds of documents, exposing a plan for an armed insurrection in South Africa titled "Operation Mayibuye"[3] (Mandela, 1994, Loc 5813). Charged and convicted in 1964 of sabotage and attempted insurrection, several ANC leaders, including Mandela, would receive life sentences and remain incarcerated on Robben Island until much later in South Africa's history (Lodge et al., 2007a, p. 337). Many remaining ANC leaders went underground or into exile. Banned and leaderless, the ANC struggled to spread its message and recruit more MK members. Despite the Soviet Union's supply of money, weaponry, strategic advice, and training, the organization would soon go into a near-revolutionary hibernation (Jeffery, 2019, Loc 239, 957).

The Faustian bargain the ANC had struck with the Soviet Union decades before did, however, bear some fruit for the wilting organization during this time. In fact, the ANC's international benefactor would prove much more effective in its "Moscow-directed campaign to isolate the NP government and have the ANC acknowledged as the 'sole authentic' representative of black South Africans." The United Nations' (UN) intergovernmental bodies, like the General Assembly, would prove essential in this Soviet strategy. From 1962, this body would frequently pass resolutions "almost unanimously asking member states to impose diplomatic sanctions and to bar South African ships and planes from their territories," while the UN Security Council approved an arms embargo on the country (Lodge et al., 2007a, p. 340).

An earlier opportunity at racial integration in South Africa brought the UN very close to substantively mitigating the evolution of apartheid. A visit during 1961 by the secretary-general of the UN, Dag Hammarskjöld saw him discussing the idea of apartheid as a "competitive alternative" to racial integration with the South African prime minister and staunch apartheid proponent, Hendrik Frensch Verwoerd. According to historians, a unique opportunity presented itself at this time "to develop a plan that could gather sufficient international support for the 'decolonization' of South Africa." Disastrously, Hammarskjöld would be killed some months later in a plane accident in what is today Zambia, silencing further conciliatory talks and leaving the South African government without an advocate of significant international stature (Lodge et al., 2007a, p. 340). During this year, Verwoerd would sever ties with the Commonwealth of Nations, thereby declaring South Africa a truly independent Republic. Inflexible in the opinion that the abandoning of apartheid would see minority race groups expelled or "absorbed" while ensuing experiments with socialism or state capitalism would destroy the economy, Verwoerd was stabbed to death in the House of Assembly prior to making a policy speech, leaving behind uncertainty in terms of the administration of the apartheid ideology (Lodge et al., 2007a, p. 341).

Appointed by Verwoerd as Minister of Justice in 1961, John Vorster held the opinion that state security, as the country's highest priority, eclipsed personal freedoms (Lodge et al., 2007a, p. 340). Under his patronage, the SAP gained excessive powers to counter what Vorster considered a growing revolution—a cause-and-effect that would perpetuate itself—as the levels of coercive prisoner interrogation by the SAP steadily increased, the security forces learned more about the Soviets' close involvement with the ANC, causing the need, real or perceived, for more stringent controls. Considering the extent of these police powers, Cawthra writes (1993, p. 15):

> New laws defined offences such as "subversion" and "terrorism" very broadly to facilitate prosecutions. Police were given powers to detain people without trial initially for 12 days, then for 90 days, then 180, and finally, with the passage of the Terrorism Act in 1967, for as long as they wanted. Torture of detainees became routine and contributed to the 68 deaths in political detention by 1990.

When Vorster became prime minister in 1966, a new intelligence agency—the Bureau of State Security (abbreviated reminiscently of a Bond villain as BOSS)—would form, together with the SAP, the crux of the intelligence community.

By the mid-1960s, despite stringent laws controlling internal unrest and banned political opposition with imprisoned leadership, new trouble was brewing for the South African government in its neighboring countries. Acting as geographical buffers from the recently independent African states further north, African nationalist movements—financed by the Soviet Union—started engaging in armed warfare in the Portuguese colonies of Angola and Mozambique, white minority-ruled Rhodesia (today Zimbabwe), and in South West Africa (SWA; today Namibia), which South Africa had administered after seizing it from Germany in 1915. The NP saw in these developments a threat to the security of the state and began deploying substantial numbers of soldiers and police officers to the country's borders and beyond (Marks, 2005, p. 37). It also introduced the South African Defence Amendment Act in 1967, which compelled all white males older than 18 to serve in the military or police service. In this way, the South African state taught men of all races in the country that violence was a legitimate and necessary tool with which either to suppress one's enemies or to defend one's freedom. As one former SAP officer recalled of this period (Faull, 2010, Loc 999):

> The [police] organisation sends you to the border. . . . There it teaches you one thing only, and that is to kill other people. Now you come back into a sane society, but that killing machine doesn't have an off switch. . . . [After border duty] you were expected to go back to your office like nothing had happened. Many things went wrong as a result of that.

With the SAP's counterinsurgency units rotating between policing South Africa and fighting a war until 1989, "many of the strategies and tactics learnt in the war were transferred to South Africa's townships—a culture of counter-insurgency was instilled" (Cawthra, 1993, p. 17).

Although this decade would be seen as the time of "Grand Apartheid," the NP fell short of achieving its aim to advance Afrikaners beyond the economic status of English-speaking whites. Although apartheid preferentially employed Afrikaans-speaking whites throughout the public service (including the police), provided disproportionate education, financial, and welfare support to whites, and protected white employment in general, English-speaking whites remained better off. In fact, comprising less than 10% of the population, by 1965 English-speaking whites controlled 90% of companies in the mining and manufacturing sectors and 80% of all financial businesses. Afrikaners, who represented 60% of whites, "earned only 45% of the total white income" (Lodge et al., 2007a, p. 345). During this time, the SAP, whose workforce consisted mainly of Afrikaners, struggled to meet its personnel requirements with a remuneration structure that simply could not compete with that of the private sector. As a result, for most of the twentieth century, between one-third and one-half of the SAP's officers were black. The government's fear that increased wages in the police force would trigger higher remuneration claims in other public sectors further exacerbated this dilemma and would hold the SAP in a loop of generally recruiting quantity over quality for decades to come (Cawthra, 1993, p. 16).

Apartheid under pressure—reforms, armed insurgencies, uprisings, and broadened policing (1970s)

In 1970, black representation in parliament (through white politicians) and black citizenship in South Africa ended with the passing of the Bantu Homelands

Citizenship Act (Act 26 of 1970). Blacks were assumed to "belong" to their homeland, though these were not recognized states (Nattrass, 2017).

Strike actions in SWA from 1971 to 1972 spilled over into a series of labor protests in Durban during 1973, heralding a tumultuous decade of political and economic predicaments for the apartheid government. With British-controlled textile plants especially targeted by these wage dispute strikes, public compassion seemed to vest in the disgruntled workers. The Afrikaans newspaper *Rapport* reported, for example, that "most whites silently sympathized with the workers with their low wages." Even the SAP's most senior official in the area stated that, although the strikes were illegal, the police "have nothing against people asking for higher wages" (Shubane et al., 2007, p. 356). Such sentiments would set the tone for comprehensive government reforms in South African workplaces during this decade. Shubane and colleagues reflect (2007, p. 356):

> By the second half of the 1970s the government began to urge employers to relax the curbs on blacks doing skilled jobs or black supervisors giving instructions to whites. It also tried to set an example. Already in 1975 the Defence Force [SADF] had announced that black soldiers would enjoy the same status as whites of equal rank, and that whites would have to take orders from blacks. By the end of the 1970s more and more companies in the private sector had accepted the principle of merit-based promotion. A 1979 survey revealed that blacks supervised whites in one-fifth of a sample of companies.

With these labor reforms, the government apparently aimed to placate black workers who, according to the NP's strategic thinking, would choose economic well-being over political demands. In case they did not choose money over freedom, the government continued to strengthen both the SAP and the SADF with more powers and resources. This form of "political schizophrenia" would ultimately prove very costly for an embattled apartheid government (Shubane et al., 2007, p. 358).

Internationally, pressure against the apartheid government was steadily increasing. In 1973, the UN's General Assembly had adopted an International Convention for the Suppression and Punishment of the Crime of Apartheid. Signed into force by about 30 countries by 1976, this convention stipulated apartheid "as a crime against humanity," comparable to Nazi atrocities (United Nations, 2019). In 1974, the General Assembly excluded the country from participation, while granting observer status to the ANC. By November, the South African ambassador to the UN informed the UN Security Council that South Africa is turning away from its discriminatory policies through all measures at its disposal. This attempt at damage control would prove easier said than done, because no feasible plan existed that could bridge the divide between what blacks were demanding and whites were willing to concede (Shubane et al., 2007, p. 357).

By 1975, the armed conflicts raging in South Africa's neighboring countries were reaching fruition, while the formation of new governments hostile to the Republic became imminent. The Portuguese government had been toppled by a coup in 1974, and its replacement soon declared its intention to hand over the Portuguese colonies to guerrilla groups. By mid-1975, Mozambique was handed over to Frelimo, a left-wing movement, but in Angola civil war loomed as three movements slugged it out for supremacy. With the Soviets meddling in the outcome by supplying arms and Cuban advisors to one group—the MPLA—the country "promptly became the site of one of the proxy wars the Soviet Union and the United States fought in distant corners of the Third World" (Shubane et al., 2007, p. 359). Encouraged by the CIA, South African troops covertly crossed into Angola in August of 1975 to assist the American funded, anti-Marxist movement active in the country. When a major FNLA attack on the MPLA failed shortly afterwards, however, the Organization of African Unity hurriedly recognized the MPLA as the legitimate government of Angola. The U.S. Congress responded by severing all aid to its allies—the anti-Marxist movements. A third-party victim to this about-face, South Africa was now committed to cross-border warfare, which became increasingly conventional as the Soviet Union and Cuba channeled arms and troops to the region. This meant an enormous increase of the Republic's military budget during a time when its annual economic growth was dropping from an average of 4.5% to 1.65% (Shubane et al., 2007, pp. 358–361). Combined with the 1977 UN Security Council's arms embargo on the country, the government's apartheid policy by then had become not only ideologically indefensible to the international community but also economically unsustainable.

During this period, the SAP began describing itself as a "professional" organization (van der Spuy, 2005; Faull, 2013). As in the U.S.A. and U.K. at the time, the strategy was to present the police as skilled, politically neutral professionals (Chackerian, 1974, p. 142). With the new rhetoric came the introduction of new technologies, the raising of educational requirements for recruits, and a focus on the "science" of policing. Police degree programs were introduced by the University of South Africa. Yet, despite its positive aspirations, the SAP's ability to police the apartheid state would soon be questioned.

By 1976, the ANC's revolutionary message and armed struggle had been subdued by the state for well over a decade. During its dormancy, its remaining leadership would commit in 1969 to a Soviet-inspired National Democratic Revolution (NDR) at a conference in Morogo, Tanzania. Despite the ANC's agreement with the primary goal of the NDR to move South Africa gradually from a free-market system to socialism and then communism, Soviet funding had all but dried up during the 1970s. In 1965, the ANC and SACP had collectively received $670,000, but, by 1973, as the Soviets expected more return on their investments from insurgencies in the Portuguese colonies, the ANC/SACP allocation had shrunk to $200,000 (Jeffery, 2019, Loc 239, 424, 436, 1000).

Contributing to its woes, the ANC now also had growing political rivals. The Black Consciousness Movement (BCM) started in the late 1960s and, by the mid-1970s, had a strong following among the township youth, where it encouraged them to "overcome the sense of inferiority that apartheid had helped to generate, and take the lead in demanding an end to racial discrimination" (Jeffery, 2019, Loc 251). In 1975, in an effort to resist apartheid's homeland policy and facilitate liberation without violence, Zulu chief Mangosuthu Buthelezi established the Inkatha Yenkululeko Yesizwe (Inkatha). The organization quickly gained widespread support, especially in the [Kwa-Zulu] Natal province and large urban townships like Soweto. By 1979, it claimed 300,000 members, much higher than the membership of any other liberation movement at the time (Jeffery, 2019, Loc 914, 1151).

In 1975, the government's policies of separate race development were so well established that the country was spending 15 times more on the education of each white child than on the education of each black child (Thomas, 1996, p. 330). Against this background, some 20,000 Soweto schoolchildren, inspired by the BC movement, took to the streets on June 16, 1976, to protest against a government decree that Afrikaans had to be used as one of the teaching languages in secondary schools. The events that followed would have far-reaching consequences for the beleaguered apartheid government and its executive branches. *Financial Mail* journalist John-Kane Berman recalled this day (cited in Shubane et al., 2007, p. 364):

> Newspaper photographs and several eyewitness accounts suggest that the marching students were good humored, high spirited and excited. Some were giving the clenched fist "Black Power" salute. Others were carrying placards bearing slogans "Down with Afrikaans," "We are not Boers," and "If we must do Afrikaans Vorster must do Zulu." Police vehicles rushed to the scene. . . . The pupils taunted them and they responded with teargas. . . . Apparently no order from the police to the marchers to disperse was heard, and a senior police officer admitted at the time that no warning shots had been fired either. The first child to be killed was evidently a thirteen-year-old schoolboy Hector Petersen, apparently by a bullet fired at him directly from behind. Several other youngsters were also shot dead. Then, in the words of one newspaper, "All hell broke loose."

Unlike previous clashes between protesters and police, images of the Soweto bloodshed were broadcast on televisions around the country and the world. In turn, violent demonstrations spread across South Africa for more than a year. A journalist flying over Soweto at the time reported he felt like "a war correspondent flying over a city after a bomb attack" (Shubane et al., 2007, p. 364). The SAP responded in accordance with their 1960s modus operandi—through decisive and fierce action the security of the state must be maintained at all costs. When the uprisings finally dwindled by October 1977, between

600 and 700 people had been killed and thousands more arrested or detained (Shubane et al., 2007, p. 365). From March 1976 to December 1977, more than 20 people had died in police custody; among them, in September 1977, was the BC leader Steve Biko, who died from injuries inflicted by members of the SAP's Security Branch while Biko was detained. With Biko's much publicized death threatening renewed uprisings, the government throttled the BC movement in October by banning 17 of its affiliate organizations. In the most "severe act of political suppression since the outlawing of the ANC and the PAC in 1960," two dissident newspapers were also closed down (Jeffery, 2019, Loc 1103, 1115).

In the aftermath of these uprisings, many inadequacies of the apartheid state's policing were exposed. Clearly, while the SAP was focused on delivering law and order in white areas, black townships had increasingly become unregulated, violent, and lawless (Posel, 2011, p. 348). Soon a commission of enquiry would conclude that "so far as manpower, equipment and mental attitude was concerned, [the police] . . . were completely unprepared for such a mass demonstration" (Cillie Report, cited in Marks, 2005, p. 42). Consequently, it was decided that mobile riot units were needed to respond rapidly and efficiently to mass protest actions. Although no initial selection criteria existed for these units, their members received some training in crowd control techniques, which became more specialized with assistance from countries like Israel (Cawthra, 1993, pp. 18–19). A national Reaction Unit was established in 1979 that could be deployed anywhere in the country where local riot units were unable to control an unrest situation. By then, policy directed all police to undergo crowd control and counterinsurgency training. The size of the police force would also grow during this time, and, by 1984, its manpower had increased by almost 30% (Marks, 2005, p. 44).

By 1978, P. W. Botha had replaced John Vorster as Prime Minister, with the latter forced to resign amid a slush-fund scandal. A firm believer in the "total onslaught" on practically every facet of society by generally subversive and, more directly, communist elements, Botha and the SADF were convinced that threats could only be countered with a "total strategy." In implementing such an encompassing tactic against Marxist insurgents, SADF generals would follow a 20% military and 80% socioeconomic approach, by which the hearts and minds of the black population had to be won prior to military intervention. This did not, however, prevent the SADF from bombing ANC facilities in Mozambique and Lesotho, or carrying out targeted assassinations during this period (Davies and O'Meara, 1985; Jacobs, 2006, p. 36). Indeed, the new government clearly preferred the SADF over the SAP as the guardian of state security. The former Minister of Defence, Botha had rapidly strengthened the military, which now consisted of a moderately sized, permanent force and large annual intakes of conscripts (Shubane et al., 2007, pp. 367–368). The annual budget increases of the SADF far exceeded that of the SAP, while the patrolling of borders gradually shifted to the military. The intelligence community also underwent noteworthy changes, with BOSS replaced by the National Intelligence Service (NIS), while SADF intelligence personnel played a more prominent role within a new body—Military Intelligence.

Among the numerous consequences of the Soweto unrests, the revival of the ANC's waning popularity would prove one of the most significant. To be sure, an opinion poll conducted by the Arnold Bergstraesser Institute of West Germany during 1978 showed that Buthelezi—the Inkatha leader—enjoyed 43% of black support in Soweto, Durban, and Pretoria, compared to the 21% for Mandela (Jeffery, 2019, Loc 1187). The Soviet Union revitalized and accelerated its support to the ANC, and, in October 1978, facilitated an ANC/SACP leadership visit to Vietnam to glean information from the local strategists' "people's war" against the U.S.A. (Jeffery, 2019, Loc 256). The lessons learned from this visit would be adopted by the ANC in 1979 and steadily implemented in its South African war of the people. The objective of this strategy would not be

> to defeat the South African security forces on the ground, for this was an impossible task. Instead, it aimed to make the country "ungovernable" and to generate a level of violence and economic malaise that in time would put enormous pressures on the NP government to jettison the already crumbling apartheid system.
>
> (Jeffery, 2019, Loc 269)

Concluding a troubled decade, an all-South African team produced its first nuclear device in 1979 as a deterrent to the buildup of Cuban troops in Angola during the mid-1970s. In total, the country would hold six such devices that could reach targets by either plane or projectile. By the start of the next decade, the South African government would follow a policy of "strategic ambiguity" and would neither confirm nor deny that it had such a destructive capability (Shubane et al., 2007, p. 362).

A litany of unbridled and sustained horror:[4] mass protest actions, border wars, international pressure, reforms, and riot policing in an ungovernable South Africa (1980s)

To ensure the security of the state, the NP government embarked on a destabilization campaign in 1981 that would see it pressuring its neighbors through military action to abandon Marxist policies and drive ANC guerrillas from the region. A 1982 meeting between the head of the CIA and Botha, orchestrated by NIS, concluded in consensus on the creation of a new safety corridor of neighboring states through military coercion (Shubane et al., 2007, p. 370).

Meanwhile, the ANC was deploying its own subversive operation to weaken its political rivals. Through hurried Soviet promotion it had already gained recognition from various Western countries as the sole representative of the South African people by 1978. When its 1979 plan to unify apartheid opposition groups under ANC patronage was rejected by Inkatha, a propaganda campaign was conceived to discredit political competitors. In 1980, the ANC's secretary-general denounced Buthelezi as "a sell-out, an enemy of the people, a collaborator," while the SACP criticized Inkatha for its "divisive influence" on the revolution. To gain populist support, the ANC called for the *Year of the Freedom Charter* and distributed copies of the 1955 document among mainly high school and university students. A global movement to release Mandela from prison was also started which would ultimately "build up a world-wide recognition of him as a martyr to Pretoria's intransigence." In parallel, the ANC continued with its revived armed struggle, and, by early 1983, had carried out some 70 bomb attacks, of which 15 occurred in city precincts and shopping centers. In May 1983, the ANC exploded a car bomb in central Pretoria, killing 19 people and injuring a further 219, most of whom were civilians (Jeffery, 2019, Loc 1012, 1267, 1324).

In January 1984, on recommendations from a SACP societal assessment, the ANC called for a people's war resting on "four pillars of revolution": the continuous underground status of the ANC, mass protest actions, MK's armed offensive, and international isolation of South Africa. In following these strategies, the ANC believed that, similar to the Vietnamese model, the country would become progressively ungovernable (Jeffery, 2019, Loc 1442).

Cognizant of the ANC's strategy, the NP government introduced further reforms, with some more precipitous than others. One such transformation was the 1984 tricameral parliament with separate houses for whites, coloreds, and Indians. Omitted from participation, blacks were regulated to their homeland administrations, albeit under new local governments or councils with more authority than before. Excluded yet again, black resentment was palpable, and when these councils abruptly raised the costs of rent and electricity, their action was met with open revolt in September 1984 in the Sebokeng Township, south of Johannesburg, which led to the murder of the council's chairman and mayor (Shubane et al., p. 370). Protests would soon spread across the country, coordinated by the United Democratic Front (UDF), an organization launched in 1983 with its executive committee consisting almost entirely of underground ANC political members (Jeffery, 2019, Loc 1396). The uprisings were much better organized than the 1976 BC movement protests: "Clergy, students, teachers, lecturers, business people, women's groups and workers were mobilized . . . there was better strategic thinking and a far more successful appeal to international opinion." In describing the protest actions in black settlements, Shubane and colleagues write (2007, p. 383):

> In many townships crowds attacked the houses of black town councilors, forcing their resignation and eventually the collapse of the system of local government. They also burned government buildings, shops and liquor

> outlets. Activists firebombed the houses of black policemen to drive them out of the townships and killed many suspected informers, sometimes by the so-called necklace—lighting a petrol-soaked rubber tyre placed around the victim's neck. This grisly form of reprisal soon led to the police's sources of information drying up.

A black SAP officer deployed to protect a councilor's vacant home at the time recalled such a murder of a local woman who was in a romantic relationship with one of his colleagues (Faull, 2010, Loc 2517):

> A [police] brother of mine kissed her, so they burnt her alive. They used a tyre and petrol and burnt her alive. Then they sang and stoned her. I saw it first hand. They took her out of the houses to a soccer field. . . . She was burning. We shot and they ran away, but she was still burning. She died helplessly.

The sheer magnitude of the countrywide protest actions necessitated practical adaptations to public-order policing. From 1984, the SAP and SADF participated jointly in township operations, "signaling that the SAP alone was no longer able to contain the intensifying resistance. . . . In the townships of South Africa, this was indeed now civil war" (Marks, 2005, p. 48). On June 16, 1985—the anniversary of the Soweto uprising—bombs were detonated in Durban and Johannesburg, and, by July 20, the government declared a state of emergency in several magisterial districts (Shubane et al., 2007, p. 383). Under this partial emergency, SAP and SADF members had the absolute authority to arrest, interrogate, search homes, and seize possessions. Day or night curfews could be implemented by the commissioner of police, who could close any building or business property and even control media reports. New offenses were created that

> effectively criminalised any activity that could be construed as political. It also empowered the police to act against collective groups of people in almost any setting—students at school, workers in a factory, mourners at funerals, and participants in peaceful demonstrations (Marks, 2005, p. 49).

Members of the security forces were also indemnified from criminal and civil litigation under the emergency, if their actions were considered in "good faith," defined as acting "under the command, direction or instruction of a superior officer" (Marks, 2005, p. 49).

Two days after the state of emergency declaration, the ANC issued a call to "render South Africa ungovernable," but the severe measures in use by the security forces would gradually extirpate the protest leadership, and, with 8,000 UDF members already detained by November 1985, further rolling protests were thwarted (Shubane et al., 2007, p. 384). Conversely, under the auspices of the people's war, political killings, intimidation, and the propaganda campaign would accelerate. The escalating violence was blamed on Inkatha and the NP government, a tactic garnering international sympathy for the ANC. As this strategy against Inkatha intensified, a low-range civil war was raging between these rivals, especially in the [KwaZulu] Natal province, where the death toll from political violence would increase from 340 in 1987, to 900 in 1988, and to 2,400 by 1989. In an attempt to address the killings and propaganda, Buthelezi released a press statement in January 1988 accusing the ANC's agenda of attempting to create a single-party Marxist state as the root cause for the violence in the country (Jeffery, 2019, Loc 278, 1793, 1826). Nevertheless, the damage to Inkatha's credibility and support during this time had been done.

The partial state of emergency was finally lifted in March 1986, but the reprieve was short-lived. A national state of emergency would follow on June 12, 1986 and would remain in place until February 1, 1990. In yet another uprising aftershock for South Africa, more than 20,000 people had been detained during the 1984 to 1986 period (Shubane et al., 2007, p. 384). The number of necklace executions had also risen to around 300 in 1986. By then, many of those killed were not only suspected informers of the security forces but also supporters of the ANC's political rivals, like Inkatha. Necklace killings were further used to intimidate as black policemen and other "enemies of the people" were similarly murdered in this gruesome manner (Jeffery, 2019, Loc 1568).

During 1986 and early 1987, Soviet military equipment and other assistance to the MPLA government in Angola reached $1 billion in value. A major offensive was planned to gain military control of southern Angola from the anti-Marxist group Unita. From this geographical vantage point, Namibia could be attacked at various places along its border, while MK camps could set up far enough south to launch infiltrations into South Africa. The offensive saw a buildup of between 50,000 to 60,000 Cuban and MPLA soldiers, supported by 500 tanks, 100 jet fighters, and 60 attack helicopters. Three months into the attack, some 3,000 South African troops supported by Unita soldiers forced the mammoth military column into retreat at a decisive battle at the Lomba River, killing between 7,000 and 10,000 MPLA soldiers. Desperately twisting this outcome in the media to save face, the Soviet Union was nonetheless humiliated and soon indicated to the U.S. Reagan administration its readiness to relinquish southern African operations. With the U.S.-Soviet negotiations set to commence in May 1988, and South African and Cuban withdrawal imminent, the NP government was reeling from the mounting costs of its cross-border warfare which already consumed 18% of the country's budget (Jeffery, 2019, Loc 1979, 1991, 2014). It would be the human cost—the increasing number of white conscript fatalities—that proved ultimately to have the farthest-reaching consequences for the government (Shubane et al., 2007, p. 390), with its voters rapidly losing their appetite for covert military operations.

The ongoing border war had gradually left the SAP "understaffed, under-resourced and poorly paid, making it even less equipped to counter a protracted people's war within the country" Jeffery, 2019, Loc 2014). Already stretched thin during the partial state of emergency, the SAP tried to include as many policing partners as it could in its suppression of mass protest actions. By 1985, homeland police forces were being trained in riot control and counter-insurgency measures. Township councils were granted their own police, and municipal police forces were soon being utilized in protest control operations. By 1986, "special constables" were inducted who, after receiving three six-week training courses, would be "used to supplement the SAP where they were overstretched." During the same year, a newly created centralized Riot Unit became the first public-order policing component in the SAP. Equipped with proper anti-riot gear and operating like a military unit, but with blurred accountability and supervision, it would soon become known as one of the more ruthless security components in the country (Marks, 2005, p. 51). The different strategic approaches of the SAP, SADF, and NIS regarding the suppression of mass uprisings were coordinated by the National Security Management System (NSMS) in 1986. Now, with domestic threats the priority, the police would "reassert their role in strategic intelligence and hence in influencing state strategy." Subsequently, more government focus was placed on strengthening the SAP as it received state funding generally allocated to the SADF (Cawthra, 1993, p. 32). Under the auspices of the NSMS, the security forces had a firm grasp on the situation in the country and were operationally geared to enact the NP government's counterrevolutionary strategies. Nevertheless, gathering external influences would have the second in command of the NIS recall this period as follows: "Nowhere was the situation out of hand, but it was clear that politically and morally we were losing our grip" (Shubane et al., 2007, p. 386).

Internationally, the sporadic condemnations of apartheid were coalescing into a deluge of disapproval. British investments in the country had halved between 1980 and 1986, while the Commonwealth introduced a ban on all South African agricultural and manufacturing exports and on new loans and investments. However, the most detrimental sanction was not inspired by anti-apartheid activism but by a major creditor's fiscal risk analyses. In August 1985, Chase Manhattan Bank abruptly announced it would not extend its short-term loans to the South African government. With the U.S. Congress also barring, among other restrictive actions, all new investments and loans to South Africa, the country would soon default on its international debt (Shubane et al., 2007, p. 384). In response to these international restrictions, the government would over time develop "sanction busting" strategies—representing a global web of corruption, money laundering, and assassinations that would be hard to quell once apartheid had ended (van Vuuren 2006, 2017).

In dire need of foreign capital, the NP government decided on more reforms to recover investor confidence. By removing parts of the apartheid system it deemed "unnecessary," the government granted full residential rights to urban blacks and rescinded the racial sex laws in 1985. The loathed pass laws were repealed in June of 1986. The disappointing "crossing the Rubicon" speech by Botha in August 1985 (see Shubane et al., 2007, p. 385; Jeffery, 2019, Loc 1932) did not only lead to additional sanctions against the country but had also publicly underscored the impasse between the government and the ANC. In response, an eclectic array of groups consisting of Afrikaner academics, editors, politicians, journalists, clerics, lawyers, and judges would conduct some 70 independent meetings with the ANC at several venues outside the country between 1985 and 1989. The Soviet Union also played its part in the de-demonizing of the ANC in the minds of white South Africans by organizing a conference between senior ANC leaders and Soviet and South African academics in West Germany in October 1988. In the ensuing discussions, the Moscow delegation had the South African academics believe that Soviet support for the ANC was waning and that, devoid of underlying communist ideologies, a rational political settlement could be reached with the organization (Jeffery, 2019, Loc 2280, 2300). Conversely, in June 1989 at a SACP congress chaired by the ANC's Thabo Mbeki,[5] amid "animated talk of 'building East Germany in Africa,'" it was decided that the armed struggle would continue until socialism could be instilled in South Africa (Shubane et al., 2007, p. 394).

In January 1988, Botha suffered a stroke and, by February, was replaced by the former Minister of Education, F. W. De Klerk. When the Berlin wall fell in November 1989, De Klerk saw the disintegration of communism as a "God-sent opportunity" to convince the NP's voter base of the diminished threat the ANC and its SACP ally would now present to the "free market, property rights and other investment-friendly policies." On February 2, 1990, he unbanned the ANC, SACP, and other liberation organizations and had announced to the parliament the need to commence with all-party negotiations toward a new constitution for the country (Shubane et al., 2007, pp. 394–395). Preempting this NP strategy, the SACP had already stressed the importance of negotiations as a main component in the people's war in April during its party congress in Cuba. It had also warned that the ANC should not be pushed into talks with the government "before it was strong enough to back its basic demands with sufficient power on the ground" (Jeffery, 2019, Loc 2357, 2370).

Prolonged political violence, intensified police killings, and a hurried election to democracy (1990–1994)

Nelson Mandela was released from prison on February 11, 1990. In step with ANC-SACP strategy, he told a Cape Town crowd on his release that sanctions, protest actions, and the armed struggle must continue, and that negotiations would not commence until all of the ANC's preconditions were met. "Now is the time to intensify the struggle on all fronts!" he announced. By April, the struggle had certainly strengthened, with seemingly uncontrolled violence in the country leaving close to 1,400 people dead, close to the total number of political killings for the whole of 1989. Prior to the start of negotiations between the parties, De Klerk had required a commitment to peace from the ANC, but this was consistently rejected with the SACP's general secretary stating at the time that, "[t]he armed struggle is going to continue . . . until we have obtained our objective" (Jeffery, 2019, Loc 2580).

Wary of their influence on his predecessor, De Klerk seldomly consulted with the commanders of the security forces. However, in early 1990 he would share with the SAP and SADF their new respective roles of fighting crime and protecting all South Africans. By then, inter-government rumors emerged of clandestine, rogue assassination units operating within the security community. Unit C-10 (or Vlakplaas, in reference to its location outside Pretoria) was one such secret police unit specializing in political murders. Project Coast was another—albeit more bizarre—covert unit that would allegedly use chemical and biological weaponry in the assassination of activists (Shubane et al., 2007, pp. 396–397).

With political violence now a strategy toward making the country ungovernable, its prevalence would assist the ANC prior to and amid the negotiations with the NP government. Throughout the period between 1984 and 1994, during which some 20,500 people were victims of political murders, the ANC had denied and deflected any report and accusation of direct involvement in the killings. In turn, it blamed the NP government or Inkatha (now the Inkatha Freedom Party or IFP) for the violence and later even the actions of a sinister Third Force. When evidence broke that the SAP had funded Inkatha as a countermeasure to the ANC, some media outlets would, for the remainder of the pre-election period, similarly point to real or imagined clandestine units operating within the security forces, all the while eroding the black support reflected in polls for De Klerk. The Truth and Reconciliation Commission (TRC) would later find that both the state's security and the liberation movements had participated in the violence (TRC, 1998, Volume 2, p. 302). At the time, however, a defensive De Klerk quickly removed both the ministers of Defence and Police (Shubane et al., 2007, pp. 397, 399).

In September 1991, the NP government and 27 political parties, including the ANC and IFP, signed a National Peace Accord (NPA) that proposed a decrease in political violence through codes of conduct, focused procedures, and new mechanisms. The NPA also established a commission of inquiry that, in accordance with the Prevention of Public Violence and Intimidation Act (1991), was tasked with finding solutions for the prevention of public violence and intimidation (Marks, 2005, pp. 58–59). While the NPA's recommendations progressively changed the role of the SAP in the policing of protest actions from suppressor to negotiator, attacks on the police increased rapidly. In 1985, 51 policemen were killed on duty. By the end of 1990, the police death toll stood at 107. By 1991, 145 officers had died, mostly ambushed while responding to fake emergency calls. During the first six months of 1993, 109 police officers were killed and 1,700 injured, indicating the degree to which the police were targeted during the period of ruthless political violence (Jeffery, 2019, Loc 2066, 2609, 2621, 3861).

In February 1991, the NP government announced that all apartheid laws would be abolished, opening the way to a negotiated political settlement. Previous attempts at such negotiations (the Groote Schuur Accord in May 1990, the Pretoria Minute in August 1990, and the D. F. Malan Accord in February 1991) had not brought the parties much closer in reaching consensus. The Convention for a Democratic South Africa (CODESA) opened on December 20, 1991, and had almost all delegations sign a Declaration of Intent to "a united, democratic, non-racial and non-sexist state in which sovereign authority is exercised over the whole of its territory." By mid-May 1992, a truculent ANC had walked out of CODESA after failing to reach agreement with the NP Government on substantial issues like minority protection, and had started a "rolling mass action" to influence the negotiations by means of public demands. Through diplomatic pressure, and in the interest of saving the country's rapidly declining economy, negotiations resumed in April 1993 under a substitute name: the Multiparty Negotiation Process (MNP). In the ensuing discussions, NP negotiators were unsuccessful in securing a multiparty government. Eventually, they agreed on a temporary government of national unity (GNU) for the country, a concept acknowledged by the head NP negotiator to be "a forced agreement of cooperation between opponents and not between two parties who have found common ground." With the NP conceding several minority-protection issues, some opposition within the party now preferred a permanent GNU, but the will of De Klerk and his negotiators persevered and the new government would remain provisional. By now, the IFP had abandoned the negotiations and rejected the principle of an elected constitution-making body, preferring a form of confederate government instead (Lodge et al., 2007b, pp. 403, 404, 406).

In the wake of the assassination of Chris Hani, the SACP's general secretary and ANC leader, by a Polish immigrant in April 1993, both De Klerk and Mandela appealed to the country for calm. As violence followed their pleas, De Klerk capitulated in accepting an election date prior to the existence of an agreed constitution. With the MNP setting the premature election date of April 27, 1994, the ANC had succeeded to increase pressure on the remaining constitution negotiations. The IFP and Conservative Party walked out in protest. Subsequently, after the interim constitution's text was approved in November 1993, ANC negotiators would comment to journalists their surprise at the easy concessions gained from the NP government in terms of important matters (Jeffery, 2019, Loc 3761, 3866–3878, 4176).

In anticipation of the election, the MNP appointed a Transitional Executive Council (TEC) that would "facilitate the transition to democratic government at all levels of government" and an Independent Electoral Commission (IEC) that would ensure that the election was free and fair, adjudicate disputes, and supervise the conduct of parties during the campaign" (Lodge et al., 2007b, p. 407). With checks and balances in place, the stage could now be set for the country's first democratic election. In December 1993, the government restored South African citizenship to residents of the homelands, allowing all people over 18 to vote. The IFP had also agreed to enter the election, albeit forced by circumstances, and within less than a week before the poll (Harris, 2010).

With seemingly uncontrolled violence persisting across South Africa, doubt over the country's ability to avoid bloodshed abounded. In the run-up to the election, both the British and Portuguese governments announced emergency plans to evacuate their citizens if necessary, while the South African Chamber of Business cautioned its sizable membership to prepare for a "worst case scenario" (Jeffery, 2019, Loc 4539). The IEC's ability to ensure free and fair elections was also questioned, with a political analyst warning at the time, "Breathtakingly routine incompetence combined with thinly veiled arrogance characterise the 'Independent' Electoral Commission—a dangerous mixture which could make our April election a mess, if not a disaster" (Donald Simpson cited in Jeffery, 2019, Loc 4597).

Many concerns around the election cut to the core of democratic voting, albeit ultimately discarded by international monitors. Wholesale intimidation in some black townships, with assaults on, and even murders of, members of opposition parties to the ANC during campaigning, had unsettled political parties to such an extent that they stopped entering a total of 165 "no-go" areas across the country (see Jeffery, 2019, Loc 4610, 4621, 4633, 4645, 4656; Lodge et al., 2007b, p. 408).

Lacking sufficient time for compilation, the election was conducted without voter rolls. Where voters could not produce identity documents, the IEC could issue temporary voting cards. With over 3.5 million cards finally issued, it was "obviously a system liable for abuse since many applicants were unable to provide proof that they were at least 18 years old" (Lodge et al., 2007b, p. 408). Facing serious logistical problems, the IEC extended the election for an additional day. Under a cloud of political party and media accusations of voter and vote counting fraud (see Jeffery, 2019, Loc 4784–4886), the results were announced on May 6: "[A]t the provincial level, the ANC won outright control of six provinces, the NP winning in the Western Cape and the IFP in KwaZulu-Natal. [The] ANC's national votes came from blacks, with only 0.5% from whites." The widest spread of minority votes vested in the NP with 49% white, 30% colored, 14% black, and 7% Indian (Lodge et al., 2007b, p. 408). With the head of the IEC stating that the results were "beyond review or appeal," international monitoring agencies stated that despite some apparent flaws, the election was conducted "in a free and fair manner" (Jeffery, 2019, Loc 4896). Through centuries of strife, strategy and a fair amount of political subterfuge, minority rule had now been officially replaced by a majority governing order.

Conclusion

The story of policing, order, and social control in South Africa is intertwined with the immense societal changes within the country, accelerated during the last three decades of apartheid rule. It is also an account bespattered with violence. Not dissimilar in its origin and initial functioning from many other colonial police forces, the South African police were strikingly different in that members would, as police officers, fight and perish in all of the country's wars—official or clandestine—and also in international conflicts.

This militant aspect of the country's policing would steadily be honed to lethal effectiveness during the 1960s to 1980s and would culminate with the police being a key element in the "total strategy" response to a "total communist onslaught." During this period, the SAP would be responsible for an estimated 3,000 people killed, whose deaths during protest actions or in cross-border warfare were all considered lawful by the government. Less officially endorsed were the rogue assassination units within the SAP's ranks, which, due to a lack of cooperation from the SAP leadership, and the tampering with evidence, would only be fully investigated after the 1994 election (Shubane et al., 2007, p. 397).

Placed in context, by then the SAP believed it was waging a war against a communist offensive in which "all aspects of life could be viewed through the prism of security, [and] all opposition could be seen as part of an omnipresent threat" (Cawthra, 1993, p. 39). In a climate of anarchy rooted in seemingly unbridled political violence, it was understood that the end justified the means. The police's adversaries—communist-affiliated activists who in 1984 had called for "a people's war" in the country—had come to use horrifying tactics like the "necklacing" of presumed "enemies of the people," which, incredulously, seemed condoned by the internationally backed liberation movements in the country. In an interview with the *Sunday Times* in London in 1986, the ANC secretary-general commented that "'collaborators with the enemy had to be eliminated," adding, "Whatever the people decide to use to eliminate these enemy elements is their decision. If they decide to use necklacing, we support it" (Jeffery, 2019, Loc 1555).

Between 1984 and 1994, about 20,000 people were killed in the conflicts raging within the country's borders. Although staggering, this number represents a mere fraction of the death tolls documented in other ethnic conflicts across the globe. Contributing largely to this was a judicial system that remained mainly independent, fostering the growing realization among police officers, especially in the 1980s, that illegal actions could be met with negative consequences. More importantly, among all the race groups within the country, no support existed for racial killings. Ultimately, although the apartheid government, the ANC, and Inkatha all contributed to the atrocities during this time, "the death toll never rose so high that a settlement became impossible" (Shubane et al., 2007, p. 398).

Established in 1996, the Truth and Reconciliation Commission (TRC) was tasked with the investigation of the different forms of human rights violations within the country from 1960 to democratic rule. It would also grant amnesty to those perpetrators who made full disclosures, while promoting reconciliation and unity among South Africans. With almost all of its members considered "tacit or overt ANC supporters," the TRC's composition would not favor its objectivity. By the time its function concluded, it had reviewed some 21,300 victim statements that recorded close to 38,000 "gross violations of human rights." Police members had asked for amnesty for the unlawful deaths of 150 people. Ultimately, the TRC did not find evidence of a Third Force, as fervently suggested by the ANC as a cause of the political violence in the 1980s and 1990s, but it did discover "that security force operatives, with the connivance of senior officers, had engaged in violence and killings." It also found the ANC accountable for violations in blurring the line between "soft" and "hard" targets (Shubane et al., 2007, p. 397; Lodge et al., 2007b, pp. 414–415).

For all the euphoria that the 1994 election had created, the country found itself in a new political dispensation "without the first and most fundamental criterion for a functioning democracy: The realistic prospect of the ruling party being voted out of power" (Jeffery, 2019, Loc 5011). Socially, the twentieth century had seemingly taught a portion of South African men of all races that violence was a legitimate means to express authority, fight injustice, defend one's honor, and be a man. In this "new" South Africa, violent crime remained alarmingly high. The 1985 call by the ANC to make the country ungovernable and the brutal responses by the state had further fostered a culture of violence often amplified in townships, where unemployment was high, socioeconomic conditions dire, and state policing remained sparse.

As discussed further in Chapter 3, the ostensible inability of the new government to address this legacy of violence meaningfully, enforce laws effectively, and bring criminals to justice in a timely manner would soon compel significant reforms to the South African police—transformations that, decades later, are still in progress.

Notes

1 Over time, the compound word "Khoisan" would be used by scholars to refer to either or both of the Bushmen and Khoikhoi groups (Elphick and Malherbe, 1989, p. 3).

2 For a period of six months prior to his departure from South Africa, Mandela worked closely with the SACP leaders to such an extent that he was referred to as "a closet member of the central committee of the SACP" (Jeffery, 2019, Loc 964).

3 According to Jeffery, Operation Mayibuye was conceived during discussions between the ANC and Soviet representatives in Moscow (2019, Loc 952).

4 Wording in a 1992 International Freedom Foundation report, describing the ANC's actions in its training camps (Jeffery, 2019, Loc 1301).

5 Deputy president to Mandela (1994–1999) and president of South Africa from 1999 to 2008.

References

Bittner, E. (1970). *The Functions of Police in Modern Society*. Rockville, MD: National Institute of Mental Health, Center for Studies of Crime and Delinquency.
Bittner, E. (1991). The functions of police in modern society. In C. Klockars and S. Mastrofski (eds.), *Thinking about Police: Contemporary Readings*, 2nd ed. Boston: McGraw-Hill.
Bonner, P. (2011). South African society and culture, 1910–1948. In R. Ross, A.K. Mager, and B. Nasson (eds.), *The Cambridge History of South Africa*, Vol. 2. Cambridge: Cambridge University Press.
Brits, J.P. and Spies, B., as adapted by A. Grundlingh (2007a). A new South Africa in the making. In H. Giliomee and B. Mbenga (eds.), *New History of South Africa*. Cape Town, South Africa: Tafelberg Publishers.
Brits, J.P. and Spies, B., as adapted by A. Grundlingh (2007b). A turbulent decade. In H. Giliomee and B. Mbenga (eds.), *New History of South Africa*. Cape Town, South Africa: Tafelberg Publishers.
Brogden, M. (1989). Origins of the South African Police: Institutional versus structural approaches. *Acta Juridica*: 1–19.
Callinicos, L., B. Nasson, H. Giliomee, and A. Grundlingh (2007). Gold and war. In H. Giliomee and B. Mbenga (eds.), *New History of South Africa*. Cape Town, South Africa: Tafelberg Publishers.
Cawthra, G. (1993). *Policing South Africa: The SAP and the Transition from Apartheid*. London, U.K.: Zed Books Ltd.
Chackerian, R. (1974). Police professionalism and citizen evaluations: A preliminary look. *Public Administration Review*, 34(2): 141–148.
Comaroff, J. (2013). Forward. In W. Garriott (ed.), *Policing and Contemporary Governance: The Anthropology of Police in Practice*. New York: Palgrave MacMillan.
Coutsoukis, P. (2004). Photius.com. *South Africa Early Development*. Available at: www.photius.com/countries/south_africa/society/south_africa_society_early_development.html (Accessed 27 November 2019).
Couzens, T. (2013). *South African Battles*. Jeppestown, South Africa: Jonathan Ball Publishers.
Davies, R. (1976). Mining capital, the state and unskilled white workers in South Africa, 1901–1913 Author(s). *Journal of Southern African Studies*, 3(1): 41–69.
Davies, R. and D. O'Meara (1985). Total strategy in Southern Africa: An analysis of South African regional policy since 1978. *Journal of Southern African Studies*, 11(2): 183–211.
Diamond, J. (1997). *Guns, Germs and Steel*. New York: Random House. Kindle Edition.
Dubow, S. (2011). South Africa and South Africans: Nationality, belonging, citizenship. In R. Ross, A.K. Mager, and B. Nasson (eds.), *The Cambridge History of South Africa*, Vol. 2. Cambridge: Cambridge University Press.
Elphick, R., M. Adhikari, C. Saunders, M. Shain, R. Mendelsohn, and H. Giliomee (2007). Diverse communities. In H. Giliomee and B. Mbenga (eds.), *New History of South Africa*. Cape Town, South Africa: Tafelberg Publishers.
Elphick, R. and H. Giliomee (1989). The origins and entrenchment of European dominance at the Cape, 1652-c.1840. In R. Elphick and H. Giliomee (eds.), *The Shaping of South African Society, 1652–1840*. Middletown, CT: Wesleyan University Press.
Elphick, R. and V.C. Malherbe (1989). The Khoisan to 1828. In R. Elphick and H. Giliomee (eds.), *The Shaping of South African Society, 1652–1840*. Middletown, CT: Wesleyan University Press.

Elphick, R., R. Shell, A. Giliomee, A. Manson, and H. Giliomee (2007). A new society from three continents. In H. Giliomee and B. Mbenga (eds.), *New History of South Africa*. Cape Town, South Africa: Tafelberg Publishers.

English-Ingles.com. (2019). *Etymology-of-Police*. Available at: http://english-ingles.com/en/etymology-of-police/ (Accessed 1 December 2019).

Etherington, N., P. Harries, and B.K. Mbenga (2009). From colonial hegemonies to imperial conquest, 1840–1880. In C. Hamilton, B.K. Mbenga, and R. Ross (eds.), *The Cambridge History of South Africa*, Vol. 1. Cambridge: Cambridge University Press.

Farwell, B. (2009). *The Great Boer War*. Barnsley, South Yorkshire, U.K.: Pen & Sword Books Ltd.

Faull, A. (2010). *Behind the Badge: The Untold Stories of South Africa's Police Service Members*. Cape Town, South Africa: Penguin Random House South Africa. Kindle Edition.

Faull, A. (2013). Towards a new professionalism for the South African Police Service. *South African Review of Sociology*, 44(2): 18–35.

Feit, E. (1972). Generational conflict and African nationalism in South Africa: The African National Congress, 1949–1959. *The International Journal of African Historical Studies*, 5(2): 181–202.

Freund, B. (2011). South Africa: The union years, 1910–1948: Political and economic foundations. In R. Ross, A.K. Mager, and B. Nasson (eds.), *The Cambridge History of South Africa*, Vol. 2. Cambridge: Cambridge University Press.

Glaser, C. (1998). Swines, hazels and the dirty dozen: Masculinity, territoriality and the youth gangs of soweto, 1960–1976. *Journal of Southern African Studies*, 24(4): 719–736.

Goldstein, H. (1975). *Police Corruption: Perspective on Its Nature and Control*. Washington, DC: Police Foundation.

Grundlingh, A. and H. Giliomee (2007). The great depression and war. In H. Giliomee, and B. Mbenga (eds.), *New History of South Africa*. Cape Town, South Africa: Tafelberg Publishers.

Hall, S. (2009). Farming communities of the second millennium: Internal frontiers, identity, continuity and change. In C. Hamilton, B.K. Mbenga, and R. Ross (eds.), *The Cambridge History of South Africa*, Vol. 1. Cambridge: Cambridge University Press.

Harris, P. (2010). *Birth: The Conspiracy to Stop the '94 Elections*. Umuzi: Roggebaai.

Huffman, T.N. (2009). Mapungube and great Zimbabwe: The origin and spread of social complexity in southern Africa. *Journal of Anthropological Archaeology*, 28: 37–55.

Jacobs, C.J. (2006). The forward defence strategy of the South African Dfence Force (SADF), 1978–1989. *Journal for Contemporary History*, 31(3 December): 23–41.

Jeffery, A. (2019). *People's War: New Light on the Struggle for South Africa*. Jeppestown, South Africa: Jonathan Ball Publishers. Kindle Edition.

Karis, T. and G. Carter (1977). *From Protest to Challenge: A documentary History of African Politics in South Africa 1882–1964*. Stanford: Hoover Institution Press.

Klockars, C.B. (1985). *The Idea of Police*. Newbury Park, CA: Sage Publications.

Klockars, C.B. and S. Kutnjak Ivković (2004). Measuring police integrity. In A.R. Piquero, J.R. Greene, and M.J. Hickman (eds.), *Police Integrity and Ethics*. Belmont, CA: Wadsworth Publishing, pp. 1.3–1.20.

Klockars, C.B., S. Kutnjak Ivković, and M.R. Haberfeld (2006). *Enhancing Police Integrity*. New York: Springer.

Klockars, C.B., S. Kutnjak Ivković, W.E. Harver, and M.R. Haberfeld (2000). *The Measurement of Police Integrity*. Research in Brief. U.S. Department of Justice, Office of Justice Programs, National Institute of Justice. Washington, DC: Government Printing Office.

Kynoch, G. (2003). Apartheid nostalgia: Personal security concerns in South African townships. *South African Crime Quarterly*, 5: 7–10.

Landau, P.S. (2009). Transformations in consciousness. In C. Hamilton, B.K. Mbenga, and R. Ross (eds.), *The Cambridge History of South Africa*, Vol. 1. Cambridge: Cambridge University Press.

Lee, R. (2011). Death "on the move": Funerals, entrepreneurs and the rural-urban nexus in South Africa. *The Journal of the International African Institute*, 81(2): 226–247. South Africa.

Legassick, M. and R. Ross (2009). From slave economy to settler capitalism: The cape colony and its extensions, 1800–1854. In C. Hamilton, B.K. Mbenga, and R. Ross (eds.), *The Cambridge History of South Africa*, Vol. 1. Cambridge: Cambridge University Press.

Lodge, T. (2011a). Resistance and reform, 1973–1994. In R. Ross, A.K. Mager, and B. Nasson (eds.), *The Cambridge History of South Africa*, Vol. 2. Cambridge: Cambridge University Press.

Lodge, T. (2011b). *Sharpeville: An Apartheid Massacre and Its Consequences*. Oxford: Oxford University Press.

Lodge, T., M. Shain, R. Mendelsohn, K. Moodley, B. Pogrund, H. Barrell, and H. Giliomee (2007a). The Afrikaner nationalists in power. In H. Giliomee and B. Mbenga (eds.), *New History of South Africa*. Cape Town, South Africa: Tafelberg Publishers.

Lodge, T., V. van der Vliet, D. Welsh, J. Myburgh, W. Gumede, L. Schlemmer, M. Bot, H. Adam, and H. Giliomee (2007b). Towards an inclusive democracy. In H. Giliomee and B. Mbenga (eds.), *New History of South Africa*. Cape Town, South Africa: Tafelberg Publishers.

Mager, A.K., and M. Mulaudzi (2011). Popular responses to apartheid: 1948-c. 1975. In R. Ross, A.K. Mager, and B. Nasson (eds.), *The Cambridge History of South Africa*, Vol. 2. Cambridge: Cambridge University Press, pp. 369–408.

Mandela, N. (1994). *Long Walk to Freedom: The Autobiography of Nelson Mandela*. Boston, MA: Little, Brown Book Group. Kindle Edition.

Manson, A., B. Mbenga, J. Peires, and H. Giliomee (2007). Consolidation and expansion of the colonial presence. In H. Giliomee and B. Mbenga (eds.), *New History of South Africa*. Cape Town, South Africa: Tafelberg Publishers.

Marks, M. (2005). *Transforming the Robocops: Changing Police in South Africa*. Scottsville, South Africa: University of KwaZulu-Natal Press.

Marks, S. (2011a). Class, culture, and consciousness in South Africa, 1880–1899. In R. Ross, A.K. Mager, and B. Nasson (eds.), *The Cambridge History of South Africa*, Vol. 2. Cambridge: Cambridge University Press.

Marks, S. (2011b). War and union, 1899–1910. In R. Ross, A.K. Mager, and B. Nasson (eds.), *The Cambridge History of South Africa*, Vol. 2. Cambridge: Cambridge University Press.

Merriam-Webster Online Dictionary (2019). *Definition of Police Power*. Available at: www.merriam-webster.com/dictionary/police%20power (Accessed 1 December 2019).

Nasson, B. (1991). Bobbies to Boers: Police, people and social control in Cape Town. In D.M. Anderson and D. Killingray (eds.), *Policing the Empire: Government, Authority and Control, 1830–1940*. Machester and New York: Manchester University Press.

Nattrass, G. (2017). *A Short History of South Africa*. London: Biteback Publishing. Kindle Edition.

Nattrass, N. and J. Seekings (2011). The economy and poverty in the twentieth century. In R. Ross, A.K. Mager, and B. Nasson (eds.), *The Cambridge History of South Africa*, Vol. 2. Cambridge: Cambridge University Press.

Peires, J.B. (1989). The British and the Cape 1814–1834. In R. Elphick and H. Giliomee (eds.), *The Shaping of South African Society, 1652–1840*. Middletown, CT: Wesleyan University Press.

Posel, D. (2011). The apartheid project, 1948–1970. In R. Ross, A.K. Mager, and B. Nasson (eds.), *The Cambridge History of South Africa*, Vol. 2. Cambridge: Cambridge University Press.

Ross, R. (2009). Khoesan and immigrants: The emergence of colonial society in the Cape, 1500–1800. In C. Hamilton, B.K. Mbenga and R. Ross (eds.), *The Cambridge History of South Africa*, Vol. 1. Cambridge: Cambridge University Press.

Shear, K.S. (1998). *Constituting a State in South Africa: The Dialectics of Policing, 1900–1939*. PhD Dissertation. Evanston, IL: Northwestern University.

Shubane, K., H. Barrell, A. Seegers, and H. Giliomee (2007). In search of a new order. In H. Giliomee and B. Mbenga (eds.), *New History of South Africa*. Cape Town, South Africa: Tafelberg Publishers.

Smalberger, S.M. (1976). The role of the diamond-mining industry in the development of the pass-law system in South Africa. *The International Journal of African Historical Studies*, 9(3): 419–434.

Steyn, M. (2007). The Mapungubwe gold graves revisited. *The South African Archaeological Bulletin*, 62(186): 140–146.

Thomas, D. (1996). Education across generations in South Africa. *The American Economic Review*, 86(2): 330–334.

Trapido, S. (2011). Imperialism, settler identities, and colonial capitalism: The hundred-year origins of the 1899 South African Warp. In R. Ross, A.K. Mager, and B. Nasson (eds.), *The Cambridge History of South Africa*, Vol. 2. Cambridge: Cambridge University Press.

Truth and Reconciliation Commission (TRC) (1998). *Truth and Reconciliation Commission of South Africa Report*, Vol. 1–5. Cape Town: Juta & Co Ltd.

United Nations (2019). *Treaty Collection: Chapter IV: Human Rights, 7: International Convention on the Suppression and Punishment of the Crime of Apartheid*. Available at: https://treaties.un.org/Pages/ViewDetails.aspx?src=TREATY&mtdsg_no=IV-7&chapter=4&clang=_en (Accessed 12 March 2019).

Van Der Spuy, E. (2005). *South African Police Reform in the 1990's: Internal Processes and External Influences*. Doctoral Thesis. Cape Town: University of Cape Town.

van Heyningen, E. (2008). Costly mythologies: The concentration camps of the South African War in Afrikaner historiography. *Journal of Southern African Studies*, 34(3): 495–513.

Vansina, J. (1995). "Bantu expansion" reconceptualized: New linguistic evidence and "The Bantu expansion". *The Journal of African History*, 2: 173–195.

van Vuuren, H. (2006). *Apartheid Grand Corruption Assessing the Scale of Crimes of Profit in South Africa from 1976 to 1994*. Pretoria: Institute for Security Studies.

van Vuuren, H. (2017). *Apartheid Guns and Money*. Auckland Park: Jacana Media.

Wright, J. (2009). Turbulent times: Political transformations in the North and East, 1760's-1830s. In C. Hamilton, B.K. Mbenga and R. Ross (eds.), *The Cambridge History of South Africa*, Vol. 1. Cambridge: Cambridge University Press.

Chapter 3

Transition to a police service

From enforcer to protector

Introduction

Core to the principle of democratic governance is the understanding that the state carries out its administrative duties in the name of the people it governs and in a manner consistent with democratic values. As seen in Chapter 4, the values of democratic governance in South Africa are enshrined in its Constitution as "one, sovereign, democratic state founded on human dignity, the achievement of equality and the advancement of human rights and freedoms" (Constitution, 1996, Section 1(a)). We have noted in the previous chapter (Chapter 2) that policing by the state is defined by a functional paradox, that its core function/role, in the words of Egon Bittner (1991), is a "mechanism for the distribution of non-negotiable coercive force" (p. 48).

Because of its history, South Africa's majority population continues to approach the exercise of police authority with deep reserve and suspicion, even as the organization attempts to transform itself from the enforcer of minority rule to serving the interests of the entire population. Subsequently, this transformation has been an enduring challenge. As Comaroff (2013) notes, referring to a key axiom of Bittner's (1991) observation, "[P]opulations are only policeable to the degree, and in the domains, that they themselves want to be policed" (p. xvii). This adage is applicable in reference to both the laws to be enforced and in the manner of enforcement. According to the fourth dimension of the organizational theory of police integrity, if the society at large has relaxed attitudes toward government misconduct in general, it is more likely that such attitudes will similarly apply to the police (Kutnjak Ivković, 2015). However, as discussed later, South Africans consistently voice concern regarding government corruption and a strong desire for fair, impartial, and effective administration of justice.

It is common—and politically salient—to split the South African story into two chapters. In the first, Europeans settle, seize land and resources from indigenous people, and establish a racist, illegitimate state based on abuse and exploitation, much of it supported by the police. In the second, freedom fighters, along with a shift in global power heralded by the collapse of the Union of

Soviet Socialist Republics (USSR), force the white government into a peaceful transition and usher in a new democratic order in which all citizens enjoy equal rights and services, including those provided by the police. In this story, the first chapter is titled *colonial and apartheid South Africa* (Chapter 2), while the latter is *the new South Africa* (Chapter 3) As common and deceptively useful for some purposes as this dichotomized narrative may be, it is ultimately a pernicious oversimplification of reality.

In this chapter, we provide a brief history of South Africa's democratic era, sketching shifts and continuities in politics, economics, violence, and policing. We show that, while much has changed in the country since the early 1990s, South Africa remains shaped by its past. A thorough understanding of this inertia is necessary to appreciate the challenge of developing a professional, democratic police service in an unequal, multicultural, and predominantly violent transitional setting.

Policing the transition

During the tumultuous 1970s, the National Party government's (NP) apartheid policy—propagating that races would develop best when kept separate—was starting to crumble "as a result of the growing predominance of the manufacturing sector with its demand for skilled labor, the demographic weakness of whites, the collapse of the resource base of the black 'reserves' and the rapid urbanization of blacks" (Grundlingh and Giliomee, 2007, p. 307). Emerging during this time as the dominant apartheid opposition in the country and aided by the USSR, the banned African National Congress (ANC) would apply lessons gleaned from Vietnam strategists in formulating its revolutionary "people's war." As the organization aimed to make the country "ungovernable," a heightened level of violence would permeate protest actions in the country (Jeffery, 2019, p. 269).

Amid the ensuing societal chaos of the 1980s, the NP began shifting its posture in a final effort to retain power. In a decade that saw unprecedented mass protests, two progressively escalating states of emergency, the occupation of townships by the South African Defense Force and SAP, and the arrest and killings of thousands by security forces, the NP initiated a number of social, administrative, and ideological reforms in an attempt to quell uprisings and regain control of the country (Morris and Padayachee, 1988). These transitions provided coloreds and Indians with new political rights, scrapped the detested pass book system and laws, and directed new welfare services to the black population. By then, probably too little, too late, these reforms proved inadequate. With pressure mounting from the Afrikaans-speaking, international, and business communities, and with the economy in steady decline, the government entered into secretive bilateral negotiations with the exiled ANC leadership (Rantete and Giliomee, 1992).

In February 1990, Nelson Mandela was released after 37 years in prison, and various liberation movements—including the ANC—were unbanned. The new NP president, F. W. De Klerk, proclaimed that "the season of violence is over" and instructed the SAP to end its politically-rooted role as apartheid enforcer and, instead, focus on "typical crime situations" (Cawthra, 1993, p. 38). Formal negotiations began, and in 1991, an agreement was reached regarding the multiparty National Peace Accord (NPA). The accord was intended to reduce the then endemic levels of violence, in part by introducing violence monitors, police monitors, a police advisory board, and a police code of conduct (Rauch, 2000). Illustrating the significance of the Accord to policing, Kinnes quotes a SAP gang-unit officer (Kinnes, 2017, p. 214):

> [W]e came from a culture of policing where it was not offensive and a violation of any human rights structure for us to engage in violence with our suspects. I still remember . . . we had to sign the Peace Accord where we undertook not to use violence. That was a huge orientation process for us, as if one day you can use violence and the next day you can't. [Violence] was almost an accepted method.

In 1992, the Convention for a Democratic South Africa (CODESA) brought key actors together to plot a way toward an "undivided nation" with "equal opportunities and social justice for all South Africans" (CODESA, 1991).

The government freed up funds to hire 10,000 new police officers over a 12-month period and improve police salaries by up to 80% (Cawthra, 1993, p. 39). Although most police adapted to the change, some within the SAP would continue to resist the coming transition by stoking violence between the ANC and Inkatha Freedom Party (IFP). Reflecting on the role of the SAP in the deluge of political killings in the early 1990s, Lodge and colleagues write (2007b, p. 397):

> There were some 20,000 political fatalities between 1984 and 1994, 70% of them between February 1990 and April 1994. Between 14,000 and 16,000 of the 20,500 fatalities involved clashes between supporters of different predominantly black organizations, mostly between supporters of the Inkatha Freedom Party (IFP) and those of the ANC. But the media put by far the greatest emphasis on rogue organizations existing within the security forces. The police killed an estimated 3,000 people in riots or fighting guerrillas in actions deemed lawful by the state. Members or ex-members of the police asked the Truth and Reconciliation Commission (TRC) for amnesty for the unlawful death of some 150 people.

While the police had killed people during the transition, police killings, both on and off-duty, reached a troubling level. In 1993, 176 police were killed, a rate of 326 murders per 100,000 citizens (Goldstone, 1994; Faull, 2018a), significantly higher than the already extreme national murder rate at the time of approximately 78 murders per 100,000 citizens (Kriegler and Shaw, 2016).

In 1991, whites made up 94% of commissioned officer ranks in the SAP. Two percent of commissioned officers were Indian, and the remainder colored and black (Cawthra, 1993, p. 79). Although formally abolished in the mid-1980s, segregated training facilities were finally integrated only in the early 1990s. Roughly 20,000 black officers were employed in the ten homeland police forces which, much like the SAP, were undemocratic and politically captured (Cawthra, 1993). Their recruitment and training standards, however, were lower than those of the SAP, and "incompetence, callousness and corruption" were commonly found within their ranks (Bruce, 2002). Notwithstanding this, by 1995 the 11 police forces would be merged, together with various "freedom fighters," to form the South African Police Service (SAPS).

Amid the violence and uncertainty of the early 1990s, it was unclear whether South Africa would manage its transition to a fully-fledged democracy without stumbling into civil war. Through numerous negotiations, concessions, ambiguous settlements, and political cozening, the country escaped this fate and the haphazard 1994 elections (see Chapter 2) ushered in a period of aspiration, hope, and fear.

Socioeconomics of transition

Despite economic stagnation in the 1980s, South Africa's per capita gross domestic product (GDP) in 1994 accounted for half the GDP of sub-Saharan Africa (Nattrass and Seekings, 2011, p. 518). Although less unequal than its 1970 peak (Simkins, 2011, p. 511), South Africa remained one of the most unequal countries in the world in the early 1990s, as its Gini index showed (Atkinson et al., 2017), with whites earning nearly ten times more than blacks (Hirsch, 2005, p. 2). Fifty-three percent of black people lived in poverty in 1989 (Fallon and Perreira da Silva, 1994, pp. 39–42), and most resided in rural areas (Simkins, 2011, p. 510). Unemployment stood at 14% (Leibbrandt et al., 2018). Centuries of racist rule had created a decidedly uneven society.

Having allied during the 1920s with the South African Communist Party (SACP), the early ANC embraced both African nationalism and Marxism. In fact, one of its most hallowed documents, the Freedom Charter, drafted by prominent SACP member Lionel Bernstein, included a list of basic human rights (Lodge et al., 2007a, p. 329), as well as demands such as "The national wealth of our country . . . shall be restored to the people; The mineral wealth . . . the banks and monopoly industry shall be transferred to the ownership of the people" (Freedom Charter, 1955). Already the ideological acolyte of the USSR, the ANC would commit to a National Democratic Revolution (NDR) at a conference in Morogo, Tanzania, in 1969. This economic strategy had as a primary goal to move South Africa from a free-market system to socialism and, ultimately, communism. In step with this vision, on his release from prison, Mandela reiterated that nationalization was the foundation of the ANC.

By 1992, the failure of Soviet communism compelled the ANC to follow a more conventional economic approach. Undeterred, however, the SACP continued to push for economic "growth through distribution." After its policy conference that year, the ANC's market orientated members, led by Thabo Mbeki, had not only shifted the organization's focus away from nationalization but had the ANC commit to trade liberalization in partnership with the World Bank and the International Monetary Fund. With an "outraged . . . ANC's Left" (Lodge et al., 2007b, p. 424), the economic policy debate would continue within the organization as it incrementally refined the NDR's implementation during its five national conferences held in the post-1994 era. For example, a Strategy and Tactics document from its 2017 conference described the ANC as "the vanguard of the NDR . . . the disciplined force of the Left," while calls for land expropriation without compensation (EWC) continued, suggesting that nationalism had remained the cornerstone of the ANC (Jeffery, 2019, pp. 424, 5162, 5233).

Since the early 1990s, access to basic services has drastically improved for most South Africans (Statistics South Africa, 2019a). However, the result of the ANC's economic policies has been a continuation of a stratified society comprised of a minority middle-class/elite and a majority working and unemployed poor, but with a shift in racial composition (Seekings and Nattrass, 2005, p. 376; Nattrass and Seekings, 2011, p. 519). While inequality has declined between race groups, it remains stark. In 2017, 64% of the black population lived in poverty, compared to 1% of the white population (Statistics South Africa, 2018). Inequality has increased within race groups as some have benefited from the new nation and others have not. This has prevented a reduction in national inequality (Leibbrandt et al., 2010). In fact, South Africa consistently ranks among the top three countries worldwide based on the degree of inequality in the distribution of family income (CIA Factbook, 2019; World Bank, 2019). Despite the expectation that the income gap will decrease over time, South Africa's Gini index scores, calculated every few years, suggest that initial improvements in equality in the early 2000s have been followed by a period of high inequality (Figure 3.1).

On the back of a growing economy in the mid-1990s, the Mandela and Mbeki administrations introduced a range of progressive poverty alleviation and social policies, initially as part of what was called the Reconstruction and Development Program (RDP). These included the rolling out of housing, education, water, and electricity to those without; a public infrastructure development program that supported skills development and basic employment; and the introduction of cash grants for the young, elderly, and disabled (Leibbrandt et al., 2010, 2018). Under president Mandela's successor, Thabo Mbeki, the RDP program was replaced by the Growth, Employment, and Redistribution (GEAR) strategy in 1996, marking a shift in the ANC's socialist ideals to

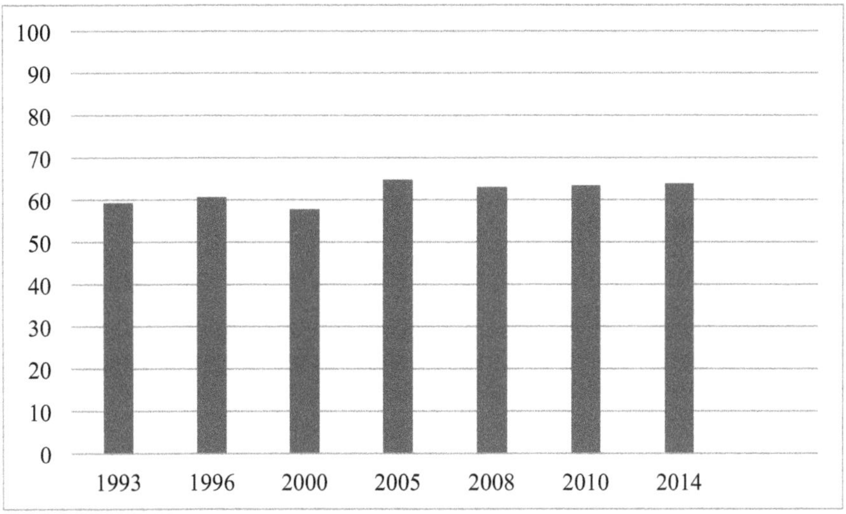

Figure 3.1 South Africa's Gini Index
Source: The World Bank (2019)

policies focused "on 'sound money,' export orientated growth, privatization of public services, public finance austerity, labor market flexibility, and a reliance on foreign investment as a key for prosperity" (Lodge et al., 2007b, p. 425).

While the number of those employed grew, the quantity of job-seekers far exceeded them, and unemployment soared. Following a global recession, economic growth slowed considerably in the 2010s, and government debt increased. In 2018, 30.4% of black people were unemployed compared to 7.6% of whites (Statistics South Africa, 2018a). Since 1994, unemployment has consistently been the issue people believe the government should most urgently address (Plus 94 Research, 2018).

Whereas the government has failed to address widespread poverty and inequality, it has managed to develop a small but growing black middle class, albeit defined by income rather than wealth. In 2016, Orthofer showed that 10% of South Africans owned 90–95% of the country's wealth, while the top 10% of income earners received 55–60% of all income (Orthofer, 2016). The next 40% of the population shared 30–35% of all income, but only owned 5–10 % of all wealth. This reveals there is no middle class with regard to wealth in the country: The middle 40% of the wealth distribution is almost as asset-poor as the bottom 50%. In 2016, 47% of employed South Africans earned less than R3,500 (US$233) per month (National Minimum Wage Panel, 2016), while SAPS constables earned four times that amount. Ninety-five percent of South Africans earn less than R11,000 (US$733) a month after tax deductions (SALDRU, 2019), meaning that the lowest SAPS ranks are part of the more "affluent" 5% of the country (Faull, 2018b).

In this context—with widespread poverty, unemployment, and a skewed distribution of wealth and opportunity—a job in the SAPS is extremely attractive to many South Africans. This is not for the occupation's honor or esteem, but because the agency is a large employer requiring only a high-school certificate yet offering relatively good salaries and an avenue out of poverty. Subsequently, in 2018, the SAPS received over 500,000 applications for just 3,500 trainee posts (Malingo, 2018).

Crime and violence

South Africa is one of the most violent countries in the world, with a murder rate surpassed only by a handful of Latin-American countries (UNODC, 2019). In 2019, it recorded 36.4 murders per 100,000 residents, averaging 406 murders per week (Crime Hub, 2019). Although murder declined by half between 1993 and 2011, by 2019 it had increased by 35% from its 2011–2012 low (Figures 3.2 and 3.3). Hsiao and colleagues (2018) estimated that the lack in addressing this level of violence costs the country 5% of its GDP. These trends are the product of violence that has become entrenched in South African homes and social relations, itself born of the state and interpersonal violence of the preceding century (see Chapter 2; Glaser, 1998,

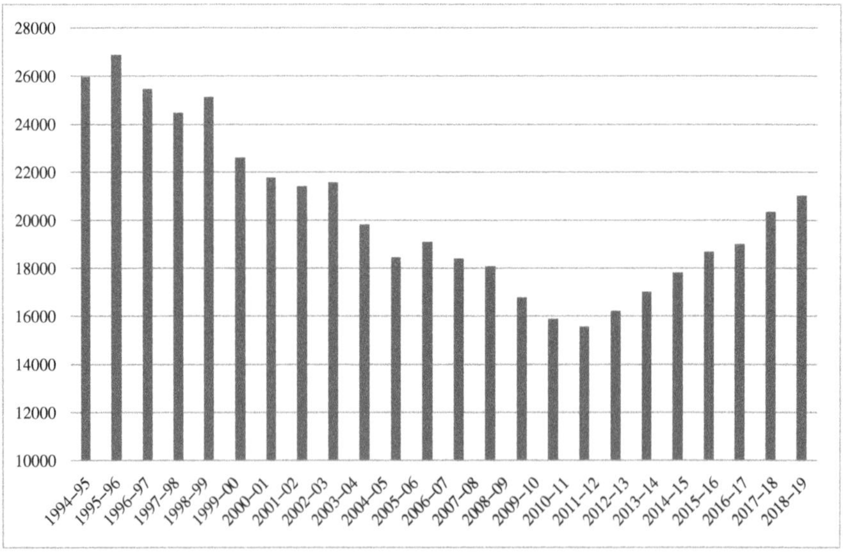

Figure 3.2 Murder in South Africa, 1994–2019

Source: ISS Crime Hub, 2019

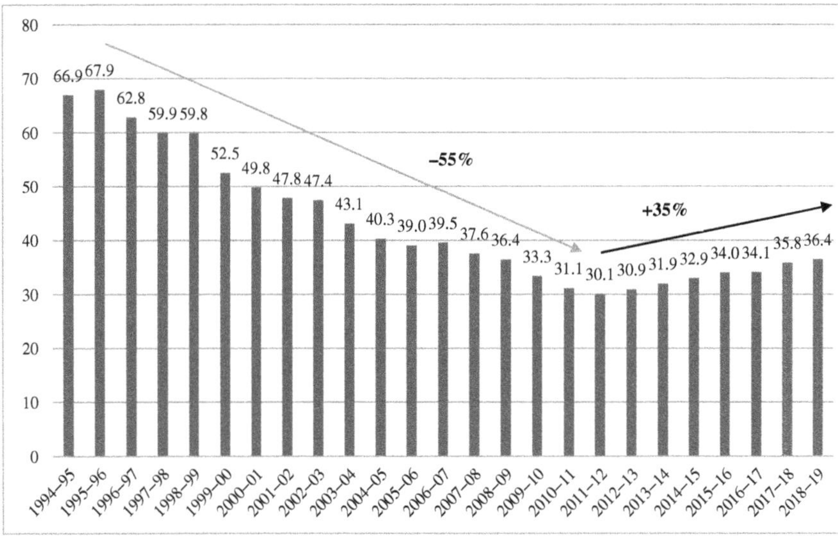

Figure 3.3 Murder rate per 100,000 residents, 1994–2019
Source: ISS Crime Hub, 2019

2000; Kynoch, 2016), the subjugation and shaming of the black population, notions of race and gender superiority, communities accustomed to policing themselves in the absence of the state, and ruthless political violence during the 1980s and 1990s (Posel, 2009; Shaw, 2002; Cooper-Knock, 2016; Super, 2017; Chapter 2).

The colonial and apartheid states strategically used violence to control population groups while providing most with only limited crime-related, professional policing. Steinberg (2008, pp. 64, 98) suggests that without a legitimate police presence, informal security providers emerged in black communities and residents grew accustomed to using violence to manage conflict. In a similar vein, apartheid's forced relocation of black, colored, and Indian residents dissolved established community structures and allowed gangs to seize the monopoly of force and order in apartheid's race-segregated settlements (e.g., Pinnock, 2016; Jensen, 2008; Standing, 2006). The apartheid state framed the quintessential violent offender as "African" and victims as white women (Super, 2010, p. 166). In reality, however, most violence occurred between black men and in the less common instances in which whites were victims, their assailants were usually other white men (Super, 2010, p. 167). Violence was also part of resistance politics. Anti-apartheid groups, including the ANC, stoked violence within township communities (Kynoch, 2016), while the police covertly did the same at times (Cawthra, 1993), without being able to steer it fully or reign it in when it no longer served their purposes.

In the early 1990s, as the old state and its criminal justice architecture were dissolved and the new government formed, existing gangs and other criminal elements outside of South Africa benefited from the opening of borders, a disorderly state, and the entrenched societal violence, by establishing new economies of coercion and social control (e.g., Shaw, 2002, 2017; Kinnes, 2017; Standing, 2006). Shaw (2017) notes that the country's contemporary "underworld" is a product of the "upperworld"—the former being a place for the many who remain excluded from the latter.

Kynoch (2016) notes that, since the late 1990s, the state has been preoccupied with crime and violence. Where the ANC had pre-1994 framed township problems as products of the apartheid state, it has become inclined to present township problems as generated by townships, or, in Samara's analysis, to see townships as problems to be managed (Samara, 2011 p. 53). Additionally, Kynoch (2016) notes that township communities have increasingly and intentionally reproduced disorder in an effort to attract state attention to their chronic dysfunction, such as by burning trains or shutting down major thoroughfares. In 2019, the SAPS recorded 2,898 incidents of public violence compared to 1,024 incidents in 2006 (Figure 3.4). In 2019 the SAPS policed 17,980 "unrest" crowd incidents and 55,582 "peaceful" incidents, describing "violent industrial and service delivery-related protest actions" as a threat to domestic stability (SAPS, 2019, p. 53).

In a shift in state violence, Super (2013) notes that criminal justice sanctions have become increasingly extreme after 1994, rather than tolerant. South Africa has some of the harshest minimum sentence legislation in the world, and between 2000 and 2018 had the most rapid increase in life sentences of all countries measured by Penal Reform International (Penal Reform International,

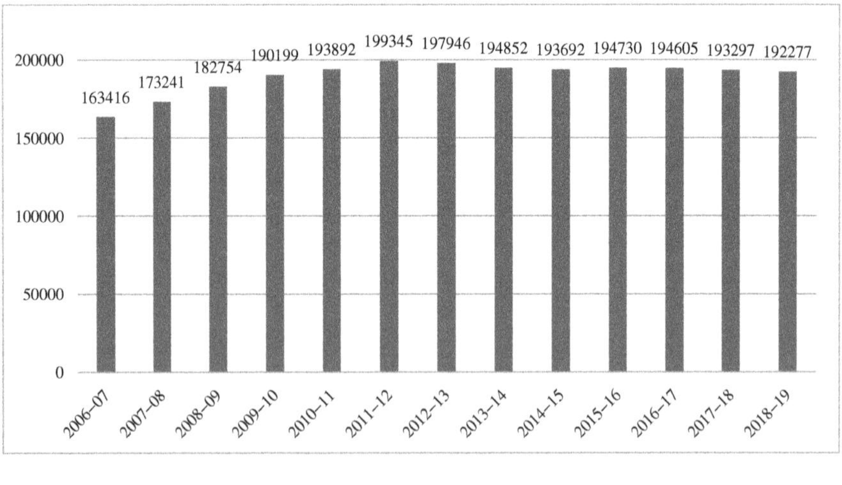

Figure 3.4 Total SAPS personnel, 2006–2019

2018). It also had the third-highest number of people serving life sentences and the second-highest ratio of prisoners serving life sentences among those countries on record. Super (2013, p. 108) sees the use of mass incarceration as a "repressive social control apparatus of the modernizing South African state." Harsh sentences and mass incarceration, she notes, coupled with a state discourse of "rehabilitation," signal both a willingness to punish and a pretense to care for the punished, walking a fine line between the liberation ideals of pre-1994 ANC and the challenges of governing a young, violent democracy.

Super's analysis of the post-apartheid prison echoes Hornberger's (2011) study of efforts to teach SAPS officers about human rights and to embed these in the organization during the early 2000s. Hornberger suggests that, by hiding behind a public image of respectful, human rights-based policing while police abuse remains fairly common, "human rights as a state practice is actually in itself a form of violence" (Hornberger, 2011, p. 152).

Together, then, Super and Hornberger suggest that the language of "human rights" and "rehabilitation" function as smoke screens behind which sinister and punitive police practices continue. As Super (2013, p. 130) points out, without such rhetoric the government could not justify locking up so many citizens. As such, Super views shifts in the country's governance of crime and institutions of justice as not simply a response to crime but rather "a highly politicized process of social transformation which began in the 1970s" (Super, 2013, p. 136).

South Africa's violence remains rooted in the lives of its citizens, having formed a cycle that policing alone cannot disrupt. This is, in part, because the country's children learn from early ages that violence is a legitimate means to solve problems and command respect. In a 22-year study of children born in Soweto in 1990, 99% confirmed their experience with violence by the age of 18 (Richter et al., 2018). Interviews with 3,515 children aged 10–17 in the North-West province found that 56.3% had experienced physical abuse, primarily by caregivers and teachers (Meinck et al., 2016). A 2015 national victim study of 15- to 17-year-olds found that 33% had been hit, beaten, or kicked by an adult, 20% had been sexually abused, 23% had witnessed violence between household members, and 16% had been attacked with a weapon (Artz et al., 2016). The results of a panel study of almost 5,000 14- to 22-year-olds (Govender, 2018) and a separate study exploring the life histories of repeat violent offenders (Gould, 2015) suggest that these experiences predispose children to future violent victimization and perpetration. A 2018 study of the cost of South Africa's violence found that interpersonal violence could be reduced by 16% if children did not witness family violence, and drug abuse could be reduced by 14% if sexual violence against children were prevented, among other outcomes (Hsiao et al., 2018).

A 2010 government-commissioned report into the violent nature of crime found that South Africans generally believed that violence was a necessary and justified conflict resolution tool (CSVR, 2010). The study of the Centre for the Study of Violence and Reconciliation (CSVR) and other studies have shown that violence has historically been, and remains, key to particular notions of what it means to be a successful man in the country (Breckenridge, 1998; Pinnock, 2016; Ratele, 2008), something of particular importance when so many men find themselves shamed by life without employment. Unemployment feeds shame, and shame is connected to violence (e.g., Gilligan, 1996, 2000; Bruce, 2006; Gqola, 2015). Advances in neuroscience have shown that chronic stress of the type brought on by unemployment and emasculation depletes dopamine, stimulates depression, and promotes aggression (Sapolsy, 2016). In addition, it seems that expressing aggression or punishing others reduces stress and stimulates states of satisfaction. One consequence of this is that, despite an extremely high incidence of violence against women and children in South Africa, most violence occurs between men (e.g., Seedat et al., 2009; SAPS, 2019; Statistics South Africa, 2019b; Bruce, 2020).

Although violence mostly occurs between people who know each other (e.g., SAPS, 2019; Statistics South Africa, 2019b), South Africans remain highly fearful of violence from strangers (e.g., Statistics South Africa, 2019b; Collins, 2013). In this arena, aggravated robbery in streets, homes, and businesses, as well as "gangsterism," are highly feared, as is burglary (Statistics South Africa, 2018b, 2019b). While acquaintance and stranger violence/crime are pervasive across the country (e.g., Abt and Winship, 2016; Weisburd, 2015), South Africa's violence clusters in particular places and times and disproportionately affects poor black and colored people (Stats SA, 2019b; Breetzke and Edelstein, 2019; Breetzke, 2018). In 2018–2019, 20% of all murders were recorded in just 2.7% of the country's police precincts, with roughly 50% of murders occurring in 13% of precincts. Within these (and other) areas, most murders occur over weekends and are often associated with alcohol use (SAPS, 2019; Western Cape Department of Health, 2018).

Although the SAPS alone is unlikely to end South Africa's violence, the police remain crucial to its decline. This could most effectively be achieved by targeting the places where harm is most predictable. However, this has not happened. Instead, SAPS officers struggle under what many perceive as an unrealistic workload to realize their impossible crime prevention mandate (Manning, 2005; Burger, 2006; Faull, 2018a). For example, detectives working in the country's most violent precincts can each be responsible for more than 100 case files (Perkins, 2018; Faull, 2018a). This workload, coupled with the introduction of a CompStat-like performance monitoring system in 2002, has resulted in a number of data-manipulation scandals (e.g., Bruce, 2010), the practices of which have been observed by researchers embedded in the SAPS (e.g., Vigneswaran and Hornberger, 2010; Faull, 2016, 2018a). In October 2019, another such scandal was exposed, this time alleging the manipulation of performance data and the premature closing of cases by the Independent Police Investigative Directorate (IPID) (Viewfinder, 2019).

Growing from the ubiquity of violence in South Africa is an economy of hitmen for hire. Assassinations have been an increasing feature in the country, deployed to remove political (de Haas, 2016), business, and romantic obstacles, and to communicate signals of power and authority (Shaw and Thomas, 2015). A study by Kim Thomas counted 159 assassinations in 2017 alone, a significant increase from 46 in 2012 (Thomas, 2018, p. 3). Shaw and Thomas argue that these killings mark a shift from the ideological assassinations carried out during apartheid to murders for hire in the new order. This includes the assassination of those deemed threatening to corrupt politicians, such as whistleblowers, political rivals, and members of the criminal justice system (Shaw, 2017; Thomas, 2018).

In 2019, Gallup's Law and Order Index, which measures citizen perceptions and experiences of crime, ranked South Africa at 137 out of 142 countries (Gallup, 2019). According to the poll, based on 152,000 interviews worldwide, 69% of people worldwide felt safe walking in their area of residence at night, and 68% had confidence in their police. In contrast, only 31% and 59% of South Africans felt the same, respectively. In the same year, the Global Peace Index ranked South Africa at 127 out of 163 countries and territories based on measures of "peacefulness" (Institute for Economics and Peace, 2019). In addition to violent crime, South Africans suffer high rates of nonviolent property crime and other offenses such as fraud and cybercrime.

Policing, policy, and politics

According to Brogden and Shearing (1993, p. 40), apartheid's police used violence strategically and gratuitously, regularly bending and breaking the SAP's formal rules (Brogden and Shearing, 1993, p. 40). Against this background, the ANC of the early 1990s viewed crime and violence as a product of abuse, inequality, and political oppression (Super, 2013), and perceived the apartheid police as militant, secretive, unaccountable, racist, and violent (Haysom, 1993). "Criminals" were perceived as victims of an unjust state, and in turn, a democratically elected government and legitimate police service would end their victimization, crime, and violence. It sought to create a politically neutral, accountable, and community-oriented police service. While decentralized, provincial policing was considered as a promising new model, centralization was ultimately pursued to promote national control and prevent potential secession (Brogden, 1994; van der Spuy, 2005).

In 1995, the South African Police Service (SAPS) was formed, and reference to a police "force" fell from political parlance. The SAPS Act (1995)—the law enabling its formation—mandated it to prevent, investigate, and combat crime and secure the safety of the inhabitants of the Republic. Policy frameworks and strategies such as community policing, many of them borrowed from the global North (e.g., van der Spuy, 2005; Steinberg, 2011), were adopted to guide the new SAPS mission, service standards, and operational practices. The SAPS Act prescribed the formation of Community Policing Forums (CPFs) at each police station, intended to instill local community oversight of police activities

(Pelser et al., 2002).[1] A new "change management" division pursued "demilitarization" and broader reform through a range of activities, including changing police rank titles, salutations, uniform, coat of arms, and vehicle color, and renaming the most infamous police stations (Rauch, 2000; Faull, 2018a).

George Fivaz, a career SAP officer, was appointed as the first National Commissioner of the SAPS. His priority was building legitimacy while preventing crime, addressing discrimination, improving discipline and morale, and establishing an organizational culture that respected human rights (Rauch, 2000). Two years into his tenure, with public anxiety about crime pervasive (StatsSA, 1999), the government declared crime prevention a national priority and in 1996 launched the National Crime Prevention Strategy (NCPS) (NCPS, 1996). The NCPS highlighted corruption in the criminal justice system as an impediment to its reform and efficacy (Parliamentary Monitoring Group, 1998). Despite its early reforms, the SAPS reported in 1997 that a "decisive break from the past" remained necessary in order to transform "the nature and style of policing" in South Africa, so that it would be "professional, transparent, accountable" and respectful of citizen rights (SAPS, 1997, p. 2).

As desired by the government, the early SAPS positioned itself as just one part of the more holistic NCPS (SAPS, 1997). The NCPS sought to address crime through the whole of government; it was an initiative for the whole of society that was focused on effective criminal justice, environmental design, public values, community mobilization, and transnational cooperation (NCPS, 1996). The strategy was laudable, but never successfully implemented (Rauch, 2004). Although overall violence (as measured by murder rates) continued to decline (see Chapter 1), violent and property crimes soared and received significant media attention, causing panic among the more affluent and a spike in skilled emigration. By the turn of the century, crime was perceived to be increasing and feelings of safety had plummeted (Burton et al., 2004).

Aligned with the NCPS, and in an effort to address corruption in the SAPS, the Anti-Corruption Unit (ACU) was established in 1996. Although increasingly successful in arresting and convicting corrupt officers (e.g., Burger and Grobler, 2017; Newham and Faull, 2011), the unit was nevertheless closed down in 2002 (see Chapters 5 and 8). Under mounting pressure from the public and frequently embarrassed by the pervasiveness of crime in the country, the SAPS in 2001 announced a new National Crime Combatting Strategy. Kinnes (2017, p. 229) notes that this unilateral police initiative seemed to signal to the government that the SAPS had given community policing a chance, and that it now would revert to what it knew best—tough law enforcement. This was followed by a decade-long surge in the SAPS's budget and staff complement. From 2001 to 2011, its personnel grew in numbers from 120,000 to 200,000 (Bruce, 2013), while public perceptions of safety continued to decline (Pharoah, 2008; Statistics South Africa, 2011) and anti-crime rhetoric became increasingly militant (Faull and Rose, 2012). Trust in the police improved slightly during this period, but this progress came from such a low starting point that most people (51%) still mistrusted police in 2011 (Chingwete, 2016, p. 6).

Victim surveys in the late 1990s and early 2000s did not reveal corruption in the SAPS as a significant problem (e.g., Statistics South Africa, 1999; Pharoah, 2008). Although traffic-related bribery was the most commonly experienced form of government corruption, it only affected 2% of the respondents, and the SAPS almost never carry out traffic enforcement. Rather, the much smaller metropolitan police services and provincial traffic authorities were likely to blame. However, this does not mean that SAPS officers were not abusing their authority.

In 1998, Nelson Mandela handed over the presidency to his deputy, Thabo Mbeki, who in 1999 established the Directorate for Special Operations (DSO) or "Scorpions" to tackle priority crime and corruption in the country (see Chapter 5). In 2000, George Fivaz handed over the SAPS to its first black commissioner, Jackie Selebi. Selebi, who was not a police officer, was an ANC and Mbeki loyalist. Under his leadership, the SAPS was restructured, with specialist units—including the anti-corruption unit—being disbanded or redeployed to police stations (Omar, 2007; Kinnes, 2017). With Selebi at the helm, the SAPS became increasingly embroiled in obstructionist "high policing" (Brodeur, 2010) used to spy on and outmaneuver rival factions within the ANC (Steinberg, 2012, 2014), which would ultimately contribute to the demise of the Scorpions. In 2008, Selebi was suspended and in 2010 convicted of corruption after investigation by the Scorpions. The court found that Selebi had accepted money and gifts from convicted drug smuggler and suspected criminal high-flyer Glen Agliotti. It also found Selebi to be a "liar [and a] person of low moral fiber" (State vs Selebi, 2010).

In the same year that Selebi was suspended, Mbeki was ousted from the presidency by a faction loyal to his corruption-accused deputy, Jacob Zuma. In 2005, Zuma's business associate, Shabir Shaik, was found guilty of corruption, including facilitating the payment of bribes to Jacob Zuma. The payment was intended to lubricate relations linked to South Africa's multibillion rand "arms deal"—the 1998 purchase of R30-billion-rand's worth of unnecessary military equipment—which saw numerous leading ANC officials corrupted by international arms manufacturers, eroding the integrity of the liberation movement (e.g., Feinstein, 2007; Holden, 2008). Shaik's conviction should have resulted in Zuma's prosecution, but following allegations that prosecutors had unfairly targeted him for political reasons, pro-Zuma allies within the ANC successfully lobbied to have the charges against him withdrawn and police investigations into the matter dropped (Nattrass, 2017).[2] However, Zuma's political career was not yet back on track.

In 2006, Zuma was accused of and tried for rape. The victim, Fezekile Kuzwayo, was the daughter of a former close friend. While he did not deny sleeping with her, Zuma argued that the sex was consensual. During the trial,

Fezekile was forced into witness protection after Zuma's supporters threatened her life and said they would "kill for Zuma" (Gqola, 2015). Zuma did not comment against such threats. Instead, he routinely sang the struggle song *umshini wam* ["bring me my machine gun"] outside the court, effectively celebrating violence. While in witness protection, police assigned to protect Fezekile illegally recorded the content of her meetings with both her advocate and psychologist, and communicated these to Zuma's legal team (Tlhabi, 2017). They told her not to inform her advocate about the recordings and fed her misinformation about the protection services available to her, among a litany of abuses of power (Tlhabi, 2017, pp. 1185–1192). In May 2006, the court found Zuma not guilty of the rape charge.

With Selebi gone and Zuma sworn in as president in 2009, the latter appointed another political ally, Bheki Cele, to lead the SAPS in 2010. Unlike Selebi, Cele was popular within the organization despite not being a police officer. This was partly because of his tough-on-crime rhetoric, his regular attendance at police funerals,[3] and his efforts to effectively "re-militarize" policing. In 2010, the SAPS announced that it would revert to the military ranks of apartheid's SAP, began exploring efforts to liberalize its use of lethal force (e.g., Bruce, 2011), and in 2011 introduced the heavily armed Tactical Response Team (TRT). Initially intended to respond to medium- and high-risk crimes, the TRT was soon performing daily policing activities. The risk involved in such utilization of a combat unit was starkly demonstrated in August 2012, when TRT and other officers shot and killed 34 striking mine workers and injured over a hundred more at Marikana, outside Rustenburg (Farlam, 2015).

The tragedy at Marikana did not occur under Cele, who had already been dismissed because of a 2010 finding of "improper, unlawful . . . maladministration" against him by the Public Protector (Ombudsman), and of being "unfit for office" by an independent board of inquiry (News24, 2012). The findings related to the leasing of property by the SAPS. Rather, the Marikana killings happened with Riyah Phiyega—a former social worker and businesswoman—as appointed National Commissioner. Phiyega was suspended after two years in her post, following findings by the Farlam (Marikana) Commission that she had intentionally lied under oath about police activities leading up to the slaughter of miners (Farlam, 2015). Though not sanctioned, she was also implicated in tipping off the then Western Cape Provincial Commissioner, Arnoe Lamoer, about a corruption investigation against him. Lamoer would be convicted of corruption in 2018 after accepting kickbacks from a business owner (ANA Reporter, 2019).

Phiyega was replaced by a career police officer, Khomotso Phahlane. He would serve as acting National Commissioner from October 2015 to June 2017, when he was suspended and later twice arrested on allegations of corruption and fraud (Wiener et al., 2019). His wife, a SAPS brigadier, was also arrested and charged for the crimes, which related to cars and other gifts allegedly provided to the Phahlanes by a business that received billions of rands in

SAPS contracts. The charges followed an investigation by the Independent Police Investigative Directorate (IPID), to be discussed later.

In November 2017, Zuma appointed Khehla Sitole to lead the SAPS. Although initially celebrated as a "clean" career police officer, he would be investigated in 2018 by the IPID for allegedly facilitating the laundering of money required to buy votes at the ANC's 2017 national conference. At the time of writing, the investigation was ongoing.

Zuma's presidency was plagued by disgrace, ranging from the pre-election allegations of rape and corruption against him to numerous scandals indicating high-level corruption and the "capture" and pillaging of key state institutions and parastatals by Zuma, his family, and his allies. This included the use of hundreds of millions of rands in state funds to provide "security upgrades" to Zuma's private residence at Nkandla and the promotion of sympathetic police officers to key positions from where they protected Zuma's cabal from investigation and prosecution (Pauw, 2018; Gqubule, 2017). The Zuma-led government repeatedly exonerated its president from any wrongdoing in the Nkandla matter, despite rising levels of public disaffection and protest (e.g., Figure 3.4) and a 2013 finding by the Public Protector against him. "What Nkandla signifies," wrote Thandeka Qhubule in 2017, "is how the ANC turned on its traditional constituency—the poor and downtrodden. . . . When the ANC defended Zuma and insulted [Public Protector] Madonsela, it was telling the population that it had chosen self-enrichment over being a movement of the people" (Gqubule, 2017, Loc 3680).

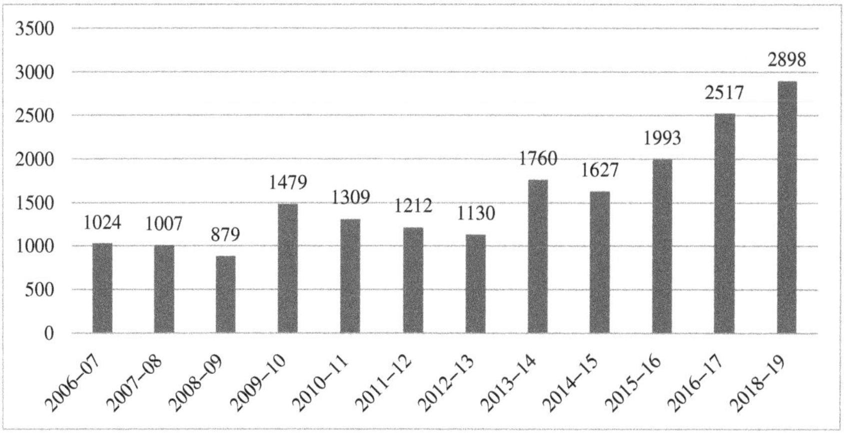

Figure 3.5 SAPS recorded incidents where police intervened in unprotected, disruptive or violent public protests, 2006–2019[1]

[1]Data for 2018 are unavailable.

Source: Crime Hub (2019)

The ANC further used its parliamentary majority to shield Zuma from successive votes of no confidence, as well as from general parliamentary scrutiny. As it became clear that the ANC was losing support under his leadership, the ANC elected Zuma's deputy, Cyril Ramaphosa, as president of the party in December 2017. Fearing that Zuma might cost the ANC the 2019 election, internal pressure mounted for him to step down as national president. He acquiesced to this in February 2018, and Ramaphosa took his place.

A period of "Ramaphoria" followed, in which, for the first time in a decade, the country seemed filled with a sense of political hope. Ramaphosa promised to clamp down on corruption, attract foreign investment, and get the state working. By December 2019, a number of initiatives to reform key institutions had been launched, including the SAPS and National Prosecuting Authority (NPA), but no significant prosecutions of allegedly corrupt politicians or business people had been recorded. Similarly, despite some efforts, Ramaphosa's government had by late 2019 yet to revive the economy. Conversely, South Africa experienced several anti-business reforms during the first year of his presidential tenure (Jeffery, 2019, Loc 5457).

Running parallel to the re-militarization of the SAPS under Zuma's government was an increase in civil claims and accusations of murder, assault, and torture against the police. In 2006–2007, 4,152 civil claims totaling R37 million (US $2,5 million) were lodged. By 2015–2016, these had increased to 16,498 claims totaling R21 billion (US $1,4 billion) (Crime Hub, 2019). In 2017–2018, a key post-apartheid accountability structure—the Independent Police Investigative Directorate (IPID)—received 5,651 criminal cases against police, 3,661 of which were for alleged assault, 217 for torture, and 436 for deaths as a result of police action (IPID, 2018, p. 38). In response, the SAPS in 2017–2018 initiated an intervention intended to curb civil claims (SAPS, 2018, p. 77). Burger has shown, however, that very few of the significant reform recommendations provided by and to the SAPS have been implemented (2015). More troubling still, in 2019, investigative journalists accused the IPID of prematurely closing cases against the police in an effort to "improve" its struggling performance against predefined indicators (Viewfinder, 2019).

Policing the police

The Independent Complaints Directorate (ICD)

South Africa's 1996 Constitution required the establishment of "an independent police complaints body . . . [to] investigate any alleged misconduct of, or offense committed by, a member of the police service" (Constitution, 1996, p. 106). In 1997, the Independent Complaints Directorate (ICD) was created as the country's first exclusive police complaints body. At its inception, it had 36 staff in three regional offices. Chapter 10 of the SAPS Act (1995) required that all deaths in police custody or resulting from police action be investigated by this body. It also allowed the ICD to investigate other complaints, ranging from corruption to torture, and poor service delivery to failure to comply with the Domestic Violence Act.

Independence was key to the vision of the ICD and its potential to pierce through the SAPS code of silence. ICD investigators were granted the same powers as police detectives. Within three years of its establishment, however, researchers were asking what the ICD could do to compel the police subjects of its investigations to answer investigator questions (Bruce et al., 2000). Moreover, the ICD could not force the SAPS to take disciplinary action against officers, while the National Prosecuting Authority (NPA) did not need to elaborate its decisions not to prosecute ICD cases. Instead, the ICD could only make recommendations to the SAPS, which could, and often did, ignore these without consequence (Berg, 2013). Despite having "independent" in its name, the ICD drew its powers and mandate from the SAPS Act and reported to the Minister of Police (then Safety and Security). This was problematic because, in the face of widespread fear of crime, the SAPS and its mandate were the Minister's priority, not that of the ICD.

Complaints reported to the ICD increased rapidly, from 1,999 in 1997–1998 to 5,675 by 2001–2002 (Table 3.1). This was probably due to increased awareness of the Directorate rather than augmented abuses. Most complaints related to minor misconduct. The most common categories of criminal complaint included assault, malicious damage to property, corruption, and theft (Table 3.1). To cope with its workload, and after only two years of operations,

Table 3.1 Complaints received by the ICD, 1997–2010

Year	*Complaints*
1997–1998	1,999
1998–1999	2,874
1999–2000	4,380
2000–2001	5,225
2001–2002	5,675
2002–2003	4,443
2003–2004	5,882
2004–2005	5,790
2005–2006	5,119
2006–2007	5,412
2007–2008	5,822
2008–2009	6,119
2009–2010	6,337
2010–2011	5,869
2011–2012	4,923

Source: ICD Annual Report 1997–1998, 1998–1999, 1999–2000, 2000–2001, 2001–2002, 2002–2003, 2003–2004, 2004–2005, 2005–2006, 2006–2007, 2007–2008, 2008–2009, 2009–2010, 2011–2012

the ICD announced that it would no longer receive certain types of complaints, including those that could, but had not yet been reported to the SAPS (ICD, 1999, p. 25). Nevertheless, the ICD continued to struggle (Bruce, 2006). In 2005, the ICD established an Anti-Corruption Command (ACC) to focus on corruption complaints. The ACC was given only four investigators, who by 2010 had received over 600 complaints, resulting in only 22 convictions (Newham and Faull, 2011, p. 10).

In 2006, then National Commissioner Jackie Selebi called for the ICD to be scrapped, saying it had outlived its purpose (Berg, 2013). As mentioned, Selebi would subsequently be convicted of corruption. By 2011, the ICD's staff complement had tripled and it had an office in each province and a national office in Pretoria. Even so, it remained woefully underfunded and in 2010 employed only 270 staff—approximately 1.5% of the SAPS workforce—with 20% of senior management posts unfilled (ICD, 2011, p. 61; Hendricks and Musavengana, 2010). Almost all of the ICD's Annual Reports referred to insufficient resources curtailing its work (Berg, 2013). Responding to its weaknesses, in 2011 the ICD was reborn as the Independent Police Investigative Directorate (IPID).

The Independent Police Investigative Directorate (IPID)

In September 2010, the IPID bill was tabled in parliament, and in May 2011 the subsequent act was promulgated. The IPID Act (2011) marked a new chapter for the ICD. The IPID Act changed the ICD's name, emphasized crime investigation over complaint intake, empowered it through a separate legislation (rather than being a product of the SAPS Act), and revised its directive. The new mandate compelled it to investigate not only deaths in police custody or those resulting from police action but also any complaints relating to the discharge of an official firearm by a police officer; rape by a police officer, or by any person in police custody; complaints of torture, assault, or corruption, and any matter referred by the Executive Director (IPID, 2011, p. 23). The Act also compels police officials to report, in writing, such incidents to the IPID and to cooperate with investigators or face prosecution (IPID, 2011, p. 26). Also, the SAPS must institute disciplinary hearings based on IPID recommendations within 30 days of their receipt.

Tables 3.2 and 3.3 present the total count and type of case intake/complaints received by the IPID between 2012 and 2018. Assaults account for 50–70% of all cases received. Considering the relative stability in figures, it is not clear from the ICD or IPID data whether the presence of "watchdogs" (i.e., the ICD or the IPID) has actually led to changes in police behavior.

According to the IPID, 31% of deaths in police custody can be attributed to natural causes, while 29% are suicides. Alarmingly, 41% of alleged rapes by police officers occur on duty, an indication that perpetrators believe their actions will not be revealed to colleagues or that their colleagues will not report them.

Table 3.2 Case intake/complaints received by the IPID, 2012–2019

Year	*Case intake/complaints*
2012–13	6,728
2013–14	5,745
2014–15	5,879
2015–16	5,519
2016–17	7,014
2017–18	5,651
2018–19	5,829

Source: IPID Annual Report 2012–2013, 2013–2014, 2014–2015, 2015–2016, 2016–2017, 2017–2018, 2018–2019

Table 3.3 IPID complaints by category, 2016–2018

Complaint Received	*2017–2018*	*2018–2019*	*% change*
Deaths in police custody	201	214	6%
Deaths as a result of police action	436	393	-10%
Complaint of the discharge of official firearm(s)	677	770	14%
Rape by police officer	105	124	18%
Rape in police custody	9	13	44%
Torture	217	270	24%
Assault	3,661	3,835	5%
Corruption	124	108	-13%
Other criminal matter	148	35	−76%
Systemic corruption	4	1	−75%
Non-compliance with Section 29 of IPID Act	69	66	−4%
Total	**5,651**	**5,829**	**3%**

Source: IPID Annual Report, 2018–2019

Furthermore, IPID reports show that, out of 2,015 cases closed by the IPID in 2018–2019 (Table 3.4; IPID, 2019, p. 70), 86% were declined by the IPID, dismissed, or found to be unsubstantiated or undetectable. In the same year, the IPID reviewed 2,337 complaints, but recommended disciplinary action to the SAPS in only 37% of cases (IPID, 2019, p. 64). Finally, of the 2,044 criminal recommendations submitted to the National Prosecuting Authority (NPA) in 20180–2019, the IPID was awaiting responses for 72% of these by the time it published its annual report (IPID, 2018–2019, p. 63) and has decided 28% (569 cases). Of all the 569 cases in which the IPID was aware of the NPA's response, the NPA had declined to prosecute in 72% of these cases, and has prosecuted only 10% of them. At the same time, the IPID's 2018 annual report notes a concern that both the SAPS and NPA were failing to act against senior officers accused of rule violations (IPID, 2018).

Crime and safety policy, 2012–2016

Amid scandals such as the "massacre" at Marikana, and the corruption, maladministration, and dishonesty findings against successive National Commissioners, the government has continued to produce impressive safety policy

Table 3.4 Outcomes of IPID cases concluded in 2018–2019

Outcome	*Number*	*Percent*
Acquitted	33	1.6%
Alternative Dispute Resolution	2	0.1%
Convicted	26	1.2%
Corrective Counseling	6	0.3%
Declined	1,051	52.2%
Dismissed	6	0.3%
Duplicate	10	0.5%
Indeterminate	2	0.1%
Inquest	20	1.0%
Policy Referred	2	0.1%
Referred	174	8.6%
Undetected	451	22.4%
Unfounded	2	0.1%
Unsubstantiated	216	10.8%
Withdrawn	14	0.7%
TOTAL	**2,015**	**100%**

Source: IPID Annual Report 2018–2019

guidelines for the SAPS and its partners. Three of the most important are the National Development Plan (NPC, 2012), the White Paper on Policing (2016), and the White Paper on Safety and Security (2016).

To ensure a decent standard of living for all South Africans by 2030, the NDP outlines the government's overall strategy and vision for South Africa, which requires that the SAPS "demilitarize" and "professionalize" by broadly enforcing codes of conduct, upholding standards of training and skill, improving recruitment and training, and establishing a National Policing Board (The Planning Commission, 2012). Such an approach is in accordance with the recommendations found in scholarly literature about the control of police misconduct and the enhancement of police integrity (e.g., Klockars et al., 2006; Kutnjak Ivković, 2015). More challenging, however, is how to go about implementing the strategy and doing so effectively and consistently.

The White Paper on Policing expands on the NDP's vision by providing an overarching, police-specific framework for its implementation (Civilian Secretariat for Police, 2016a). It notes the SAPS is regularly criticized for abuse of force, in part because "para-military police units" are carrying out regular policing tasks, and police leaders and politicians engage in warlike rhetoric. Much like the ANC's pre-1994 envisaged police, the White Paper suggests that the SAPS should be demilitarized, community centered, rights-based, accountable, well-resourced, and made up of highly skilled officers (Civilian Secretariat, 2016a, p. 14). In fact, it relies on demilitarization to instill police conduct characterized by respect for human rights and restrained use of force.

However, the White Paper does not provide clear definitions of, or guidance on, demilitarized, professional, or democratic policing. While it stresses that "community oriented," "democratic," and "professional" policing are the cornerstones of policing in the twenty-first century, it leaves it to police leadership "to establish the benchmark for professionalism within the SAPS" (2016a, p. 44). This is a problematic approach, with recent history demonstrating that SAPS's leadership might suffer from serious integrity challenges and consequently should not be entrusted to police the police. The task of establishing the standards of professional and democratic policing is thus left in the hands of those who either may not know what these values entail or may not be willing to uphold them.

Another key policy has been the White Paper on Safety and Security (Civilian Secretariat for Police, 2016b)—the government's flagship safety-related policy—which in many ways echoes the "whole of society" approach to safety as was envisaged by the NCPS of the mid-1990s. It recommends the establishment of a national crime center to oversee the coordination of relevant activities across government departments and stresses the importance of evidence-based interventions. However, like the NCPS, little progress has been made in the White Paper's implementation.

In September 2018, in an effort to garner support for the White Paper on Safety and Security, the Civilian Secretariat for Police Service (CSPS) hosted a National Summit on Crime and Safety in Johannesburg. Speaking at the event, the Minister of Police, Bheki Cele, told delegates that crime was a product of where and how South Africans were socialized: "If people live as animals they will behave as animals," he said (Faull, 2019). He said that criminal justice institutions could achieve little until the conditions in which South Africa's poor live are addressed. Reflecting on the White Paper—a product of his own department—he said, "I like these documents, but when they are not implemented, they do not help" (Faull, 2019).

Conclusion

Although South Africa's political transition was ultimately relatively peaceful, it fell short of the high national and international expectations for a new South Africa. A country in which all people have equal opportunities and rights—a notion fought and longed for by so many— remains a distant hope. While significant gains have been made, public anger festers, and protests against the state and private sector abound (Lancaster, 2018).

Since its inception in 1995, many mechanisms have been put in place to oversee the country's police service (SAPS) in efforts to develop and support an organization of integrity (see the discussions in Chapters 4 and 5). In spite of these endeavors, the SAPS is consistently regarded by the public as among the most corrupt government organizations in South Africa and is subsequently viewed with great suspicion (e.g., Transparency International, 2019). In what Steyn refers to as "police subcultural strategies" (Steyn, 2017b), SAPS members in turn become increasingly cynical toward the public, isolate themselves from the community, and ultimately regard the public with permanent distrust (Meyer and Steyn, 2009; Steyn, 2017a).

According to Steyn,

> [p]olice officials use police subculture strategies to protect themselves from the pressure and anxiety that originate from conspicuous qualities of two interdependent but paradoxical environments within which police officials perform their duties, the police occupational environment and the police organizational environment.
>
> (Steyn, 2017b)

In the context of the organizational environment of the SAPS, he postulates that the two most salient components of the organizational environment are the ambiguity of the police role emanating in the police occupation environment ("the unique forcible 'license' police officials have over the populace") and erratic disciplinary oversight (Steyn, 2017b). The SAPS is not unique in its efforts to address this core contradiction (as repository for the distribution of non-negotiable coercive force [Bittner, 1991]), inherent in modern policing. According to Manning's summary of the four-part mandate of the "modern" London police of 1828, the police were to prevent crime without the use of force, manage public order nonviolently, use force to obtain compliance only as a last resort, and minimize and reduce conflict between the police and the public (Manning, 1977, pp. 98–99). This core tension between the police and the policed appears present in every police organization founded on the organizing principles (Reith, 1952) of the London police tradition.

As society stabilized in the wake of the disorganized 1994 elections, it became apparent that the Afrikaner nationalist movement was no more—the transfer of political power from white to black had been "abrupt, decisive and irreversible" (Lodge et al., 2007b, p. 453). By 1996, whites dominated the country's wealthiest 1% but represented less than 54% of the wealthiest top 20%, a significant decline from their 87% position in 1975. By 2000, less than 10% of the population were white, of which less than 5% were Afrikaans-speaking. Yet, according to Lodge and colleagues (2007b, p. 436), the ANC persisted in its rendition of "democratic authoritarian rule," and, by the late 1990s, the Mandela government had people outside ANC ranks "afraid of voicing criticism or organizing protest action. South Africa's liberal democracy almost suffocated at birth." Since then, the high levels of intimidation and even physical attacks at public displays of dissent to ANC actions or proposals, coupled with the organization's insistence on following through with its National Democratic Revolution (NDR) strategy, cast doubt in some circles on whether the organization does in fact rule at the behest of democratic values and principles (see Jeffery, 2019, Loc 5272–5358).

Similarly, the SAPS, formed through the merger of eleven agencies, has never settled into the role of coherent, professional, rights-respecting agency as envisaged by the country's architects. This is surely in part the result of its integrity-challenged leadership, which, since 1995, has seen only one National Commissioner complete a five-year term free of integrity violations. While the organization is home to thousands of dedicated, talented police officers who work under difficult circumstances in an often thankless profession, it has also become a home for thousands of others who simply wanted a job (Steinberg, 2008; Faull, 2018a). Many of the latter have been recruited from poor families whom they must support with their police salaries. As a result, although SAPS officers are paid relatively well in the South African context, many remain trapped in poverty, which, in turn, erodes morale and undermines organizational efficiency. It may also incline some officers to engage more readily in corruption and abuses of force (Faull, 2013, 2018a).

Although a black middle class has emerged, most South Africans remain poor, with many reliant on state grants for survival. Whether through government intent or ineptitude, the destitute remain dependent on ANC rule which, according to the former president, Zuma, in 2014—will last "till Jesus comes" (Jeffery, 2019, Loc, 432). Where the ANC has succeeded in delivering social services to millions of formally underserved citizens, it has failed in many other important ways, such as not holding its own members accountable for corruption, sexual abuse, and violence.

By 2019, few apartheid police officers implicated in human rights violations at the Truth and Reconciliation Commission (TRC) had been prosecuted, while corruption and abuses within the new SAPS and broader state have since flourished (see Chapter 5). Under President Zuma, loyal supporters were deployed to key positions within the criminal justice system, including the SAPS, to keep Zuma's opponents at bay and protect him from investigation (Steinberg, 2014; Pauw, 2018; Wiener, 2018). During the same period, in 2012, murder began to increase for the first time since 1994, and was still rising in 2019. The swearing in of Cyril Ramaphosa as president in 2018 once again brought hope to the country, especially among its citizens who still expect the ANC to bring the corrupt and criminal within its ranks to justice.

Notes

1 Although these forums have not been uniformly successful in their objectives (Minnaar, 2010).

2 The matter was not forgotten, and in March 2018 the National Prosecuting Authority announced that it would reinstate charges against Zuma. Following various appeals, in October 2019, the trial was set for February 2020.

3 Although police murders have steadily declined since 1993, they remain a priority concern and are regularly referred to by politicians seeking sympathy for, or approval from, the police. In 2018–2019, 26 officers were murdered on duty (SAPS, 2018–2019).

References

Abt, T. and T. Winship (2016). *What Works in Reducing Community Violence: A Meta-Review and Field Study for the Northern Triangle*. Bethesda: USAID.

Afrobarometer (2018). *Summary of Results: Afrobarometer Round 7 Survey in South Africa*. Available at: http://afrobarometer.org/sites/default/files/publications/Summary%20of%20results/saf_r7_sor_13112018.pdf

ANA Reported (2019). Former Western Cape top cop Arno Lamoer found guilty of corruption. *IOL*, 18 February. Available at: www.iol.co.za/news/south-africa/western-cape/former-western-cape-top-cop-arno-lamoer-found-guilty-of-corruption-13374018

Artz, L., P. Burton, C. Ward, L. Leoschut, J. Phyfer, S. Loyd, and R. Kassanjee (2016). *Research Bulletin: The Optimus Study on Child Abuse, Violence and Neglect in South Africa*. Cape Town: Centre for Justice and Crime Prevention.

Atkinson, A.B., J. Hasell, S. Morelli, and M. Roser (2017). *The Chartbook of Economic Inequality: South Africa*. Available at: www.chartbookofeconomicinequality.com/about/

Berg, J. (2013). Civilian oversight of police in South Africa: From the ICD to the IPID. *Police Practice and Research: An International Journal*, 14: 144–154.

Bittner, E. (1991). The functions of police in modern society. In C. Klockars and S. Mastrofski (eds.), *Thinking about Police: Contemporary Readings*, 2nd ed. Boston: McGraw-Hill.

Breckenridge, K. (1998). The allure of violence: Men, race and masculinity on the South African goldmines, 1900–1950. *Journal of Southern African Studies*, 24(4): 669–693. DOI: 10.1080/03057079808708596

Breetzke, G.D. (2018). The concentration of urban crime in space by race: Evidence from South Africa. *Urban Geography*, 39(8): 1195–1220. DOI: 10.1080/02723638.2018.1440127

Breetzke, G.D. and I.S. Edelstein (2019). The spatial concentration and stability of crime in a South African township. *Security Journal*, 32(1): 63–78. https://doi.org/10.1057/s41284-018-0145-2

Brodeur, J.P. (2010). *The Policing Web*. Oxford: Oxford University Press.

Brogden, M. (1994). Reforming police powers in South Africa. *Police Studies: International Review of Police Deviance*, 17(1): 25–44.

Brogden, M. and C. Shearing (1993). *Policing for a New South Africa*. London: Routledge.

Bruce, D. (2002). New Wine from an old cask? The South African Police Service and the process of transformation. Paper presented at the John Jay College of Criminal Justice, New York, Centre for the Study of Violence and Reconciliation, Cape Town, South Africa, 9 May.

Bruce, D. (2006). Racism, self-esteem and violence in South Africa: Gaps in the NCPS' explanation? *South African Crime Quarterly*, 17: 31–36.

Bruce, D. (2010). The ones in the pile were going down: The reliability of violent crime statistics. *South African Crime Quarterly*, 31: 9–18.

Bruce, D. (2011). Beyond Section 49: Control of the use of lethal force. *South African Crime Quarterly*, (36): 3–12.

Bruce, D. (2013). New blood—Implications of en masse recruitment for the South African Police Service. *SA Crime Quarterly*, 2013(43): 17–28.

Bruce, D. (2020). *A Question of Gender? Violence against Women and Men in South Africa*, Southern Africa Report, ISS: Pretoria.

Bruce, D., K. Savage, and J. de Waal (2000). A duty to answer questions? The police, the independent complaints directorate and the right to remain silent. *South African Journal on Human Rights*, 16(1): 71–96.

Burger, J. (2006). Crime combating in perspective: A strategic approach to policing and the prevention of crime in South Africa. *Acta Criminologica*, 19(2): 105–118.
Burger, J. (2015). Leading a horse to water: Assessing review mechanisms of SAPS performance. *South African Crime Quarterly*, 53: 49–58.
Burger, J. and S. Grobler (2017). *Why the SAPS Needs an Internal Anti-Corruption Unit*. Policy Brief No. 101. Pretoria: Institute for Security Studies.
Burton, P., A. Du Plessis, T. Leggett, A. Louw, D. Mistry, and H. van Vuuren (2004). *National Victims of Crime Survey: South Africa 2003*. ISS Monograph No. 101. Pretoria: Institute for Security Studies.
Cawthra, G. (1993). *Policing South Africa: The SAP and the Transition from Apartheid*. London, U.K.: Zed Books Ltd.
Chingwete, A. (2016). *In South Africa, Citizens' Trust in President, Political Institutions Drops Sharply*. Afrobarometer Dispatch No. 90. Cape Town: Afrobarometer and Institute for Justice and Reconciliation.
CIA (2019). *The World Factbook*. Available at: www.cia.gov/library/publications/the-world-factbook/rankorder/2172rank.html
Civilian Secretariat for Police (2016a). *White Paper on Policing*. Pretoria: Department of Police.
Civilian Secretariat for Police (2016b). *White Paper on Safety and Security*. Pretoria: Department of Police.
CODESA (1991). *Declaration of Intent*, 20 December. Available at: https://peacemaker.un.org/sites/peacemaker.un.org/files/ZA_911229_CodesaDeclCODESA Declaration of Intent 20 December 1991rationIntent.pdf
Collins, A. (2013). Violence is not a crime: A broader view of interventions for social safety. *South African Crime Quarterly*, 43: 29–37.
Constitution of the Republic of South Africa (1996). Act No. 108 of 1996. Available at: www.gov.za/documents/constitution-republic-south-africa-1996
Cooper-Knock, S.J. (2016). Behind closed gates: Everyday policing in Durban, South Africa. *Africa*, 86(1): 98–121.
Crime Hub (2019). *ISS Crime Hub*. Available at: https://issafrica.org/crimehub
CSVR (Centre for the Study of Violence and Reconciliation) (2010). *Why South Africa Is So Violent and What Should Be Done about It: Statement by the Centre for the Study of Violence and Reconciliation*. Available at: www.csvr.org.za/old/index.php/media-articles/news-and-events/56-latest-news/2447-press-statement-the-study-on-the-violent-nature-of-crime-91110-.html
De Haas, M. (2016). The killing fields of kzn local government elections, violence and democracy in 2016. *South African Crime Quarterly*, 57: 43–53.
Fallon, P. and L. Perreira Da Silva (1994). *South Africa: Economic Performance and Policies*. Discussion Paper 7, April. Washington, DC: World Bank Southern Africa Department.
Farlam, I.G. (2015). *Marikana Commission of Inquiry: Report on Matters of Public, National and International Concern Arising Out of the Tragic Incidents at the Lonmin Mine in Marikana, in the North West Province*. Available at: www.sahrc.org.za/home/21/files/marikana-report-1.pdf
Faull, A. (2013). Fighting for respect: Violence, masculinity and legitimacy in the SAPS. *South African Crime Quarterly*, 44: 5–14.
Faull, A. (2016). Measured governance? Policing and performance management in South Africa. *Public Administration and Development*, 36(2): 157–168.
Faull, A. (2018a). *Police Work and Identity: A South African Ethnography*. Abingdon: Routledge.

Faull, A. (2018b). Are South African police salaries fair? *ISS Today*, 27 November. Available at: https://issafrica.org/iss-today/are-south-african-police-salaries-fair

Faull, A. (2019). South Africa's presidency must drive safety. *ISS Today*, 7 January. Available at: https://issafrica.org/iss-today/south-africas-presidency-must-drive-safety

Faull, A. and B. Rose (2012). *Professionalism and the South African Police Service: What Is It and How Can It Help Build Safer Communities?* ISS Occasional Paper No 240. Pretoria: Institute for Security Studies.

Feinsten, A. (2007). *After the Party: Corruption, the ANC and South Africa's Uncertain Future.* Verso: Johannesburg.

Freedom Charter (1955). *The Freedom Charter: Adopted at the Congress of the People at Kliptown, Johannesburg, on June 25 and 26, 1955.* Available at: www.historicalpapers.wits.ac.za/inventories/inv_pdfo/AD1137/AD1137-Ea6-1-001-jpeg.pdf

Gallup (2019). *2019 Global Law and Order Report.* Available at: www.gallup.com/analytics/267869/gallup-global-law-order-report-2019.aspx

Gilligan, J. (1996). *Violence: Our Deadly Epidemic and Its Causes.* New York: GP Putman's & Sons.

Gilligan, J. (2000). *Violence: Reflections on Our Deadliest Epidemic.* London: Jessica Kingsley.

Glaser, C. (1998). Swines, hazels and the dirty dozen: Masculinity, territoriality and the youth gangs of Soweto, 1960–1976. *Journal of Southern African Studies*, 24(4): 719–736.

Glaser, C. (2000). *Bo-Tsotsi: The Youth Gangs of Soweto, 1935–1976.* Social History of Africa Series. Oxford: James Currey Ltd.

Goldstone Commission (1994). *Goldstone Commission: Final Report on Attacks on Members of the South African Police.* Available at: www.icla.up.ac.za/images/un/commissionsofinquiries/files/South%20Africa%201993%20Goldstone%20commission%20TOR%20_%20Final%20report.pdf

Gould, C. (2015). *Beaten Bad: The Life Stories of Violent Offenders.* Pretoria: Institute for Security Studies.

Govender, R. (2018). Children and violence: Victimisation and perpetration. Predictive Modelling Using the Cape Area Panel Study CAPS, presentation delivered at the Safer Western Cape Conference, November. Available at: www.westerncape.gov.za/assets/day_1_session_1_rajen_govender.pdf

Gqola, P.D. (2015). *Rape: A South African Nightmare.* Jacana: Auckland Park.

Gqubule, T. (2017). *No Longer Whispering to Power.* Johannesburg, South Africa: Jonathan Ball Publishers. Kindle Edition.

Grundlingh, A. and H. Giliomee (2007). The great depression and war. In H. Giliomee and B. Mbenga (eds.), *New History of South Africa.* Cape Town, South Africa: Tafelberg Publishers.

Haysom, F. (1993). *Policing the Transition: Transforming the Police.* African National Congress. Available at: www.anc.org.za/content/policing-transition-transforming-police.

Hendricks, C. and T. Musavengana (2010) (Eds.). *The Security Sector in Southern Africa,* IISS Monographs, No. 174. Available at: https://issat.dcaf.ch/Learn/Resource-Library2/Policy-and-Research-Papers/The-Security-Sector-in-Southern-Africa

Hirsch, A. (2005). *Season of Hope: Economic Reform under Mandela and Mbeki.* Scottsville: University of KwaZulu-Natal Press.

Holden, P. (2008). *The Arms Deal in Your Pocket.* Cape Town: Jonathan Ball.

Hornberger, J. (2011). *Policing and Human Rights: The Meaning of Violence and Justice in the Everyday Policing of Johannesburg.* Abingdon: Routledge.

Hsiao, C., D. Fry, and C.L. Ward (2018). Violence against children in South Africa: The cost of inaction to society and the economy. *BMJ Global Health*, 3(1): 1–8.
Independent Police Investigative Directorate Act 1 of 2011 (IPID) (2012). South Africa Government Gazette no. 35018. Available at: https://www.gov.za/documents/independent-police-investigative-directorate-act
Institute for Economics and Peace (2019). *Global Peace Index 2019*. Available at: http://visionofhumanity.org/app/uploads/2019/06/GPI-2019-web003.pdf
ICD (1997). *Annual Report 1996/1997*, Independent Complaints Directorate: Pretoria.
ICD (1998). *Annual Report 1997/1998*, Independent Complaints Directorate: Pretoria.
ICD (1999). *Annual Report 1998/1999*, Independent Complaints Directorate: Pretoria.
ICD (2000). *Annual Report 1999/2000*, Independent Complaints Directorate: Pretoria.
ICD (2001). *Annual Report 2000/2001*, Independent Complaints Directorate: Pretoria.
ICD (2002). *Annual Report 2001/2002*, Independent Complaints Directorate: Pretoria
ICD (2003). *Annual Report 2002/2003*, Independent Complaints Directorate: Pretoria.
ICD (2004). *Annual Report 2003/2004*, Independent Complaints Directorate: Pretoria.
ICD (2005). *Annual Report 2004/2005*, Independent Complaints Directorate: Pretoria.
ICD (2006). *Annual Report 2005/2006*, Independent Complaints Directorate: Pretoria
ICD (2007). *Annual Report 2006/2007*, Independent Complaints Directorate: Pretoria.
ICD (2008). *Annual Report 2007/08*, Independent Complaints Directorate: Pretoria.
ICD (2009). *Annual Report 2008/2009*, Independent Complaints Directorate: Pretoria.
ICD (2010). *Annual Report 2009/2010*, Independent Complaints Directorate: Pretoria.
ICD (2011). *Annual Report 2010/2011*, Independent Complaints Directorate: Pretoria.
ICD (2012). *Annual Report 2011/2012*, Independent Complaints Directorate: Pretoria.
IPID (2013). *Annual Report 2012/2013*, Independent Police Investigative Directorate: Pretoria
IPID (2014). *Annual Report 2013/2014*, Independent Police Investigative Directorate: Pretoria.
IPID (2015). *Annual Report 2014/2015*, Independent Police Investigative Directorate: Pretoria.
IPID (2016). *Annual Report 2015/16*, Independent Police Investigative Directorate: Pretoria.
IPID (2017). *Annual Report 2016/2017*, Independent Police Investigative Directorate: Pretoria.
IPID (2018). *Annual Report 2017/2018*, Independent Police Investigative Directorate: Pretoria.
IPID (2019). *Annual Report 2018/19*, Independent Police Investigative Directorate: Pretoria.
Jeffery, A. (2019). *People's War: New Light on the Struggle for South Africa*. Jeppestown, South Africa: Jonathan Ball Publishers. Kindle Edition.
Jensen, S. (2008). *Gangs, Politics & Dignity in Cape Town*. Chicago, IL: University of Chicago Press.
Kinnes, I. (2017). *Contested Governance: Police and Gang Interactions*. PhD Thesis. Cape Town: University of Cape Town.
Klockars, C.B., S. Kutnjak Ivković, and M.R. Haberfeld (2006). *Enhancing Police Integrity*. New York: Springer.
Kriegler, A. and M. Shaw (2016). *A Citizen's Guide to Crime Statistics in South Africa*. Johannesburg and Cape Town: Jonathan Ball.
Kutnjak Ivković, S. (2015). Studying Police Integrity. In Kutnjak Ivković, S. and M. Haberfeld (eds.), *Police Integrity across the World*. New York: Springer.
Kynoch, G. (2016). Apartheid's afterlives: Violence, policing and the South African state. *Journal of Southern African Studies*, 42(1): 65–78. DOI: 10.1080/03057070.2016.1087167
Lancaster, L. (2018). Unpacking discontent: Where and why protest happens in South Africa. *South African Crime Quarterly*, 64: 29–43.

Leibbrandt, M., V. Ranchod, and P. Green (2018). *Taking Stock of South African Income Inequality*. WIDER Working Paper 2018/184. United Nations University.
Leibbrandt, M., I. Woolard, A. Finn, and J. Argent (2010). *Trends in South African Income Distribution and Poverty since the Fall of Apartheid, OECD Social*. Employment and Migration Working Papers, No. 101. OECD Publishing. DOI: 10.1787/5kmms0t7p1ms-en
Lodge, T., M. Shain, R. Mendelsohn, K. Moodley, B. Pogrund, H. Barrell, and H. Giliomee (2007a). The Afrikaner nationalists in power. In H. Giliomee and B. Mbenga (eds.), *New History of South Africa*. Cape Town, South Africa: Tafelberg Publishers.
Lodge, T., V. van der Vliet, D. Welsh, J. Myburgh, W. Gumede, L. Schlemmer, M. Bot, H. Adam, and H. Giliomee (2007b). Towards an inclusive democracy. In H. Giliomee and B. Mbenga (eds.), *New History of South Africa*. Cape Town, South Africa: Tafelberg Publishers.
Malingo, B. (2018). 500 000 apply for SAPS' learnership programme which has 3 500 spaces. *The Citizen*, 20 August. Available at: https://citizen.co.za/news/south-africa/1997933/500-000-apply-for-saps-learnership-programme-which-has-3-500-spaces/
Manning, P.K. (1977). *Police Work: The Social Organization of Policing*. Cambridge, MA: The MIT Press.
Manning, P.K. (2005). The police: Mandate, strategies and appearances. Reprinted in T. Newburn (ed.), *Policing: Key Readings*. Cullompton: Willan Publishing, pp. 191–214.
Meinck, F., L.D. Cluver, M.E. Boyes, and H. Loening-Voysey (2016). Physical, emotional and sexual adolescent abuse victimisation in South Africa: Prevalence, incidence, perpetrators and locations. *Epidemiology Community Health*, 70: 910–916. DOI: 10.1136/jech-2015-205860
Meyer, M.E. and J. Steyn (2009). Nurturing isolation in the South African police service A comparison of male and female recruits. *Policing: An International Journal of Police Strategies & Management*, 32(1): 108–127.
Minnaar, A. (2010). The changing face of 'community policing' in South Africa, post-1994. *Acta Criminologica: Southern African Journal of Criminology,* Special Edition 2: 189–210.
Morris, M. and V. Padayachee (1988). State reform policy in South Africa. *Transformation,* 7: 1–26.
National Minimum Wage Panel (2016). *A National Minimum Wage for South Africa: Recommendations on Policy and Implementation*. National Minimum Wage Panel Report to the Deputy President. Available at: www.groundup.org.za/media/uploads/documents/NMWReportFinal.pdf
National Planning Commission (NPC) (2012). National Development Plan 2030: Our future—make it work, The Presidency, Pretoria,
Nattrass, G. (2017). *A Short History of South Africa*. London: Biteback Publishing. Kindle Edition.
Nattrass, N. and J. Seekings (2011). The economy and poverty in the twentieth century. In R. Ross, A.K. Mager, and B. Nasson (eds.), *The Cambridge History of South Africa*, Vol. 2. Cambridge: Cambridge University Press.
NCPS (1996). *National Crime Prevention Strategy*. Available at https://www.csvr.org.za/docs/crime/1996nationalcrime.pdf
Newham, G. and A. Faull (2011). *Protector or Predator: Tackling Police Corruption in South Africa*. Pretoria: Institute for Security Studies.
News24 (2012). Bheki Cele fire. *News24*, 12 June. Available at: www.news24.com/SouthAfrica/News/Bheki-Cele-fired-20120612

Omar, B. (2007). *SAPS' Costly Restructuring: A Review of Public Order Policing Capacity*. Pretoria: Institute for Security Studies.

Orthofer, A. (2016). *Wealth Inequality in South Africa: Evidence from Survey and Tax Data*. REDI3x3 Working Paper 15, June.

Parliamentary Monitoring Group (1998). *National Crime Prevention Strategy: Briefing by SAPS*. Minutes of Briefing to the Parliamentary Portfolio Committee on Safety and Security, 17 February. Available at: https://pmg.org.za/committee-meeting/6272/ (Accessed 1 September 2019).

Pauw, J. (2018). *The President's Keepers: Those Keeping Zuma in Power and Out of Prison*. Johannesburg: Tafelberg.

Pelser, E., J. Schnetler, and A. Louw (2002). Not Everybody's Business: Community Policing in the SAPS' Priority Areas. Institute for Security Studies Monograph No. 71, Pretoria: South Africa: Institute for Security Studies.

Penal Reform International (2018). *Life Imprisonment: A Policy Briefing*. Nottingham, U.K.: University of Nottingham.

Perkins, G. (2018). *Danger and Death: Organisational and Occupational Responses to the Murder of Police in South Africa: A Case Study*. PhD Thesis. Cape Town: University of Cape Town.

Pharoah, R. (2008). *National Victims of Crime Survey: Overview of Key Data*. ISS Occasional Paper No. 175. Pretoria: Institute for Security Studies.

Pinnock, D. (2016). *Gang Town*. Tafelberg: Cape Town.

The Planning Commission (2012). *National Development Plan*, November. Pretoria: The Planning Commission.

Plus 94 Research (2018). *Afrobarometer: South Africa Results*. Available at: www.afrobarometer.org/publications/south-africa-summary-results-2018

Posel, D. (2009). The apartheid project, 1948–1970. In R. Ross, A.-K. Mager, and B. Nasson (eds.), *The Cambridge History of South Africa*, Vol. 2. Cambridge: Cambridge University Press, pp. 319–368.

Rantete, J. and H. Giliomee (1992). Transition to democracy through transaction? Bilateral negotiations between the ANC and NP in South Africa. *African Affairs*, 91(365): 515–542.

Ratele, K. (2008). Masculinity and male mortality in South Africa. *African Safety Promotion: A Journal of Inquiry*, 6(22): 19–41.

Rauch, J. (2000). *Police Reform and South Africa's Transition*. Available at: www.csvr.org.za/docs/policing/policereformandsouth.pdf

Rauch, J. (2004). *Police Transformation and the South African TRC*. Available at: www.csvr.org.za/docs/policing/policetransformationandTRC.pdf

Reith, C. (1952). *The Blind Eye of History: A Study of the Origins of the Present Police Era*. London: Faber and Faber.

Richter, L.M., S. Mathews, J. Kagura, and E. Nonterah (2018). A longitudinal perspective on violence in the lives of South African children from the Birth to Twenty Plus cohort study in Johannesburg-Soweto. *South African Medical Journal*, 108(3): 181–186. DOI: 10.7196/SAMJ.2018.v108i3.12661

SALDRU (2019). *Income Check*. Available at: www.saldru.uct.ac.za/income-comparison-tool/

Samara, T.R. (2011). *Cape Town after Apartheid: Crime and Governance in the Divided City*. Minneapolis, MN: University of Minnesota Press.

SAPS (1997). *Annual Report 1996/1997*.

SAPS (2018). *Annual Report 2017/2018*.

SAPS (2019). *Crime Situation in the Republic of South Africa: Twelve (12) months (April to March 2018_19*. Available at: www.saps.gov.za/services/april_to_march2018_19_presentation.pdf

Seedat, M., A. van Niekerk, R. Jewjes, S. Suffla, and K. Ratele (2009). Violence and injuries in South Africa: Prioritising an agenda for prevention. *The Lancet*, 374: 1011–1022.

Seekings, J. and N. Nattrass (2005). *Class, Race and Inequality in South Africa*. New Haven, CT: Yale University Press.

Shaw, M. (2002). *Crime and Policing in Post-Apartheid South Africa: Transforming under Fire*. Indiana: Indiana University Press.

Shaw, M. (2017). *Hitmen for Hire: Exposing South Africa's Underworld*. Johannesburg: Jonathan Ball Publishers. Kindle Edition.

Shaw, M. and K. Thomas (2016). The commercialisation of assassination: "Hits" and contract killing in South Africa, 2000–2015. *African Affairs*, 116(465): 597–620.

Simkins, C. (2011). The evolution of the South African population in the twentieth century. In R. Ross, A.K. Mager, and B. Nasson (eds.), *The Cambridge History of South Africa*, Vol. 2. Cambridge: Cambridge University Press.

Standing, A. (2006). *Organised Crime: A Study from the Cape Flats*. Pretoria: Institute for Security Studies.

State vs. Selebi (2010). *(25/09) [2010] ZAGPJHC 53*, 5 July. Available at: www.saflii.org/za/cases/ZAGPJHC/2010/53.pdf

Statistics South Africa (1999). *Victims of Crime Survey 1998*. Pretoria: Statistics South Africa.

Statistics South Africa (2011). *Victims of Crime Survey 2010/11*. Pretoria: Statistics South Africa.

Statistics South Africa (2018a). *General Household Survey*. Statistical Release P0318. Pretoria: Statistics South Africa.

Statistics South Africa (2018b). *Victims of Crime Survey 2017/18: Statistical Release P0341*. Pretoria: Stats SA.

Statistics South Africa (2019a). *Inequality Trends in South Africa: A Multidimensional Diagnostic of Inequality*. Pretoria: Statistics South Africa. Available at http://www.statssa.gov.za/publications/Report-03-10-19/Report-03-10-192017.pdf

Statistics South Africa (2019b). *P0341: Victims of Crime: Governance, Public Safety and Justice Survey*. Pretoria: Statistics South Africa. Available at http://www.statssa.gov.za/publications/Report-03-10-19/Report-03-10-192017.pdf

Steinberg, J. (2008). *Thin Blue: The Unwritten Rules of Policing South Africa*. Johannesburg: Jonathan Ball.

Steinberg, J. (2011). Security and disappointment: Policing, freedom and Xenophobia in South Africa. *British Journal of Criminology*, 52(2): 345–360.

Steinberg, J. (2012). Establishing police authority and civilian compliance in post-apartheid Johannesburg: An argument from the work of Egon Bittner. *Policing and Society*, 22(4): 481–495.

Steinberg, J. (2014). Policing, state power, and the transition from apartheid to democracy: A new perspective. *African Affairs*, 113(451): 173–191.

Steyn, J. (2017a). Police sub-culture; to what extent does it prepare police officials to police in a complex society? *SAPS Research Colloquium, CSIR*, 7–9 February 2017. Available at: https://www.saps.gov.za/resource_centre/publications/jean_steyn_saps_research_colloquium.pdf

Steyn, J. (2017b). Lilac? Ten years of blue nurtured nature. *Police Practice and Research*, 18(3): 259–274.

Super, G. (2010). The spectacle of crime in the "new" South Africa: A historical perspective (1976-2004). *British Journal of Criminology*, 50(2): 165–184.
Super, G.J. (2013). *Governing through Crime in South Africa: The Politics of Race and Class in Neoliberalizing Regimes*. Oxford: Routledge.
Super, G.J. (2017). What's in a name and why it matters: A historical analysis of the relationship between state authority, vigilantism and penal power in South Africa. *Theoretical Criminology*, 21(4): 512–531.
Thomas, K. (2018). *The Rule of the Gun: Hits and Assassinations in South Africa*. University of Cape Town & Global Initiative against Transnational and Organised Crime. Available at: https://assassinationwitness.org.za/wp-content/uploads/2018/03/The-rule-of-the-gun_Assassination-Witness.pdf
Tlhabi, R. (2017). *Khwezi: The Remarkable Story of Fezekile Ntsukela Kuzwayo*. Johannesburg: Jonathan Ball Publishers. Kindle Edition.
Transparency International. Afrobarometer, Global Corruption Barometer 2019. Citizens Views and Experiences of Corruption.
UNODC (2019). *Global Study on Homicide 2019*. Vienna, Austria. Available at https://www.unodc.org/documents/data-and-analysis/gsh/Booklet1.pdf
Van Der Spuy, E. (2005). *South African Police Reform in the 1990's: Internal Processes and External Influences*. Doctoral Thesis. Cape Town: University of Cape Town.
Viewfinder (2019). Viewfinder: IPID's cover-up of police brutality in SA. *Daily Maverick*, 7 October. Available at: www.dailymaverick.co.za/article/2019-10-07-viewfinder-ipids-cover-up-of-police-brutality-in-sa/
Vigneswaran, D. and J. Hornberger (eds.) (2010). *Beyond "Good Cop"/"Bad Cop": Understanding Informality and Police Corruption in South Africa*. Johannesburg: Open Society Reports & University of the Witwatersrand.
Weisburd, D. (2015). The law of crime concentration and the criminology of place. *Criminology*, 53(2): 133–157.
Western Cape Department of Health (2018). *Western Cape Injury Mortality Profile 2016*. Available at: www.westerncape.gov.za/assets/departments/health/mortality_profile_2016.pdf
Wiener, M. (2018). *Ministry of Crime: An Underworld Explored*. Johannesburg, South Africa: Pan Macmillan. Kindle Edition.
Wiener, M., J. Chabalala, and K. Cowan (2019). Phahlane arrested on new charges linked to R86m blue lights tender. *News 24*, 1 March. Available at: www.news24.com/South-Africa/News/breaking-phahlane-arrested-on-new-charges-linked-to-r86m-blue-lights-tender-20190301
World Bank (2019). *Gini Index*. Available at: https://data.worldbank.org/indicator/SI.POV.GINI?end=2014&locations=ZA&start=1993&view=chart

Chapter 4

South African policing through normative and legal boundaries

Introduction

Although frontline police officers are allowed wide discretion in many everyday operations, their work and actions are guided, regulated, and limited by official rules. In situations in which officers can use their discretion, they may select actions among several legally available options—issue a ticket, make an arrest, warn a citizen, or simply do nothing.[1] Once they decide on an action (e.g., make an arrest), legal rules may exist to guide subsequent procedures (e.g., how to carry out an arrest). Legal rules may also govern the general conduct of officers, be it in their professional capacity (e.g., rules prohibiting the use of torture to obtain a confession) or in their personal lives (e.g., rules prohibiting secondary employment on days off).

The conduct of police officers is typically regulated by three types of legal rules. First, a country's constitutional framework may contain norms that either directly or indirectly regulate police conduct. The constitution may contain norms requiring of public servants in general, or police officers in particular, to carry out their duties with integrity and respect. In addition, the constitution may indirectly limit the discretion of officers by guaranteeing citizens their rights (e.g., the right not to be searched or arrested without a warrant issued by a judge; the right to have an attorney present during police questioning). Second, countries typically have legal statutes that provide more detailed rules regulating all areas of citizen activities. Some of these, when articulated in criminal codes and similar statutes, may prohibit certain behaviors, either for all citizens (e.g., murder, theft) or for public servants only (e.g., acceptance of gifts, violating citizens' procedural rights). Other legal rules, when formulated in criminal codes or similar documents, may outline citizens' rights and introduce limits on police behavior when conducting investigations, collecting evidence, or questioning citizens. Third, countries may develop specific rules, both in their legal codes and internal agency ordinances, dealing exclusively with the police and outlining their rights and responsibilities. In countries with decentralized police systems (e.g., Germany, United States), individual police agencies would have the right to create such rules, providing for a potential dissonance of official agency rules. In centralized police systems, the same official rules apply to all police agencies and officers within the system (e.g., Croatia, Poland, France).

In South Africa, police conduct is regulated through a complex set of legal rules. The discussion commences with the country's constitutional norms, followed by an analysis of the criminal substantive law and common law, as well as specific SAPS rules regulating the conduct of police officers. The chapter also provides a discussion about the first dimension of the theory of police integrity (Klockars and Kutnjak Ivković, 2004) and the relevant extant research. It includes empirical analyses of the survey data we have collected that analyze SAPS officers' knowledge of official rules and the seriousness with which they evaluate various forms of police misconduct.

Legal framework

South Africa's legal system can be best described as a mixed legal system, combining elements of the civil-law tradition inherited from the Dutch, the common-law tradition inherited from the British, and customary-law traditions derived from indigenous Africans. Four sources of South African law (constitutional law, statutory law, case law, and the amalgam of Roman-Dutch law with British common law) provide the foundation of police rights and duties, as well as the bases for regulating police conduct through the promulgation of rules and regulations by the SAPS management.

The South African Constitution

Following the country's April 1994 democratic election, conducted under a hurriedly agreed upon interim constitution, the Constituent Assembly (CA) met in May to finalize a new constitutional text within two years. Soon, interests of political factions within the Assembly would encumber negotiations over a range of constitutional issues, including the relationship between the national and provincial governments, official languages, protection of property, the restriction of rights during states of emergency, and the death penalty. With the deadline fast approaching, political tradecraft had the primary negotiators reach consensus on this important document "partly by fudging matters, by ambiguous clauses or qualifying principles by means of 'escape' clauses" (Lodge et al., 2007, pp. 410–411). The day before the deadline, four political parties voted in favor, with the National Party (NP) and the African National Congress (ANC) as the main negotiators leading the process. Another influential political organization—the Inkatha Freedom Party (IFP)—shunned these proceedings. After Constitutional Court approval, the document was signed into law by President Mandela as the Constitution of the Republic of South Africa, Act 108 of 1996 (Lodge et al., 2007, p. 411). Despite negotiation strategies by the key political parties that threatened to derail the conclusion of this document, the Constitution would ultimately be widely recognized as one of the most progressive constitutions of any modern state, with legal scholar Cass Sunstein praising it as "the most admirable constitution in the history of the world" (Sunstein, 2001).

Although it has been amended 17 times since its enactment, the basic tenets—the guarantee of human rights and the establishment of the South African government and its structures—are well preserved. The Constitution (1996) contains references to broad principles as well as specific provisions regarding the rights of citizens and limitations on the powers of governmental agencies and its employees. It guarantees citizens the protection of their human rights by the government and its agents, which, in turn, imposes restrictions on the conduct of police officers. Although the Constitution positions the SAPS as central to its success, mandating that it "protect and secure" all inhabitants of the country and "uphold and enforce the law" (Constitution, 1996, section 205[3]), only three of its chapters have direct bearing on the integrity of the organization by establishing normative and legal boundaries.

Chapter 1, Founding Provisions, lays out the broadest and most comprehensive principles, which include the provision that the Republic of South Africa is one sovereign, democratic state founded on "[h]uman dignity, the achievement of equality and the advancement of human rights and freedoms" (Constitution, 1996, section 1(a)). In turn, these principles inform all other constitutional provisions as well as substantive and procedural law and the rules and regulations of the SAPS. These principles also contain the broad provisions that require police officers to respect citizens' human rights in general and the achievement of equality in particular. In a country that has experienced centuries of state-sponsored oppression (see Chapter 2 of this book) and is striving to establish a non-discriminatory and democratic regime (see Chapter 3 of this book), embracing such key principles very early and including them in this most fundamental legal document has been of pivotal importance.

Chapter 2 of the Constitution (1996) establishes the Bill of Rights, reaffirming freedom, equality, and dignity as the basic values protected by the Constitution (section 7.2). Equality is an overarching principle in the Constitution, rooted in several of its chapters. For example, Chapter 2, section 9(1), states: "Everyone is equal before the law and has the right to equal protection and benefit of the law" (Constitution, 1996). Some constitutional norms (Constitution, 1996, Chapter 2, section 9) directly and explicitly prohibit direct and indirect discrimination by the state based on "race, gender, sex, pregnancy, marital status, ethnic or social origin, colour, sexual orientation, age, disability, religion, conscience, belief, culture, language and birth." As state employees, the prohibition of discrimination outlined in the Constitution applies directly to police officers and their conduct.

Several chapters of the Constitution establish further foundations for the development of rules regulating the work of the SAPS officers. The Bill of

Rights contains specific provisions describing other protected rights such as life, freedom of religion, freedom of expression, and freedom of assembly and demonstration (Chapter 2), values that would typically be included in the constitutions of established democracies. In addition to establishing citizen rights, these provisions also create the boundaries of proper conduct for government officials, including the police. Chapter 2 also contains the norms directly affecting frontline police officers' routine duties, including search and seizure activities. The first set of rules outlined in section 12 (Chapter 2, 1996) states:

> Everyone has the right to freedom and security of the person, which includes the right—(a) not to be deprived of freedom arbitrarily or without just cause; (b) not to be detained without trial; (c) to be free from all forms of violence from either public or private sources; (d) not to be tortured in any way; and, (e) not to be treated or punished in a cruel, inhuman or degrading way.

These norms indirectly prohibit arbitrary arrest by the police (i.e., without legally valid reason) and the abuse of force (Constitution, 1996, Chapter 2, section 12).

Chapter 2, section 35 of the Constitution (1996) establishes the rights of arrested, detained, and accused persons. These include the rights not to have their person or home searched, their property searched, or their possessions seized (Constitution, 1996). This section also establishes specific rights of arrested, detained, and accused persons to include the right to remain silent, the right to be informed of such a right, and to be informed of the consequences of not exercising such a right (Constitution, 1996). All of these rights, which constitute a substantial development in the country's democratization, apply directly and unequivocally to the police. Paragraph 5 of Chapter 2, section 35 states that "[e]vidence obtained in a manner that violates any right in the Bill of Rights must be excluded if the admission of that evidence would render the trial unfair or otherwise be detrimental to the administration of justice" (Constitution, 1996). While the exclusionary rule—the prohibition of the use of illegally obtained evidence in court proceedings—affects court cases and, potentially, their outcomes, it can also be used to guide proper officer training and to justify the disciplining of officers who obtain evidence illegally. Similarly, failure to act against violations of the Bill of Rights would reflect negatively on the integrity of the organization as a whole.

While Chapter 2 establishes the rights of persons on the one hand and limits police powers on the other, Chapter 11 is focused solely on the "security services"—including the police service, defense force, and intelligence services. Chapter 11, section 199, paragraph 1, authorizes the establishment of a single national police service. Earlier provisions of this chapter (Constitution, 1996, section 199, paragraph 5) already required that the security services "must act, and must teach and require their members to act, in accordance with the

Constitution and the law." While outlining the tasks of the police service, its organizational structure, and oversight, the Constitution further provides for the creation of legal statutes and other acts to regulate police work in more detail (e.g., "National legislation must establish the powers and functions of the police service and must enable the police service to discharge its responsibilities effectively, taking into account the requirements of the provinces," Constitution, 1996, Chapter 11, section 205, paragraph 2).

Criminal law

As inhabitants of the Republic, police officers are governed not only by the same legal responsibilities as other citizens but also by provisions specifically related to the conduct of public officials. In this section, we identify some of the activities that, if carried out by police officers, can result in criminal responsibility.

The criminal law of the Republic of South Africa, which is not codified, is derived primarily from Dutch civil law and substantially influenced by English common law (Kemp et al., 2015). While the influence of Dutch civil law and English common law traditions have ebbed and flowed in relationship to one another over time, the rules of criminal procedure and evidence have remained guided by the English law of the time (Kemp et al., 2015). In addition, many common-law crimes have been supplemented, and in some cases replaced, by "Acts" of Parliament, creating a complex criminal law system.

Among the common-law offenses of general application are those constituting crimes against human life and person (e.g., murder and attempted murder, culpable homicide, common assault and assault with intent to do grievous bodily harm), property crimes (e.g., theft, robbery, housebreaking), and drug offenses (e.g., using or possessing a prohibited substance, dealing in prohibited substances). Sexual crimes, in particular, mostly prohibited under common law, are stellar examples of the mixed system of common law crimes and statutory offenses in South Africa. The Criminal Law (Sexual Offenses and Related Matters) Amendment Act 32 of 2007 codified or consolidated sexual crimes under South African law (Snyman, 2014). The objectives of the Act include repealing common-law sexual crimes, such as rape and incest, and replacing them with statutory versions (Kemp et al., 2015), while also introducing a number of new crimes that did not exist in common law.

There are other criminal offenses within South African jurisprudence that, although applicable to the general public, may also have special application to the police, such as fraud, corruption, perjury, and related integrity crimes. The crime of corruption, traced to the common-law offense of bribery, is today encompassed by the Prevention and Combating of Corrupt Activities Act 12 of 2004 (Corrupt Activities Act). Under the Act, corruption is divided into a general crime of corruption, resembling the common-law crime of bribery, and specific forms of corruption (Kemp et al., 2015). Section 4(1) of the Act is directed in particular to the corruption of public officers. Although more specific in its definition of the offense, section 4(1) essentially defines such corruption as the unauthorized or improper inducement to either do or not do anything (Prevention and Combating of Corrupt Activities Act 12 of 2004).

In addition, section 34 of the Corrupt Activities Act makes it an offense when a person of authority does not report violations of the Act to a police officer. This raises the question of when and to whom a police officer must report such a violation. The extent of perjury by SAPS officers is not clear. Certainly, perjury has occurred frequently enough in the American context to have received the label of "testilying," and, considering that two former National Commissioners were found to have lied in court (Selebi) or to a Commission of Inquiry (Phiyega; see Chapters 3 and 6), it would not be surprising if perjury also occurs among lower ranks. Under South African law, this common-law offense involves making a false statement under oath or affirmation in the course of a judicial hearing. It is also a crime under section 9 of the Justice of the Peace and Commissioners of Oaths Act 16 of 1963 to knowingly make a false statement in an affidavit, affirmation, or solemn or attested declaration, other than testimony under oath during court proceedings (Kemp et al., 2015). As noted above, criminal acts by police officers not only constitute violations of law but also violations of the rules and regulations of the SAPS—the subject of the next section.

The South African Police Service Act

Derived from the mandate in Chapter 14, section 214 of the Interim Constitution (1993), the South African parliament enacted legislation establishing a national police service—the South African Police Service Act of 1995 (Police Service Act). The wording of the Act is sufficiently nebulous to complicate defining specific actions that constitute police misconduct. While the Constitution (1996) directly establishes human rights and indirectly limits police powers, the Police Service Act (1995, Chapter 5, section 13, paragraph 1) requires, for example, that the police follow constitutional and other legal norms in the performance of their duties and protection of citizen rights:

> Subject to the Constitution and with due regard to the fundamental rights of every person, a member may exercise such powers and shall perform such duties and functions as are by law conferred on or assigned to a police official.[2]

Left to interpretation, any actions by the police in contravention of these broad legal limits could likely constitute police misconduct. Fortunately, the Police Service Act (1995) includes additional rules regulating proper police conduct. To begin, when officers use force—a defining feature of police powers and functions (Bittner, 1970)—they are required to use only the minimum amount necessary (South African Police Service [SAPS] Act (No. 68), 1995, Chapter 5, section 13, paragraph 3b). Any amount of force used beyond the minimum required in a given situation constitutes a violation of official rules and is consequently viewed as police misconduct. This has been called a central contradiction in democratic policing—that their use of force in defense of rules and rights puts the police at risk of breaking rules and harming rights (Bittner, 1970; Young, 1991; Manning, 2010). In a country struggling to rid itself of a legacy of violence (see Chapters 2 and 3), regulating the police use of force is of particular relevance.

Beyond these specific references, Chapter 8, Regulations, confers broad authority to the Minister for Safety and Security (presently, the Minister of Police) to "make regulations" regarding "conduct by members that will constitute misconduct" as well as "the issue of a code of conduct for the Service and the upholding thereof." Subject to interpretation, the SAPS Act (1995) incorporates additional norms that could be interpreted further to regulate police conduct. In particular, Chapter 9, section 34, lists situations that may herald an inquiry (e.g., absent without leave for over a month, damage or loss of state property, and questionable suitability of property purchased for the state), which implies that such behavior potentially also violates official rules.

Another example in which the SAPS Act (1995) regulates police (mis)conduct in a secondary way is found in Chapter 9, section 36, which authorizes the National Commissioner to dismiss an officer convicted of a crime and sentenced to "imprisonment without a possibility of a fine." The implication is clear: Police misconduct includes any behavior resulting in such a sentence. Any crime carrying a less severe punishment would be addressed by the SAPS (SAPS Act, 1995, Chapter 9, section 43), but would not automatically result in dismissal. Committing a serious crime, striking, inducing other police officers to strike, or conspiring to strike against the police agency and police administration could also lead to dismissal (SAPS Act, 1995, Chapter 9, section 41), suggesting that such behaviors constitute further violations of official rules.[3]

The SAPS Act (1995) prescribes that the National Commissioner establish disciplinary procedures (Chapter 9, section 40) or inquiries (Chapter 9, section 34) to investigate potential misconduct. Clearly, such procedures will be described in separate legal acts (i.e., the South African Police Service [SAPS] Disciplinary Regulations, 2006). The SAPS Act (1995) imposes an obligation on a police officer who becomes aware of an offense committed to inform his or her commanding officer as soon as possible. A contravention of this provision constitutes a violation of the Act. Where obliged to perform an official duty, the Act requires officers to perform it in a manner that is reasonable under the circumstances. Chapter 8 also obliges officers to obey lawful orders, with the caveat that a member must not obey a patently unlawful order or instruction. Similarly, the Constitution (1996, Chapter 11, section 199, paragraph 6) contains a provision directly prohibiting security officers (including the police) from obeying an order that is illegal.

The SAPS Code of Conduct and the SAPS Discipline Regulations

The third level of laws, rules, and regulations addressing expectations for the conduct of police consists of those promulgated by the organization itself, as permitted by the SAPS Act 68 of 1995 (Chapter 8 (24)(1)(g)). These rules exist as the South African Police Service Code of Conduct and the South African Police Service Discipline Regulations. One of the initial reform focuses—a uniform set of internal rules in terms of police conduct (Bruce, 2002)—appears after 20 years to be well established in the SAPS. Since 1997, every member is expected to sign a standardized Code of Conduct that has them warrant, among several behavioral and ethical matters, that they shall "act with integrity," "work towards preventing any form of corruption and to bring the perpetrators thereof to justice," exercise their conferred powers "in a responsible and controlled manner," and, more generally, "contribute to the reconstruction and development of, and reconciliation" of the country (SAPS Code of Conduct, 1997).

The Code of Conduct is given force through the application of the SAPS Disciplinary Regulations (SDR) (2006, section 20[q]), defining the contravention of "any prescribed Code of Conduct for the Service or the Public Service, whichever may be applicable to him or her" as constituting misconduct. This application of the SDR to the Code of Conduct is observable in Table 4.1 (discussed later), in which every case scenario presented can be constituted as a Code of Conduct violation in contravention of Section 20(q) of the SDR. For example, Scenario 6, an assault of a suspect in custody, can be interpreted to violate the principles of "achieve[ing] a safe and secure environment," by "uphold[ing] and protect[ing] the fundamental rights of every person," and "act[ing] impartially, courteously, respectfully, in an accountable manner." Scenario 1 presents another example of a violation of the Code of Conduct wherein SAPS officers are directed to "undertake to work towards preventing any form of corruption." As such, the SDR creates the potential to transform the Code of Conduct from a set of broad principles to a set of rules, whenever section 20(q) of the SDR is applied in such a rigid manner.

Internally, the SAPS regulates the behavior of their members through a set of disciplinary regulations. The SDR of 2006 was repealed and replaced by the South African Police Service Discipline Regulations of 2016 (Department of Police, 2016). As our data were collected under the terms of the 2006 regulations, knowledge and awareness of the rules by SAPS members are relevant only to those rules in effect at that time. This limitation, however, does not affect the role of internal rules, the awareness of such rules, and the assessments by members of the organization regarding the seriousness of such rule violation, aspects all relevant to the first dimension of the theory of police integrity.

The SDR prohibits a range of possible misconduct by SAPS members and civilian personnel alike (2006, regulation 20), stating "the support [of]

Table 4.1 Scenario descriptions and official rules

Scenario description	*Statute articles*	*Official SAPS rules*
Scenario 1: A police officer is widely liked in the community. Local merchants and restaurant owners regularly show their appreciation for his attention by giving him gifts of food, cigarettes, and other items of small value.	Prevention and Combating of Corrupt Activities Act 12 of 2004	Section 20(h) Unauthorized Compensation
Scenario 2: A police officer is aware that there is an issued warrant of arrest for a longtime friend of his. Although he sees his friend frequently over a period of more than a week and warns his friend of its existence, he does not arrest him.	Prevention and Combating of Corrupt Activities Act 12 of 2004	Sections 20(q) Violate Code of Conduct 20(z) Statutory offense
Scenario 3: A police officer discovers a burglary of a hardware store. The display cases are smashed, and many items have obviously been taken. While searching the store, he takes an expensive pocketknife and slips it into his pocket. He reports that the knife has been stolen during the burglary.	Common law theft & Section 9 of the Justice of the Peace and Commissioners of Oaths Act 16 of 1963	Sections 20(q) Violate Code of Conduct 20(w) False statement 20(x) Falsifies record 20(z) Statutory offense
Scenario 4: An officer who was severely beaten by a person resisting arrest, has just returned to duty. On patrol, the officer approaches a person standing in a dimly lit alley. Suddenly, the person throws a gym bag at the officer and begins to run away. The officer fatally shoots the person, striking him in the back. It was later determined that the person was unarmed.	SA Constitution, Chapter 2, Bill of Rights, Paragraph 11, Life. Criminal Procedure Act, Section 49(2). Culpable homicide, possibly common law murder	Sections 20(a) Contravenes legal obligation 20(e) Endangers life 20(q) Violate Code of Conduct 20(z) Statutory offense
Scenario 5: A police officer is scheduled to work during coming holidays. The supervisor offers to give him these days off, if he agrees to run some personal errands for the supervisor. Evaluate the supervisor's behavior.	Common law fraud	Sections 20(f) Prejudices discipline 20(p) Improper conduct 20(q) Violate Code of Conduct 20(r) Incites unlawful conduct 20(t) Victimized employee

Scenario 6: In responding with her male partner to a fight in a bar, a young female officer receives a black eye from one of the male combatants. The man is arrested, handcuffed, and as he is led into the cells, the male member of the team punches him very hard in the kidney area saying, "hurts, doesn't it."	SA Constitution, Chapter 2, Bill of Rights, Paragraph 12, Freedom and security of the person. Crime of Common Assault	Sections 20(a) Contravenes legal obligation 20(e) Endangers life 20(q) Violate Code of Conduct 20(z) Common law crime
Scenario 7: A police officer stops a motorist for reckless driving. As the officer approaches the vehicle, the driver yells, "What the hell are you stopping me for?" The officer replies, "Because today is 'Arrest an Asshole Day.'"	SA Constitution, Chapter 2, Bill of Rights, Paragraph 10, Human Dignity	Sections 20(a) Contravenes legal obligation 20(l) Displays disrespect 20(q) Violate Code of Conduct 20(s) Displays disrespect
Scenario 8: At 2:00 A.M. a police officer, who is on duty, is driving his patrol car on a deserted road. He sees a vehicle that has been driven off the road and is stuck in a ditch. He approaches the vehicle and observes that the driver is not hurt but is obviously intoxicated. He also finds that the driver is a police officer. Instead of reporting this accident and offense, he transports the driver to his home.	This is a Common Law offense of "Defeating or Obstructing the Course of Justice".	Section 20(f) Prejudices discipline 20(p) Improper conduct 20(q) Violate Code of Conduct
Scenario 9: A police officer has a private arrangement with a local auto body shop to refer the owners of cars damaged in accidents to the shop. In exchange for each referral, he receives a payment of 5% of the repair bill from the shop owner.	Prevention and Combating of Corrupt Activities Act 12 of 2004	Sections 20(h) Unauthorized compensation 20(p) Improper conduct 20(q) Violate Code of Conduct 20(z) Statutory offense
Scenario 10: A police officer arrests two drug dealers involved in a street fight. One has a large quantity of heroin on his person. In order to charge them both with serious offenses, the officer falsely reports that the heroin was found on both men.	Section 9 of the Justice of the Peace and Commissioners of Oaths Act 16 of 1963	Sections 20(q) Violate Code of Conduct 20(w) False statement 20(x) Falsifies record 20(z) Statutory offense

(Continued)

Table 4.1 (Continued)

Scenario description	*Statute articles*	*Official SAPS rules*
Scenario 11: A police sergeant, without intervening, watches officers under his supervision repeatedly strike and kick a man arrested for child abuse. The man has previous child abuse arrests. Evaluate the Sergeant's behavior.	SA Constitution, Chapter 2, Bill of Rights, Paragraph 12, Freedom and security of the person. Crime of Common Law Assault	Sections 20(a) Contravenes legal obligation 20(e) Endangers life 20(f) Prejudices discipline 20(p) Improper conduct 20(q) Violate Code of Conduct 20(r) 20(z) Common law crime
Scenario 12: In responding to a stabbing, a police officer arrests a suspect under circumstances that suggest that the attack was racially motivated. The police officer fails to write on the official report that the attack was racially motivated.	Obstructing the course of justice	Sections 20(l) Unfair discrimination 20(q) Violate Code of Conduct 20(w) False statement 20(x) Falsifies record
Scenario 13: A police officer stops a motorist for not obeying a red traffic light. The officer agrees to accept a personal gift for half of the amount of the fine in exchange for not issuing a citation.	Common law bribery & Prevention and Combating of Corrupt Activities Act 12 of 2004	Sections 20(f) Prejudices administration 20(p) Improper conduct 20(q) Violate Code of Conduct 20(z) Statutory offense
Scenario 14: While on patrol in his patrol car, a police officer notices a group of juveniles writing graffiti on a building. The police officer decides not to intervene and instead drives on.	Obstructing the course of justice	Sections 20(a) Contravenes legal obligation 20(f) Prejudices administration 20(q) Violate Code of Conduct

constructive labour relations in the Service" as one of its purposes (regulation 3(a)). Whether this objective is met seems unclear. Newham (2005) reported that commanders have struggled to apply the SDR since its promulgation and that they believed that discipline within the agency had generally deteriorated as a result. Masuku (2005, p. 16) is evenly skeptical and identifies impending absences of official attention and resources to the disciplinary system. In 2019, on reviewing the SAPS's enforcement of its disciplinary regulations, the African Centre for Justice Reform (ACJR) concluded that "the accountability chain in SAPS is broken" (ACJR, 2019, p. 5).

Following the provision in the SAPS Act (1995) that disciplinary procedures will be regulated in a separate legal document, the SDR actually regulates the disciplinary process (2006, articles 4–22). We provide a detailed account of the disciplinary procedure in Chapter 5. In this chapter, our focus lies in determining how legal rules delineate police misconduct from proper police behavior. To that end, the SDR contains a detailed list of 26 behaviors that constitute police misconduct (2006, article 20). The list includes a heterogeneous group of actions and omissions (2006, article 20) and features examples of police corruption (e.g., "operates any money lending scheme for employees," "accepts any compensation in cash or otherwise"), mismanagement of state funds (e.g., "wilfully and negligently mismanages the finances of the State," "intentionally or negligently damages and or causes loss to State property"), and biased/discriminatory policing (e.g., "unfairly discriminates against others on the basis of race, gender, disability, sexuality or other grounds prohibited by the Constitution," "prejudices the administration, discipline or efficiency of a department, officer, or institution," "misuses his or her position . . . to promote or to prejudice the interests of any political party"). In addition, the list contains examples of perjury (e.g., "gives a false statement or evidence in the execution of his or her duties"), falsification of evidence (e.g., "falsifies records or any other documentation"), and intimidation (e.g., "intimidates or victimizes other employees"). It also contains examples of behaviors dealing with the consumption of alcohol and illegal substances (e.g., "while on duty, is under the influence of intoxicating, illegal, unauthorized, habit-forming drugs, including alcohol"), absenteeism (e.g., "absents himself or herself from work without reason or permission"), sleeping on duty (e.g., "without authorization, sleeps on duty"), and sexual misconduct (e.g., "commits an act of sexual harassment"). The rules contain explicit examples of union-related misconduct (e.g., "prevent other employees from belonging to any trade union," "participates in any unlawful labour or industrial action"), as well as general statements about the failure to comply with regulations or commit common-law or statutory offenses. Absent from the rules are examples of abuse of force as a form of police misconduct, stating instead that misconduct includes behavior that "endangers the lives of others by disregarding safety rules or regulations."

The regulations aim to ensure "a mutual understanding of misconduct and discipline among supervisors and employees" (SDR, 2006, regulation 3(c)).

Although this objective underscores the organizational theory of police integrity's proposed correlation between a police agency of high integrity and its members' ability to identify misconduct in accordance with the agency's rules and regulations (see, e.g., Klockars and Kutnjak Ivković, 2004; Klockars et al., 1997, 2000), it remains uncertain whether it is realized at the SAPS's operational level. With limited information available on the extent to which SAPS members recognize incidents of SDR-violating behavior, our 2005 research shed light on this aspect (Sauerman and Kutnjak Ivković, 2008). In measuring the levels of integrity among 379, mostly supervisory SAPS members, our results indicated a substantial majority who recognized police misconduct within a provided set of rule-violating scenarios. Disconcertingly, we also found a considerable minority of supervisors who did not identify the most severe scenario representations as violations of regulations, even though such violations would generally be subject to criminal investigation (Sauerman and Kutnjak Ivković, 2008, p. 31).

Police integrity theory and official rules

The police integrity theory (e.g., Klockars et al., 2000, 2004, 2006; Kutnjak Ivković, 2015) rests on the premise that enhancing police integrity and curtailing police misconduct are organizational and administrative responsibilities (see Chapter 1). The first dimension of the theory focuses on organizational rules and the way they are made by police agencies, communicated to police officers, and understood by officers (e.g., Klockars et al., 2006). While virtually every police agency is governed by a set of legal rules, what differs across countries, and potentially even police agencies, is "the way these official rules are made, how often they are enforced, the level of police officers' familiarity with the rules, and the extent to which police officers support them" (Kutnjak Ivković, 2015, p. 5). Klockars and colleagues (2006, p. 9) argued:

> The obligations of rulemaking require police agencies not only to develop policies and create both formal and informal rules that specify expectations of integrity but also to communicate those rules and the reasons for them to their employees. In a police agency of integrity police officers ought to know the agency's integrity relevant rules, understand the agency's rationale for them, and believe in the rightness of both.

The bottom line is that agencies of high integrity should have developed rules that outline prohibited police behavior. Yet legal rules addressing some forms of police misconduct (e.g., police corruption, perjury, planting of evidence) are easier to design than rules dealing with other police misconduct types (e.g., racial profiling, milking the clock). In addition, governments and police agencies are more likely to prohibit serious misconduct in their respective laws and official rules than they are to prohibit lesser forms of misconduct (e.g., kickbacks in comparison with the corruption of authority; Roebuck and

Barker, 1974). Police agencies and governments are also more likely to differ in how they define less serious misconduct, and even more importantly, whether they prohibit it in the first place (Klockars et al., 2006).

Developing rules and teaching them to police officers is further complicated when discrepancies exist between the rules on the books (i.e., official rules) and what seems to be allowed in practice (i.e., unofficial rules). Such divergences are common among government workers (e.g., Lipsky, 2010), including the police (Cockcroft, 2013; Reiner, 2010), who routinely develop "shortcuts" in their work environments. As such, a police agency of high integrity would need not only to develop and teach clear rules to its officers but also to enforce them consistently (see Chapter 5), thereby effectively closing the gap between official and unofficial rules. Quite possibly, when rules prohibiting misconduct are not enforced, the informal, unofficial rules that develop in the resulting vacuum allow misconduct to continue unabated (Klockars et al., 2006). Under such circumstances—some of which have been documented by ethnographers (e.g., Marks, 2005; Steinberg, 2008; Hornberger and Vigneswaran, 2009; Hornberger, 2011; Faull, 2018; Perkins, 2018)—police officers may perceive official rules as unimportant, assess that misconduct is tacitly allowed, and fear no consequences if they engage in rule-violating behavior. Alternatively, if there is a large discrepancy between official and unofficial rules, police officers may be confused about the content of the actual, endorsed rules. For example, Klockars and colleagues (2006, p. 256) reported that in one police agency, where both an official policy prohibiting the acceptance of gratuities and an informal policy tolerating such behavior were present, police officers did not know what behavior was officially expected of them. Various independent commissions that investigated police misconduct in the United States (e.g., Pennsylvania Crime Commission, 1974; Mollen Commission, 1994) found similar disjunctions between official policies and informally accepted practices.

In a comparative study involving ten countries as diverse as Croatia, Russia, South Korea, Thailand, and the United States, Kutnjak Ivković and Haberfeld (2015) found that in all countries less serious forms of misconduct were less likely to be recognized by police officers as violations of official rules. Specifically, the acceptance of gratuities, the least serious form of corruption, was much less likely to be labeled as a violation of official rules than a kickback or an opportunistic theft, both of which were evaluated as much more serious. Similarly, respondents were uniformly more likely to recognize the abuse of deadly force as a violation, compared to the verbal abuse of citizens. Despite these similarities, the study also found large differences in how likely the respondents from different countries were to say that the same behavior violated the official rules in their respective countries. Although extant literature includes a number of other comparative studies (e.g., Andreescu et al., 2012; Ekenvall, 2003; Huberts et al., 2003; Khruakham and Lee, 2013; Klockars et al., 2004; Kutnjak Ivković, 2005; Kutnjak Ivković and Shelley, 2008), none of them focused on evaluating the respondents' knowledge of official rules.

The police integrity theory also emphasizes the importance of police officers' support for official rules (e.g., Klockars et al., 2006). In police agencies of high integrity, officers are expected to be familiar with official rules, understand the rationale behind them, and believe that they are correct and necessary (Klockars et al., 2006, p. 9). As Klockars and colleagues (2006, p. 9) argue, if officers do not support the rules, the role of police administrators and managers in agencies of high integrity is to teach those officers why their support of the rules is necessary. Whereas the police integrity questionnaire does not contain questions probing the respondents' level of support for official rules, it contains questions probing the perceived seriousness of various forms of misconduct. When presenting the results of a two-year study of three agencies, Klockars and colleagues (2006, p. 9) explained how police officers learned whether misconduct was perceived as serious:

> [I]n none of the agencies was there anything but the most minimal instruction in the moral gravity of specific types of police misconduct. The way officers learned to evaluate the seriousness of various types of misbehavior was by gleaning it from the department's behavior in detecting and punishing it. If a department welcomed complaints of misconduct, investigated those complaints thoroughly, and severely disciplined officers who committed them, officers concluded that misconduct of that type was serious. If agencies ignored or discouraged complaints of certain types of misconduct, failed to investigate certain types of complaints, or failed to punish officers for certain violations, officers learned not to take those violations seriously.

Studies have consistently found that police officers' assessments of misconduct seriousness vary not only across countries (e.g., Andreescu et al., 2012; Ekenvall, 2003; Huberts et al., 2003; Johnson, 2003; Khruakham and Lee, 2013; Klockars et al., 2004; Klockars and Kutnjak Ivković, 1999; Kutnjak Ivković, 2004, 2005; Kutnjak Ivković and Haberfeld, 2015; Kutnjak Ivković and Shelley, 2008; Vito et al., 2011) but also within police agencies in decentralized police systems (e.g., Klockars et al., 2000, 2004, 2006). While the relative order of seriousness was very similar across the 30 U.S. police agencies in the first study of police integrity that Klockars and colleagues conducted (2000, 2004), that study also uncovered large differences in officer evaluations of the seriousness of corruption across the agencies. In a subsequent U.S. study, based on the police integrity questionnaire, Klockars and colleagues (2006) emphasized that, despite overall small differences in evaluations of seriousness provided by officers at three police agencies of high integrity, there were large differences across the agencies in how their respective officers evaluated the acceptance of gratuities, which authors connected to the heterogeneity of the agencies' unofficial policies on gratuities.

To illustrate potential differences in seriousness evaluations across countries, we rely once again on the ten-country comparative study conducted by Kutnjak Ivković and Haberfeld (2015). While there were large and meaningful differences in officer evaluations of misconduct seriousness across the ten countries, the respondents from these heterogeneous countries were mostly in agreement about what constitutes more serious misconduct. In particular, opportunistic theft and the abuse of deadly force were typically perceived as the most serious violations, while the acceptance of gratuities and verbal abuse of citizens were typically viewed as the least serious violations (Kutnjak Ivković and Haberfeld, 2015). However, the authors noted that the country's sociohistorical context (e.g., relaxed attitudes toward the use of excessive force in South Korea; widespread corruption in Russia) directly and strongly affected how the respondents evaluated misconduct.

Survey findings

Our survey conducted in South Africa includes responses from 868 SAPS officers. The questionnaire was designed by Klockars and colleagues (2006) and later revised by Kutnjak Ivković (2009). The respondents were asked to assess 14 hypothetical scenarios describing a range of police misconduct, including police corruption, use of excessive force, falsification of official documents, and the planting of evidence. After the respondents read the description of each scenario, they were asked to assess whether the described behavior violated official rules, how serious they thought the behavior was, their perceptions of the appropriate and expected discipline, as well as their propensity to report such behavior. For details about the methodology, please see Chapter 1.

In light of the focus of this chapter on official rules, we will discuss the results primarily as they relate to the respondents' familiarity with, and support for, such rules. The first question explored the officers' own perceptions of misconduct seriousness ("How serious do you consider this behavior to be?"). It was followed by the question about their estimates of how serious most SAPS officers would evaluate the same behavior ("How serious do most police officers in your agency consider this behavior to be?"). The respondents could select one answer for each question using a 5-point scale ranging from 1 = "not at all serious" to 5 = "definitely serious." In addition, they were also asked to assess whether the described misconduct violated the SAPS official rules ("Would this behavior be regarded as a violation of official policy in your agency?"). Possible answers ranged on a 5-point scale from 1 = "definitely not" to 5 = "definitely yes."

As shown in Table 4.1, 12 of the 14 scenarios describe clear instances of either common law or statutory crimes. Furthermore, all 14 describe violations of the SDR. In all instances, the scenarios depict behaviors that contravene the SAPS Code of Conduct in one form or another, the violation of which in itself constitutes misconduct under section 20(q) of the SDR. In addition, all scenarios depict behaviors that constitute at least one additional act of police misconduct beyond violation of the SAPS Code of Conduct. This is not to say that an officer would actually be charged with all the separate acts of misconduct listed, but that a very specific application of the description and listings of 26 instances of what the SDR identifies as misconduct could result in a charge. A challenge to accurate matching of the scenarios with specific instances of misconduct defined or described in the SDR is that all examples of misconduct are written very broadly and are thus open to interpretation, except in the most serious, and thus clearest, instances. This characteristic of the SDR is possibly what led Newham (2005) to observe that commanders have struggled to apply the SDR since its promulgation.

Knowledge of official rules

To assess their familiarity with official rules, our respondents were asked to state whether the behaviors described in the 14 hypothetical scenarios violated the official SAPS rules. With the exception of two scenarios (Scenario 7: Verbal abuse; Scenario 14: Failure to react to graffiti), the mean values, ranging between 4 and 5, indicate that the respondents were able to recognize and label the described behaviors as violations of official rules (Table 4.2). Behaviors such as opportunistic theft (Scenario 3: Theft of knife), failure to make an arrest (Scenario 2: Failure to arrest), and acceptance of a bribe (Scenario 13: Bribe from reckless motorist) had the highest mean values (between 4.58 and 4.75), suggesting that the respondents would be most likely to recognize them as rule violations (Table 4.2). On the other hand, verbally abusing citizens (Scenario 7: Verbal abuse), not reacting to graffiti (Scenario 14: Failure to react to graffiti), and covering-up a police DUI and accident (Scenario 8: Cover-up of DUI) had the lowest mean values (between 3.75 and 4.01), implying that the respondents would be least likely to recognize these as rule-violating behaviors (Table 4.2).

Aside from analyzing mean values, an exploration of the full distribution of answers for each scenario offers valuable insights (Figures 4.1 to 4.14). First, if we assume that the respondents know the official rules only if they were able to say that the behavior "definitely violates the rules" (by selecting answer "5" on the questionnaire), fewer than 50% definitely recognized verbal abuse of a citizen (Scenario 7: Verbal abuse) and the failure to react to graffiti (Scenario 14: Failure to react to graffiti) as violations of official rules. On the other hand, the majority of the respondents, between one-half and two-thirds, were able to evaluate the behaviors described in six scenarios as definite rule-violations (Scenario 1: Acceptance of gratuities; Scenario 5: Supervisor corruption; Scenario 6: Striking arrestee; Scenario 8: Cover-up of DUI; Scenario 9: Kickback; Scenario 12: Failure to report hate crime). Some of these scenarios include serious forms of corruption (Scenario 9: Kickback) and the use of excessive

Table 4.2 Violation of official rules

Scenario	*Mean*	*Rank (smallest to largest)*	*St. deviation*	*Rank (smallest to largest)*	*Rank (largest to smallest)*
Scenario 1: Acceptance of gratuities	4.26	5	1.184	11	4
Scenario 2: Failure to arrest	4.58	12	.881	3	12
Scenario 3: Theft of knife	4.75	14	.687	1	14
Scenario 4: Abuse of deadly force	4.48	9	.985	6	9
Scenario 5: Supervisor corruption	4.36	8	1.075	9	6
Scenario 6: Striking arrestee	4.35	7	1.086	10	5
Scenario 7: Verbal abuse	3.95	2	1.218	12	3
Scenario 8: Cover-up of DUI	4.01	3	1.367	13	2
Scenario 9: Kickback	4.32	6	1.054	8	7
Scenario 10: False report on drugs	4.49	10	.895	4	11
Scenario 11: Supervisor fails to stop excessive force	4.50	11	.929	5	10
Scenario 12: Failure to report hate crime	4.15	4	1.039	7	8
Scenario 13: Bribe from reckless motorist	4.68	13	.771	2	13
Scenario 14: Failure to react to graffiti	3.75	1	1.655	14	1

force (Scenario 6: Striking arrestee), all of which are directly prohibited by the official rules (Table 4.1). Finally, the overwhelming majority of the respondents, about three-quarters, could recognize the behaviors described in six scenarios as rule violations (Scenario 2: Failure to arrest; Scenario 3: Theft of knife; Scenario 4: Abuse of deadly force; Scenario 10: False report on drugs; Scenario 11: Supervisor fails to stop excessive force; Scenario 13: Bribe from reckless motorist). The behaviors most likely to be evaluated as rule-violating are not only in a direct violation of the SAPS policy but are also in a violation of the norms of criminal law (Table 4.1). These include some of the most

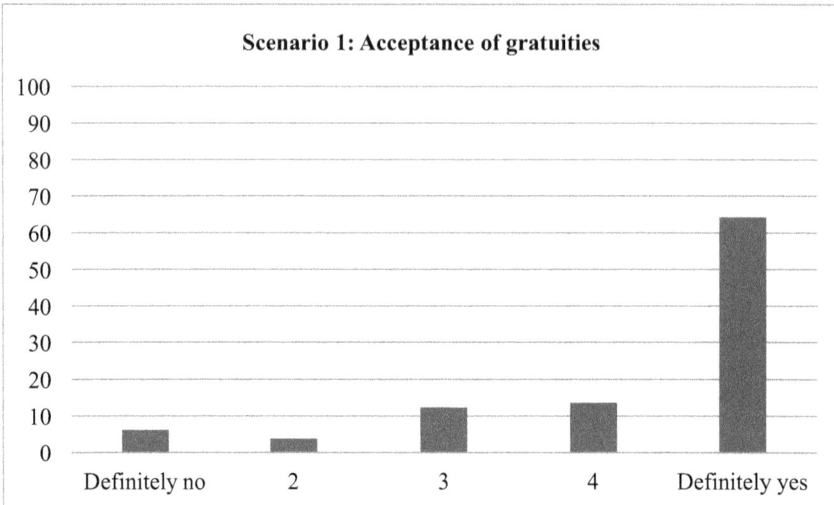

Figure 4.1 Distribution of answers about rule violations for Scenario 1

serious forms of police misconduct, such as the abuse of deadly force (Scenario 4: Abuse of deadly force), the acceptance of a bribe (Scenario 13: Bribe from reckless motorist), opportunistic theft (Scenario 3: Theft of knife), and the planting of evidence (Scenario 10: False report on drugs).

Second, if we use a less stringent approach and regard as affirmative both answers "4" and "5," the proportion of respondents familiar with the official rules expectedly increases. While there were no scenarios in which the majority of the respondents were unsure whether the described behavior was rule-violating (Figures 4.1 to 4.14), there were four scenarios (Scenario 7: Verbal abuse; Scenario 8: Cover-up of DUI; Scenario 12: Failure to report hate crime; Scenario 14: Failure to react to graffiti) in which only about one-half to three-quarters of the respondents identified the behavior as rule-violating. However, in most scenarios (10 out of 14), over three-quarters of the respondents were able to recognize the behaviors as violations of official rules. In fact, in two scenarios (Scenario 3: Theft of knife; Scenario 13: Bribe from reckless motorist), the overwhelming majority of respondents—over 90%—evaluated these behaviors as rule violations.

Third, regardless of the approach we use, there remain respondents who did not evaluate these behaviors as rule-violating (Figures 4.1 to 4.14). If we utilize the first approach, relying only on the respondents who selected "5" or "definitely yes," to determine the respondents who correctly identified the behavior in question as a violation of official rules, the percentage of respondents unsure whether the behavior in question definitely violates the official

Figure 4.2 Distribution of answers about rule violations for Scenario 2

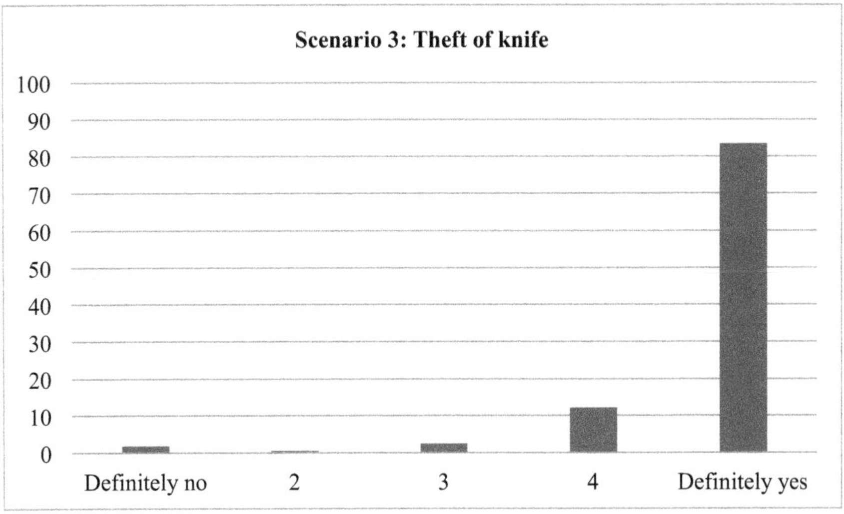

Figure 4.3 Distribution of answers about rule violations for Scenario 3

rules ranges from 17% (Scenario 3: Theft of knife) to 62% (Scenario 14: Failure to react to graffiti). In many of these scenarios (9 out of 14), there was a substantial minority of police officers—about one-third or more—who were unsure whether the described behaviors definitely violated official rules. In two

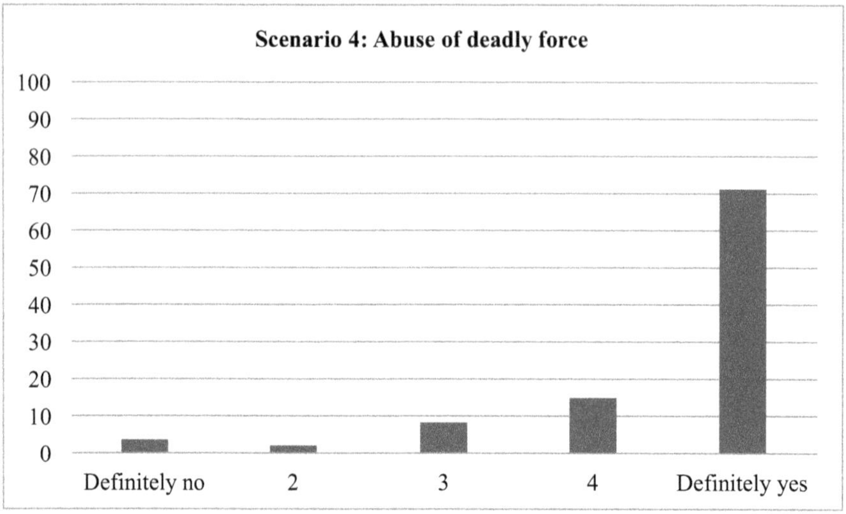

Figure 4.4 Distribution of answers about rule violations for Scenario 4

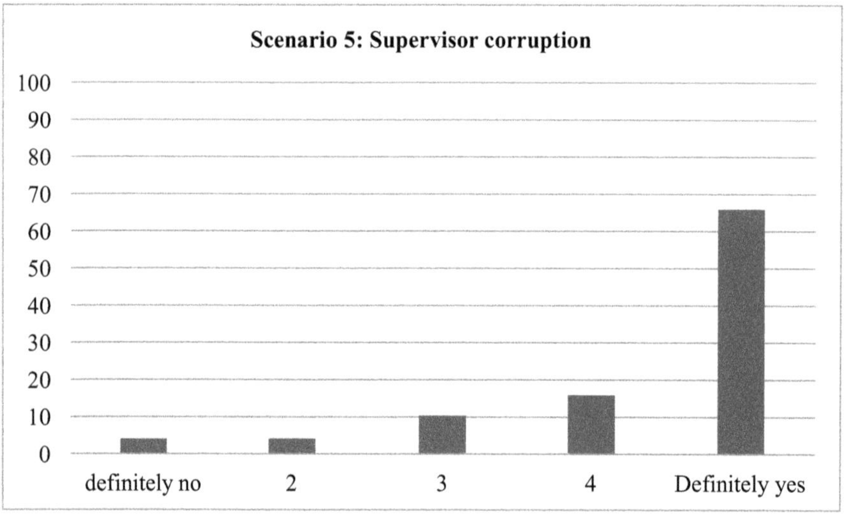

Figure 4.5 Distribution of answers about rule violations for Scenario 5

additional scenarios (Scenario 7: Verbal abuse; Scenario 14: Failure to react to graffiti), the *majority* of the respondents were unsure whether the described behaviors were definite violations of official rules. Even in the more relaxed, second approach (i.e., answers "4" and "5"), the percentage of respondents

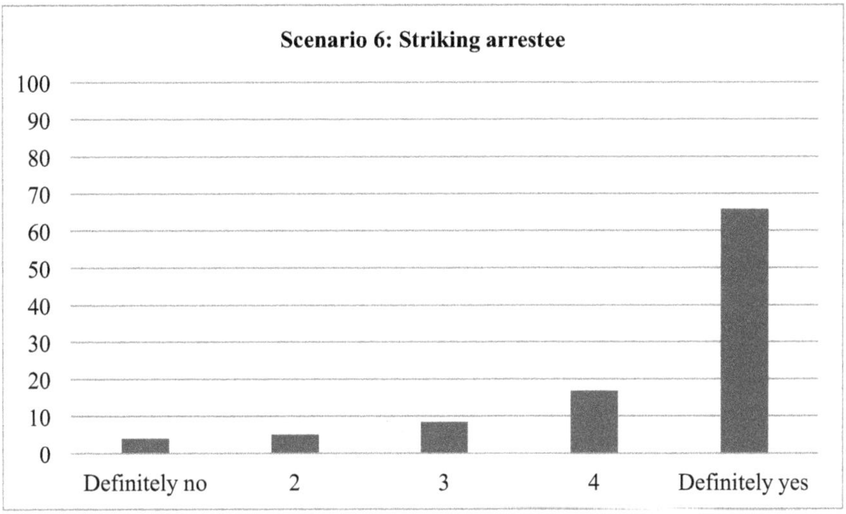

Figure 4.6 Distribution of answers about rule violations for Scenario 6

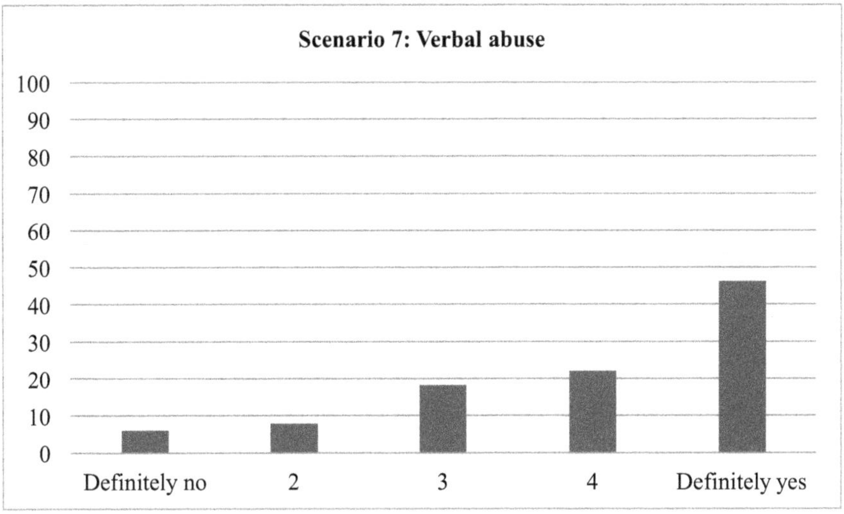

Figure 4.7 Distribution of answers about rule violations for Scenario 7

unsure whether the behavior is rule-violating varies from 9% (Scenario 13: Bribe from reckless motorist) to 38% (Scenario 14: Failure to react to graffiti).

Fourth, there is more certainty in labeling some behaviors as rule-violating than others. A visual inspection of the answer distribution for each scenario in

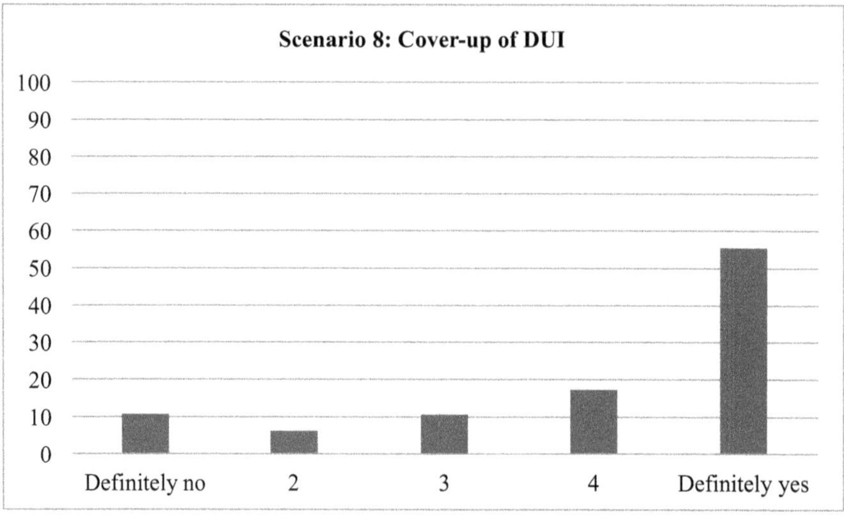

Figure 4.8 Distribution of answers about rule violations for Scenario 8

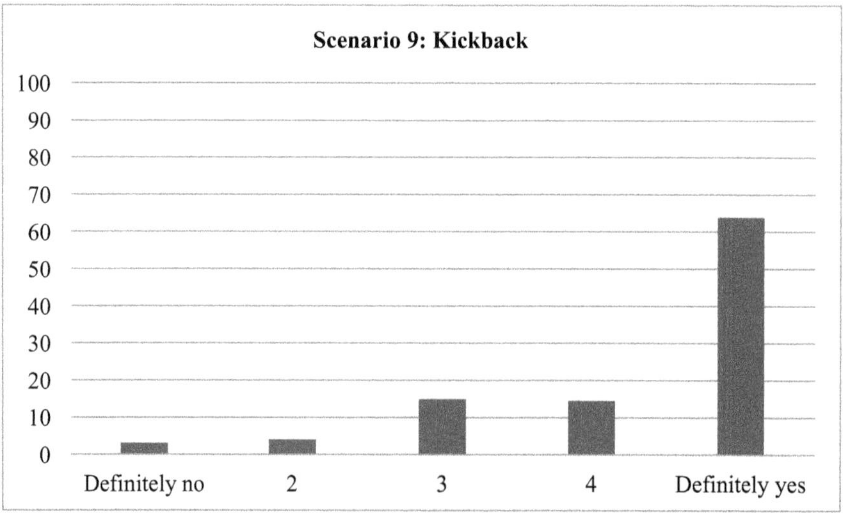

Figure 4.9 Distribution of answers about rule violations for Scenario 9

Figures 4.1 to 4.14 and the standard deviation for each scenario in Table 4.2 reveal that the respondents' views were more homogenous for opportunistic theft (Scenario 3: Theft of knife), failure to execute an arrest warrant (Scenario 2: Failure to arrest), and acceptance of a bribe from a motorist (Scenario 13: Bribe from reckless motorist) than for verbal abuse (Scenario 7: Verbal

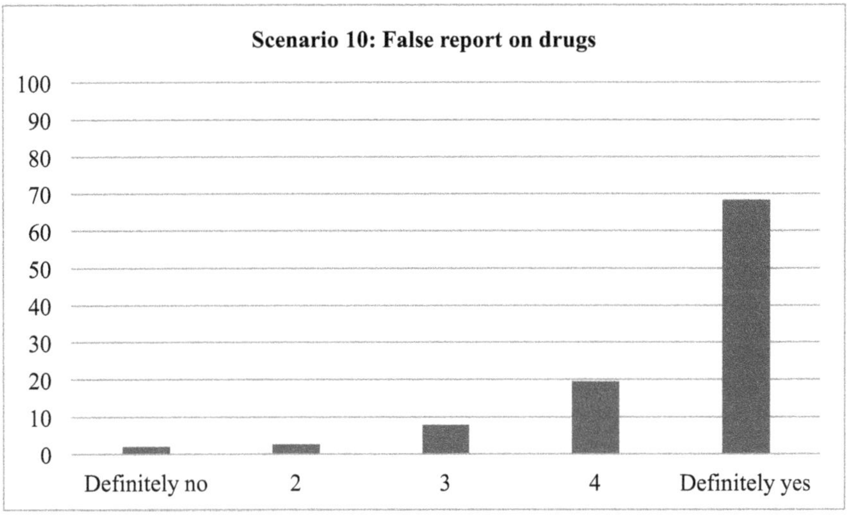

Figure 4.10 Distribution of answers about rule violations for Scenario 10

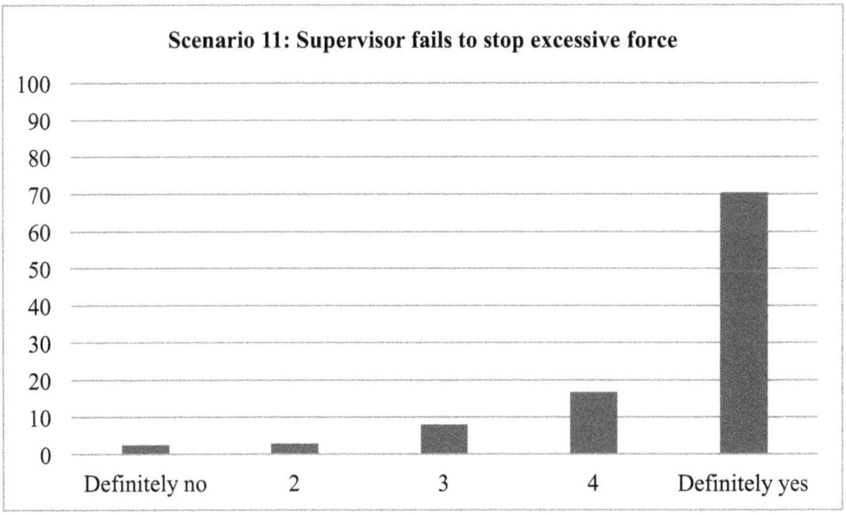

Figure 4.11 Distribution of answers about rule violations for Scenario 11

abuse), cover-up of DUI (Scenario 8: Cover-up of DUI), and failure to report a hate crime (Scenario 12: Failure to report hate crime). In fact, there is a strong negative correlation between the respondents' evaluation of behavior as rule-violating and the size of the standard deviations (Spearman's rho = −.934, $p < .001$).

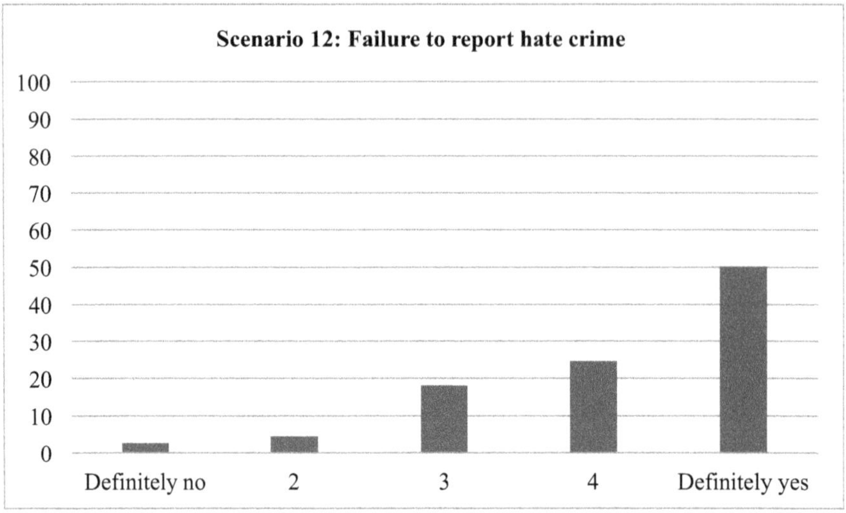

Figure 4.12 Distribution of answers about rule violations for Scenario 12

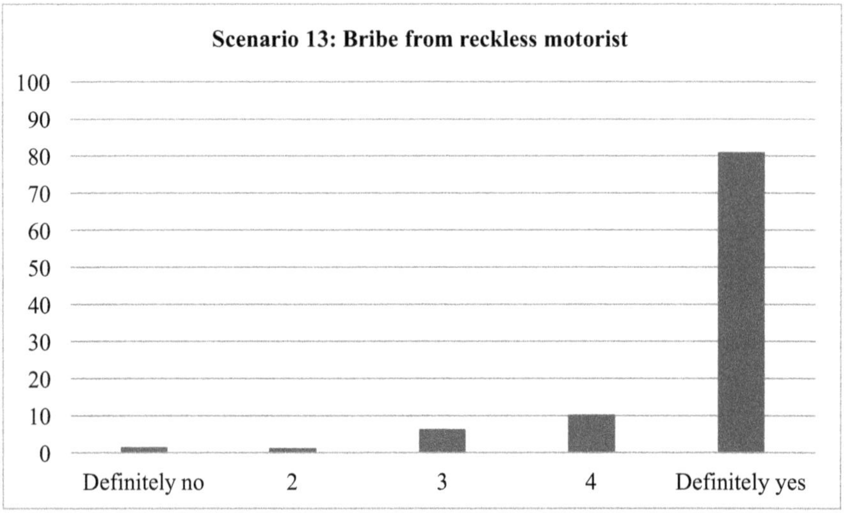

Figure 4.13 Distribution of answers about rule violations for Scenario 13

Support for official rules

One approach toward assessing how strongly police officers support official rules is examining their evaluations of misconduct seriousness (e.g., Klockars et al., 2006), presented in Table 4.3. The misconduct contained in the questionnaire

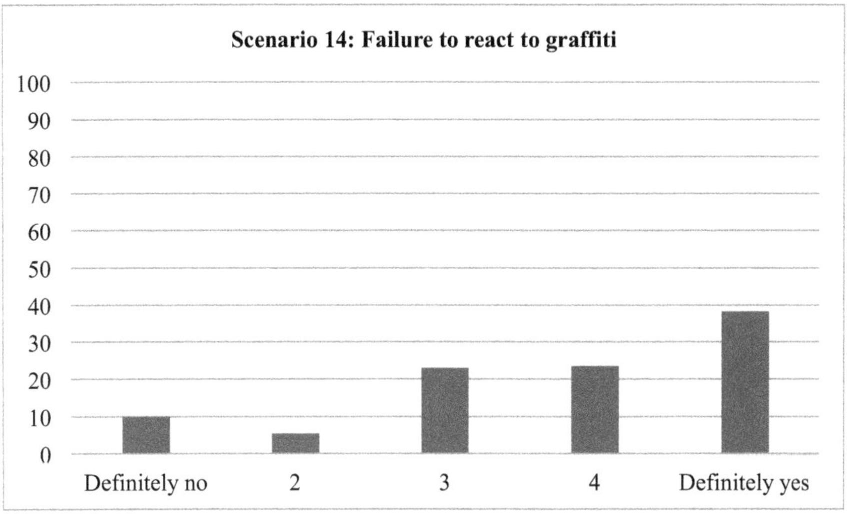

Figure 4.14 Distribution of answers about rule violations for Scenario 14

varies in terms of its seriousness. The respondents perceived four scenarios (Scenario 2: Failure to arrest; Scenario 3: Theft of knife; Scenario 4: Abuse of deadly force; Scenario 13: Bribe from reckless motorist) as the most serious; the mean values for all of these scenarios (between 4.5 and 5 on a 5-point scale) are the highest among all 14 scenarios (ranked as #11 to #14). Because the abuse of deadly force, the acceptance of a bribe, theft, and a failure to execute an arrest warrant exemplify behaviors that constitute severe violations of both the SAPS organizational rules and the criminal code (see Table 4.1), it is not surprising that officers evaluated them as the most serious and, hence, most strongly supported the rules prohibiting them.

At the other end, the respondents were least supportive of the rules prohibiting acceptance of gratuities (Scenario 1: Acceptance of gratuities), verbal abuse of citizens (Scenario 7: Verbal abuse), cover-up of police DUI and accident (Scenario 8: Cover-up of DUI), and failure to react to graffiti (Scenario 14: Failure to react to graffiti). In these four scenarios, the mean values (all below 4.0 on a 5-point scale) are the lowest among the means for all 14 scenarios (ranked #1 to #4), suggesting that the respondents evaluated these forms of misconduct as least serious and, hence, would be less supportive of the official rules prohibiting them.

The respondents also provided estimates of how seriously they thought most SAPS officers would evaluate the same behaviors (Table 4.3). Similar to their own perceptions, respondents assessed most police officers to also differentiate across the scenarios when evaluating misconduct seriousness. The relative

Table 4.3 Own and most police officers' seriousness estimates

Scenario	*Own seriousness*		*Most police officers' seriousness*		*t-test*	*Difference of means*	*Cohen's d*
	Mean	*Rank*	*Mean*	*Rank*			
Scenario 1: Acceptance of gratuities	3.98	4	3.39	2	11.64***	.59	0.43 Small-med.
Scenario 2: Failure to arrest	4.60	12	3.98	9	15.84***	.62	0.60 Medium
Scenario 3: Theft of knife	4.78	14	4.26	12	14.91***	.52	0.60 Medium
Scenario 4: Abuse of deadly force	4.57	11	4.30	13	8.00***	.27	0.27 Small
Scenario 5: Supervisor corruption	4.35	8	3.88	6.5	11.21***	.47	0.39 Small
Scenario 6: Striking arrestee	4.29	6	3.79	5	13.55***	.50	0.41 Small-med.
Scenario 7: Verbal abuse	3.88	2	3.53	3	10.94***	.35	0.26 Small
Scenario 8: Cover-up of DUI	3.94	3	3.59	4	10.69***	.35	0.25 Small
Scenario 9: Kickback	4.32	7	3.95	8	10.97***	.37	0.31 Small
Scenario 10: False report on drugs	4.52	10	4.14	11	11.75***	.38	0.37 Small
Scenario 11: Supervisor fails to stop excessive force	4.46	9	4.01	10	12.76***	.45	0.39 Small
Scenario 12: Failure to report hate crime	4.17	5	3.88	6.5	9.28***	.29	0.25 Small
Scenario 13: Bribe from reckless motorist	4.71	13	4.32	14	12.00***	.39	0.42 Small-med.
Scenario 14: Failure to react to graffiti	3.72	1	3.38	1	10.91***	.34	0.26 Small

Note: Cohen's *d* calculated at Becker, Effect size calculators, available at: www.uccs.edu/lbecker/

order of scenario seriousness—based on the rank-order of scenarios by their mean values—was very similar between the respondents' own answers and their estimates of most police officers' assessments (Spearman's rho = .928, $p < .001$). Conversely viewed, the respondents assumed that other officers would use the same internal relative order when evaluating the seriousness of the scenarios.

We also compared the mean values for the respondents' own evaluations of seriousness with the estimates of most police officers' evaluations of seriousness (Table 4.3). In all 14 scenarios, the differences between the respondents' own evaluations of seriousness and their assessments of most officers' evaluations of seriousness are positive and statistically significant. Although the differences were statistically significant in all scenarios, Cohen's *d* indicates that they were larger ("small to medium" or "medium") in only 5 out of 14 scenarios (Scenario 1: Acceptance of gratuities; Scenario 2: Failure to arrest; Scenario 3: Theft of knife; Scenario 6: Striking arrestee; Scenario 13: Bribe from reckless motorist). These scenarios do not form a homogeneous group because they vary in their severity from those evaluated as the least serious (Scenario 1: Acceptance of gratuities), those of medium seriousness (Scenario 6: Striking arrestee), to those evaluated as the most serious in the questionnaire (Scenario 2: Failure to arrest; Scenario 3: Theft of knife; Scenario 13: Bribe from reckless motorist).

Klockars and colleagues (2006, p. 26) developed a rule of thumb that, in the context of 5-point scales, the calculated differences in means should be regarded as meaningful if they reach or exceed 0.5. Applying this guideline to explore the differences in the evaluations of seriousness between our respondents' own estimates and their assessments of most SAPS officers' estimates, the inference is similar to the conclusion we reached on the basis of Cohen's *d*. Specifically, in the majority of the scenarios, the differences were statistically significant, but not meaningful; the respondents evaluated misconduct as substantively more serious than they expected most police officers would for only four scenarios (Scenario 1: Acceptance of gratuities; Scenario 2: Failure to arrest; Scenario 3: Theft of knife; Scenario 6: Striking arrestee).

Conclusion

This chapter addresses the first dimension of the Klockars and Kutnjak Ivković (2004) theory of police integrity, focusing on organizational rules and the way they are made by police agencies, communicated to, and understood by their officers. In South Africa, these rules are grounded in the country's Constitution, statutory and common law, the SAPS Code of Conduct, and the SAPS Disciplinary Regulations.

The Klockars and Kutnjak Ivković (2004) theory of police integrity rests on three basic requirements. First, a police organization must have a set of behavioral expectations written in official rules. In this regard, the SAPS has a detailed set of rules describing and identifying behaviors constituting police misconduct and creating procedures for disciplinary repercussions. Moreover, each of the scenarios presented to our sample of officers constitutes an example of police misconduct defined by the SAPS's Disciplinary Procedures (Table 4.1).

Second, in addition to establishing the official rules, members of a police agency must be aware of its rules. Our results clearly show that the majority of SAPS respondents recognized the behaviors described as violations of organizational rules (Table 4.2). However, as prior research (e.g., Klockars et al., 2004, 2006; Kutnjak Ivković and Haberfeld, 2015) suggests, the extent of such recognition varied across the different scenarios.

Third, organizational members, individually and collectively, must recognize the imperative of these rules and provide support for them. To gauge this, the questionnaire measures the degree to which respondents believe the described behaviors constitute serious violations. In all scenarios, the majority of our respondents considered the behaviors to be serious violations (Table 4.3). However, as with the question soliciting whether the described behaviors violated organizational rules, assessments of the behavior seriousness varied across the scenarios. This result is by no means surprising, because the extant police integrity research provided ample evidence of such variation in a number of other countries across the world (e.g., Andreescu et al., 2012; Ekenvall, 2003; Huberts et al., 2003; Johnson, 2003; Khruakham and Lee, 2013; Klockars et al., 2004; Klockars and Kutnjak Ivković, 1999; Kutnjak Ivković, 2004, 2005; Kutnjak Ivković and Haberfeld, 2015; Kutnjak Ivković and Shelley, 2008; Vito et al., 2011).

Finally, respondent assessments of how most SAPS officers would view the seriousness of the various behaviors lead to several observations. First, respondents expected that most of their fellow officers would evaluate the behaviors described in the questionnaire as serious violations of the official rules (Table 4.3). At the same time, they expected that their fellow officers would not evaluate all behaviors to be of the same level of seriousness and, instead, would vary in their assessments across the scenarios. Second, the respondents' personal assessments of seriousness were greater than their predictions of most officers' assessments for all scenarios. We attribute this finding to self-serving bias. Third, the relative rank ordering of the perceived seriousness of a violation by most officers is consistent with the individual respondent's own assessments of seriousness, implying that the officers believed that their fellow officers shared the same underlying views of misconduct seriousness.

At face value, our findings suggest that most of the SAPS officers in our sample have a good understanding of the organization's formal rules and generally also believe this to be true of their colleagues. However, although the establishment of the rules by the organization and support for those rules by its officers are necessary requirements for the SAPS to be deemed an agency of integrity, meeting these prerequisites is not necessarily sufficient. In addition to its familiarity with rules, a police agency of high integrity must also provide accountability for violations, and its officers must be supportive of such a system. We turn to these criteria in Chapter 5.

Notes

1 Unlike in other jurisdictions, SAPS officers almost never carry or issue tickets. These are more commonly issued by traffic and metro police in South Africa.
2 In the lingua franca of the SAPS, "member" and "official" are the preferred terms for police "officer."
3 Because they are constitutionally deemed an "essential service" without which society could not function, SAPS officers are not allowed to participate in industrial actions when on duty. Disputes must instead be referred to the Commission for Conciliation, Mediation and Arbitration (CCMA) for conciliation and arbitration.

References

ACJR (African Centre for Justice Reform) (2019). Failing to discipline in SAPS: Fostering a culture of impunity. *ACJR Factsheet 9*, February. Available at: https://acjr.org.za/resource-centre/fact-sheet-9-failing-to-discipline-v-3.pdf

Andreescu, V., D. Keeling, G. Vito, and M. Voinic (2012). Romanian and American police officers' perceptions of professional integrity and ethical behavior. *Revista Română de Sociologie*, 23(3–4): 185–207.

Bittner, E. (1970). *The Functions of the Police in Modern Society: A Review of Background Factors, Current Practices, and Possible Role Models*. Rockville, MD: National Institute of Mental Health, Center for Studies of Crime and Delinquency.

Bruce, D. (2002). *New Wine from an Old Cask? The South African Police Service and the Process of Transformation*. Paper presented at John Jay College of Criminal Justice, New York, 9 May 2002. Cape Town, South Africa, Centre for the Study of Violence and Reconciliation. Available at https://csvr.org.za/docs/policing/newwinefromold.pdf.

Cockcroft, T. (2013). *Police Culture*. New York: Taylor and Francis. Kindle Edition.

Constitution of the Republic of South Africa, Act No. 108 of 1996. Available at: www.gov.za/documents/constitution-republic-south-africa-1996

Criminal Law (Sexual Offenses and Related Matters) Amendment, Act 32 of 2007. Available at: https://justice.gov.za/legislation/acts/2007-032.pdf

Department of Police (2016). *Discipline regulations for the South African Police Service, Government Gazette No. 40389*. Available at: www.gov.za/sites/default/files/gcis_document/201611/40389rg10662gon1361.pdf

Ekenvall, B. (2003). Police attitudes toward fellow officers' misconduct: The Swedish case and a comparison with the USA and Croatia. *Journal of Scandinavian Studies in Criminology and Crime Prevention*, 3(2): 210–232.

Faull, A. (2018). *Police Work and Identity: A South African Ethnography*. Abingdon: Routledge.

Hornberger, J. (2011). *Policing and Human Rights: The Meaning of Violence and Justice in the Everyday Policing of Johannesburg*. Abingdon: Routledge.

Hornberger, J. and D. Vigneswaran (eds.) (2009). *Beyond "Good Cop"/"Bad Cop": Understanding Informality and Police Corruption in South Africa*. Johannesburg: University of the Witwatersrand.

Huberts, L., T. Lamboo, and M. Punch (2003). Police integrity in the Netherlands and the United States: Awareness and alertness. *Police Practice and Research*, 4(3): 217–232.

[Interim Constitution] Constitution of the Republic of South Africa (1993). *Act No. 200 of 1993. Gazette Number: 1994-01-28*. Available at: www.justice.gov.za/legislation/acts/1993-200.pdf

Johnson, D.T. (2003). Above the law? Police integrity in Japan. *Social Science Japan Journal*, 6(1): 19–37.

Justice of the Peace and Commissioners of Oaths Act 16 of 1963. Available at: www.justice.gov.za/legislation/acts/1963-016.pdf

Kemp, G., S. Walker, R. Palmer, D. Baqua, C. Gevers, B. Leslie, and A. Steynberg (2015). *Criminal Law in South Africa*. Cape Town, South Africa: Oxford University Press.

Khruakham, S. and J. Lee (2013). Cross-nation comparison of the intolerance to police misconduct: Findings from a Thai police cadet survey. *International Journal of Police Science & Management*, 15(3): 237–245.

Klockars, C.B. and S. Kutnjak Ivković (1999). The measurement of police delinquency. In W. Laufer and F. Adler (eds.), *The Criminology of Criminal Law. Advances in Criminological Theory*, Vol. 8. New Brunswick, NJ: Transaction, pp. 87–106.

Klockars, C.B. and S. Kutnjak Ivković (2004). Measuring police integrity. In M.J. Hickman, A.R. Piquero, and J.R. Greene (eds.), *Police Integrity and Ethics*. Belmont, CA: Wadsworth Publishing.

Klockars, C.B., S. Kutnjak Ivković, and M.R. Haberfeld (2004). The contours of police integrity. In C.B. Klockars, S. Kutnjak Ivković, and M.R. Haberfeld (eds.), *The Contours of Police Integrity*. Thousand Oaks, CA: Sage Publications, pp. 1–18.

Klockars, C.B., S. Kutnjak Ivković, and M.R. Haberfeld (2006). *Enhancing Police Integrity*. Dordrecht, The Netherlands: Springer.

Klockars, C.B., S. Kutnjak Ivković, W.E. Harver, and M.R. Haberfeld (1997). *The Measurement of Police Integrity*. Final Report Submitted to the U.S. Department of Justice, Office of Justice Programs, National Institute of Justice.

Klockars, C.B., S. Kutnjak Ivković, W.E. Harver, and M.R. Haberfeld (2000). *The Measurement of Police Integrity*. Research in Brief. U.S. Department of Justice, Office of Justice Programs, National Institute of Justice. Washington, DC: Government Printing Office.

Kutnjak Ivković, S. (2004). Evaluating the seriousness of police misconduct: A cross-cultural comparison of police officer and citizen views. *International Criminal Justice Review*, 14: 25–48.

Kutnjak Ivković, S. (2005). Police (mis)behavior: A cross-cultural study of corruption seriousness. *Policing: An International Journal of Police Strategies and Management*, 28(3): 546–566.

Kutnjak Ivković, S. (2009). The Croatian police, police integrity, and transition toward democratic policing. *Policing: An International Journal of Police Strategies and Management*, 32(3): 459–488.

Kutnjak Ivković, S. (2015). Studying police integrity. In S. Kutnjak Ivković and M. Haberfeld (eds.), *Police Integrity across the World*. New York: Springer.

Kutnjak Ivković, S. and M.R. Haberfeld (2015). A comparative perspective on police integrity. In S. Kutnjak Ivković and M. Haberfeld (eds.), *Police Integrity across the World*. New York: Springer.

Kutnjak Ivković, S. and T.O. Shelley (2008). The contours of police integrity across Eastern Europe: The case of Bosnia and Herzegovina and the Czech Republic. *International Criminal Justice Review*, 18(1): 59–82.

Lipsky, M. (2010). *Street-Level Bureaucracy: Dilemmas of the Individual in Public Services*. New York: Russell Sage Foundation.

Lodge, T., V. van der Vliet, D. Welsh, J. Myburgh, W. Gumede, L. Schlemmer, M. Bot, H. Adam, and H. Giliomee (2007). Towards an inclusive democracy. In H. Giliomee and B. Mbenga (eds.), *New History of South Africa*. Cape Town, South Africa: Tafelberg Publishers.

Manning, P.K. (2010). *Democratic Policing in a Changing World*. Boulder, CO: Paradigm.

Marks, M. (2005). *Transforming Robocops: Changing Police in South Africa*. Durban, South Africa: University of Scottsville and KwaZulu-Natal Press.

Masuku, T. (2005). Strengthening Democratic Policing in South Africa: Enhancing and Coordinating he Internal and External Accountability Systems of the South African Police. Research report written for the Centre for the Study of Violence and Reconciliation. Available at: http://www.csvr.org.za/docs/policing/strengtheningdemocraticpolicing.pdf

[Mollen Commission] New York City Commission to Investigate Allegations of Police Corruption and the Anti-Corruption Procedures of the Police Department (1994). *Commission Report*. New York: Author.

Newham, G. (2005). Strengthening police accountability in South Africa through internal systems for officer control. *South African Review of Sociology*, 36(2).

Pennsylvania Crime Commission (1974). *Report on Police Corruption and the Quality of Law Enforcement in Philadelphia*. Saint Davids, PA: Author.

Perkins, G. (2018). *Danger and Death: Organisational and Occupational Responses to the Murder of Police in South Africa: A Case Study*. Unpublished PhD Thesis. University of Cape Town.

Prevention and Combating of Corrupt Activities Act 12 of 2004. *Gazette No. 26311*. Available at: www.saps.gov.za/dpci/downloads/prevention_combating_corrupt_activities_act.pdf

Reiner, R. (2010). *The Politics of the Police*, 4th ed. Oxford: Oxford University Press.

Roebuck, J.B. and T. Barker (1974). A typology of police corruption. *Social Problems*, 21(3): 423–437.

South African Police Service Act (1995). *Act No 68 of 1995*. Government Gazette, no. 16731. Available at: https://www.gov.za/sites/default/files/gcis_document/201409/act68of1995.pdf

South African Police Service Code of Conduct (1997). Available at: https://www.saps.gov.za/about/conduct.php

South African Police Service Disciplinary Regulations (2006). Government Gazette no. 28985. Pretoria, South Africa.

Sauerman, A. and S. Kutnjak Ivković (2008). Measuring the integrity of the South African Police Service during transitional times. *Acta Criminologica*, 2: 21–39.

Snyman, C.R. (2014). Criminal law. In G. Kemp, S. Walker, R. Palmer, D. Baqua, C. Gevers, B. Leslie, and A. Steynberg (eds.), *Criminal Law in South Africa*, 6th ed. Cape Town, South Africa: Oxford University Press.

Steinberg, J. (2008). *Thin Blue: The Unwritten Rules of Policing South Africa*. Cape Town, South Africa: Jonathan Ball.

Sunstein, C.R. (2001). *Designing Democracy: What Constitutions Do*. New York: Oxford University Press.

Vito, G., S. Wolfe, G. Higgins, and W. Walsh (2011). Police integrity: Rankings of scenarios on the Klockars scale by "management cops". *Criminal Justice Review*, 36: 152–164.

Young, M. (1991). *An Inside Job: Policing and Police Culture in Britain*. Oxford: Oxford University Press.

Chapter 5

Policing the police

Internal control of police misconduct

Introduction

Establishing systems of accountability and control was central to the birth of the SAPS in 1995 and has remained essential to its identity since. The ubiquitous nature of these structures is, however, a reflection not of their enduring efficacy but of their collective failure. Unquestionably, the SAPS could be significantly more flawed, corrupt, and abusive than it is, and in this sense perhaps the systems established since 1994 have played some positive part. Nonetheless, it is far from the effective, professional, and legitimate police organization imagined by the architects of South Africa's democracy or the authors of South Africa's National Development Plan (The Planning Commission, 2012).

As described in Chapter 3, the 1990s saw the introduction of a raft of accountability mechanisms (Rauch, 2000, 2004). These included the separation of the SAPS policy and operational functions through the creation of the Civilian Secretariat for Police Services (CSPS), the formation of community police forums (CPF) to provide citizen oversight of frontline police, the establishment of the Independent Complaints Directorate (ICD) to independently receive and investigate complaints against the police, and the establishment of the SAPS's Anti-Corruption Unit (ACU), among others. In the decades since, some of these structures have been closed and reopened, while others were gradually added. What has remained fairly constant is the SAPS internal disciplinary system and regulations (introduced in Chapter 4). Yet despite these structures, as well as the National Development Plan (The Planning Commission, 2012), which compels the SAPS to improve its professionalism and integrity, the SAPS reported in 2019 that it continued to struggle with "consequence management" (SAPS, 2019 pp. 299, 305, 339), while public surveys revealed declining trust in the police (Research, 2018).

In the wake of this struggle for legitimacy, this chapter explores the SAPS's internal control mechanisms using the second dimension of the police integrity theory (e.g., Klockars and Kutnjak Ivković, 2004; Klockars et al., 2006; Kutnjak Ivković, 2015). This dimension examines the relation between the level of integrity in a police agency and organizational mechanisms intended to prevent and control integrity violations, such as codes of conduct, the investigation of alleged transgressions, integrity testing and entrapment, and disciplinary processes (Klockars et al., 2000). The chapter reviews the SAPS's code of conduct, its anti-corruption unit, and other specialized units and agencies tasked with addressing police abuses and corruption. It also incorporates our analysis of SAPS officers' views about the discipline they expect the SAPS to administer in response to various misconduct and their views about disciplinary fairness.

The SAPS code of conduct and discipline management

In 1997, the SAPS introduced a code of conduct aligned with its new values of democratic and community policing. At the time, the code was relatively progressive. It offered a strong normative framework to guide the conduct of officers working in the new organization and remained relatively unchanged in 2019 (see Chapter 4). Through it, officers must commit to a range of values and intentions, including acting with integrity, upholding the fundamental rights of every person, working impartially, courteously, honestly, respectfully, transparently, and accountably, preventing and acting against corruption, and exercising their powers in a responsible and controlled manner (SAPS, 2019). The SAPS also has a Code of Ethics and a set of official organizational values.

Although a code of conduct can support police integrity, it is by no means sufficient. Police agencies rely instead on broader systems of formal rules and discipline management to hold officers accountable (see Chapter 4). In this regard, the SAPS's Discipline Regulations (SDR, 2016) are intended to generate a common understanding of misconduct and discipline within the organization, promote acceptable conduct, provide a user-friendly disciplinary framework, and prevent arbitrary actions by supervisors in response to misconduct (Department of Police, 2016). This may appear perfectly reasonable, but applying these regulations fairly and consistently has proved challenging for the SAPS.

In an effort to ensure fairness, the SDR requires that for each case, the SAPS consider the actual or potential impact of the alleged misconduct on the work or the image of the Service, station, unit, the employee, or the colleagues; the nature of the work and responsibilities of the employee; and the circumstances under which the alleged misconduct took place (Department of Police, 2016). In the absence of effective leadership, the SDR allows for significant discretion and, thereby, potential inconsistency in the manner in which cases are addressed. As we show in this chapter, the SAPS's disciplinary data do not clearly demonstrate whether its systems are effective or not.

The 2016 SDR was drafted after the country's cabinet adoption of the National Development Plan (NDP) in 2012. The NDP represents government's meta-strategy and vision for South Africa until 2030. It calls for the reform of the SAPS, in part by linking its code of conduct more clearly to promotion and disciplinary regulations, and for officers charged with breaches of the code to vacate their posts pending the outcome of a disciplinary hearing (The Planning Commission, 2012, pp. 389–390). In theory, the SAPS's 2006 SDR, in force during the NDP's drafting, already enabled this. As such, the 2016 regulations did not introduce anything significantly new on the matter. Rather, their primary contribution was the introduction of 12 new forms of misconduct. These are not, however, any more clearly linked to the code of conduct or predictable consequences (as recommended by the NDP) than the preexisting offenses were. According to the SAPS, this is the result of the 2006 Regulations already presenting the transgression of the code of conduct as a form of misconduct, which was simply restated in the new SDR (Personal Communication, 2019). The NDP also recommended that police oversight bodies should be able to request a review of the professional ethics of individual officers or units and called for the establishment of a police board to set and regulate professional standards for the police. Neither of these recommendations had been implemented by early 2020.

The SAPS Anti-Corruption Unit and discipline management

Disciplinary regulations and procedures

The SAPS's Disciplinary Regulations of 2016 (SDR) outline the definitions and processes governing discipline management in the organization. They require the National and Provincial Commissioners to appoint disciplinary officers to develop and maintain standards for provincial disciplinary officers and employer representatives, and to appoint other officers to exercise the powers provided for in the SDR on behalf of the SAPS (Regulation 5). Through this process, station level officers are able to bring charges against colleagues.

Where charges are brought against an employee, the SDR differentiates between "less serious misconduct" (Regulation 7) and "serious misconduct" (Regulation 12). To distinguish between the two, managers must consider the impact of the alleged misconduct on the work or image of the SAPS and its associates, the nature of the work and responsibilities of the employee, and the circumstances under which the alleged misconduct took place (Regulation 5.2). The SDR also provides a list of serious offenses for which an "expeditious process" is warranted (Regulation 5.4). Almost all of these offenses are also criminal offenses and include allegations of corruption, bribery, extortion, and defeating the course of justice, as well as any act of misconduct involving an element of dishonesty.

Although the determination of "less serious" versus "serious" misconduct remains somewhat subjective, the regulations provide clear guidelines on how the two categories of offenses should be dealt with once established.

The clearest difference is that a supervisor can directly discipline an officer who admits to less serious misconduct, while serious misconduct requires that an investigation be initiated and concluded within 30 days of the complaint, followed by a disciplinary hearing if necessary. Where an accused does not admit to less serious misconduct, a disciplinary hearing must be instituted (Regulation 7). For less serious misconduct, disciplinary options increase in severity from corrective counseling (Regulation 7.1) to verbal warnings (Regulation 7.2), written warnings (Regulation 7.3), and final written warnings (Regulation 7.4).

Allegations of serious misconduct must be investigated, and the investigation completed within 30 days or as soon as possible thereafter. If the investigation confirms that the alleged offense is sufficiently serious to warrant a disciplinary hearing, the outcome of the investigation must be communicated to the relevant disciplinary officer within seven working days, in order for a disciplinary enquiry to be initiated. The accused employee must be informed of the alleged misconduct and investigation and, within 15 working days, be served with a written notice describing the alleged misconduct, his or her rights, and the details of the disciplinary hearing (Regulation 8).

In addition to the disciplinary options available for less serious police misconduct, disciplinary options for serious misconduct include additional and more serious selections. First, until the disciplinary hearing takes place, the employee may be suspended with pay for up to 60 days or transferred to another post. If the employee fails to appear for the disciplinary hearing, such employee is immediately suspended without remuneration (Regulation 9.7.a). If the employee does appear and is found guilty[1] of the alleged misconduct, the employee may be sent for counseling, given a warning or final written warning (both valid for 6 months), be suspended without salary for up to two months, or dismissed from the Service (Regulation 12.1). Whenever the latter discipline option is selected, the National, Provincial or Divisional Commissioner may impose a less severe discipline if it is deemed in the interest of the Service (Regulation 9.4).

If supervisors believe that an allegation of misconduct is so serious that it requires expeditious intervention, they must submit a report to a disciplinary officer with the rank of Brigadier or above (Regulation 9). If the latter does not agree that the matter requires expeditious processing, they may refer it back to the supervisor for normal processing. On the other hand, where they approve expeditious processing, they must notify the accused employee of the allegations and instruct them to answer in person within no more than five days. Within five days of this meeting, the senior officer must reach a finding on the matter and, if necessary, impose a discipline. All findings and discipline are subject to dispute resolution through the Safety and Security Sectoral Bargaining Council (Regulation 13).

The manner in which the SAPS reports on its disciplinary processes has changed over time, encumbering trend analyses. This is partly because the organization expanded and has subsequently become more competent and

Table 5.1 Police officers disciplined and dismissed, 1992 and 1993

	Disciplined	*Dismissed*
1992	31	2,026
1993	20	1,303

Source: SAP (1993)

complex, and also partly because its systems and priorities have changed with the political climate. For example, in 1993 the South African Police force (SAP) reported that it disciplined 31 officers and dismissed 2,026 for "misconduct" during 1992, and disciplined 20 and dismissed 1,303 during 1993 (see Table 5.1). The number of dismissals is significant at a time when the SAP employed just over 100,000 personnel, 83,000 of whom were police officers (SAP, 1993). This means that officers were dismissed at a rate of roughly 2,440 per 100,000 in 1992 and 1,570 per 100,000 in 1993. The report does not explain what had led to the dismissals.

As the SAPS matured into a more transparent, democratic police service, it began publishing more details on its disciplinary matters but also dismissed officers at a far lower rate than the SAP had—156 per 100,000 in 2019 (SAPS, 2019, pp. 34, 234). This may be driven by a range of factors, including new and necessary rights protections for officers, as well as a SAPS leadership more focused on "fighting crime" in the streets than within its own ranks. However, before this shift in disciplinary procedure and reporting, the SAPS embarked on an important period of corruption busting, when it created its Anti-Corruption Unit.

The SAPS Anti-Corruption Unit (1996–2001)

The 1996 National Crime Prevention Strategy (NCPS) identified corruption within the criminal justice system as a significant risk to the reform (Parliamentary Monitoring Group, 1998). In response, the SAPS established its first Anti-Corruption Unit (ACU). The ACU's mandate was to investigate allegations of corruption in the SAPS, raise awareness about corruption within the organization, identify and report internal dysfunctions, and maintain an information management system to support relevant investigations (Newham and Faull, 2011). The unit had a presence at the SAPS's Head Office in Pretoria, and in each province. Investigators were carefully selected and given top-secret security clearance. In an apparent demonstration of support, the National Commissioner directed that ACU members should be provided protection whenever they or their families felt threatened as a result of their work (Newham and Gomomo, 2003).

Despite various obstacles, including capacity constraints, the anonymous nature of many tip-offs, and an Anti-Corruption Act (1992) which made formal proof of corruption extremely difficult, the ACU was relatively successful during its short tenure (Newham and Gomomo, 2003). In its first year, it arrested 30 officers, which resulted in 30 convictions. By 2000, with 250 personnel, it was investigating 4,000 complaints and arresting over 1,000 officers, resulting in approximately 150 convictions (Burger and Grobler, 2017; Newham and Faull, 2011).

The rapid growth in complaints and personnel likely resulted from two parallel processes. First, the surge in corruption reports mirrored the increase in crimes reported to the SAPS, even though murder (Kriegler and Shaw, 2016) and overall victimization rates were declining at the time (Statistics South Africa, 1999; Pharoah, 2008). This phenomenon may indicate increasing public trust in and expectations of the new democratic government and its institutions at the time. With most South Africans having had little recourse to state justice or accountability structures before 1994, and with the mood in the country by the late 1990s being one of optimism and expectancy, it is likely that victims turned to the ACU believing that they might find justice. At the same time, additional resources were steadily made available to support the ACU's work, and as it grew, it was able to receive and investigate an increasing number of complaints. A second reason for the surge in complaints in the late 1990s may have something to do with the fact that the SAPS—like the country—was in a state of transition. It had been formed in 1995 through a merger of 11 police forces, all of which had been, to a greater or lesser extent, undemocratic, corrupt, and abusive (Cawthra, 1993).

However, just as the ACU was gaining significant momentum, it was shut down. This occurred under National Commissioner Jackie Selebi's leadership in 2002 as part of a disastrous organizational restructuring process (e.g., Omar, 2008). The ACU's mandate was moved to the SAPS's Organized Crime Unit and later to the Directorate for Priority Crime Investigations (DPCI—better known as the "Hawks"), both of which have been implicated in abuse of power, often in service of corrupt politicians. For example, in 2015, a court found that Hawks's head, Berning Ntlemeza, lacked integrity and had lied under oath (Sapa, 2015). In 2017, the High Court declared he did not possess the qualities necessary to hold any public office, let alone to lead the elite corruption busting unit (de Langa, 2017). Similarly, testifying in 2018 at the Zondo "State Capture" Inquiry into high-level corruption within government, Pravin Gordhan, former minister of finance, described how Hawks's officers had threatened him with arrest in an effort to muscle him out of the Treasury, where he was perceived as an obstacle to politicians wanting to plunder government funds (Chabalala, 2018). The goal was to replace him with someone more compliant to President Zuma's nefarious plans (Pauw, 2018; Basson and Du Toit, 2018). The abuse of the Hawks is discussed in more detail later and with reference to "high policing" (Brodeur, 2010) in Chapter 6. However, it should be noted that, while the capacity and integrity of the Hawks were undermined by Zuma's allies, many dedicated officers within the unit continued to carry out their work with integrity and resisted the capture of their division (Pitchford, 2016). Ntlemeza was forced out of his post in 2017, and in 2018 a former deputy national commissioner, Godfrey Lebeya, took his place. In contrast to Ntlemeza, Lebeya appears to be a dedicated police officer who believes the "integrity of the Hawks should be beyond reproach" (Burger, 2019).

Replacing the ACU with strategy

Whether it was already planning to close the ACU or it wanted to pursue a more holistic approach to integrity management, in 1999 the SAPS began developing a Service Integrity Framework (SIF) to address the risks posed to the organization by ill-disciplined officers (SAPS, 1999, 2000). Over the next decade, the SIF was renamed the Service Integrity Strategy, the Corruption and Fraud Prevention Plan, and, finally, the Anti-Corruption Strategy (ACS) and Anti-Corruption Framework. These approaches were indicative of the national government's response to corruption, which emphasized prevention, detection, investigation, and restoration, the latter implying the regular assessment of fraud prevention and control measures (Faull, 2007). The introduction of the strategies was in part compelled by an obligation that the SAPS comply with the government's Minimum Anti-Corruption Capacity Guidelines, among other imperatives. At face value, these approaches and the manner in which they are presented in the SAPS Annual Reports suggest an impressive understanding of integrity-related risks and the interventions needed to address them. However, it is unclear to what extent they were ever implemented, or whether they had any measurable effect, with successive Annual Reports hinting at repeated delays in their implementation (SAPS, 2010, p. 61, 2016).

The SAPS's various strategies sought both to coerce officials to improve their conduct by raising awareness and increasing the likelihood of sanctions for misconduct, and to educate and rehabilitate the officers who transgressed and punish those who have engaged in serious abuse (SAPS, 2011, p. 26). The strategies required provincial offices and individual stations to take a range of actions to improve frontline "control" and service delivery and to educate staff about relevant policies and consequences (SAPS, 2008, p. 66). In 2009, a communication plan was introduced to publicize the strategy within the SAPS and public at large. This included the production of pamphlets and dissemination of information to employees via email and on their monthly pay slips (SAPS, 2009, p. 51, 2013, p. 159). The communication plan also promoted the National Anti-Corruption Hotline—a toll-free telephone and web-based reporting tool managed by government's Public Service Commission (PSC). The Constitution (1996, Sections 195 and 196) tasks and empowers the PSC to investigate, monitor, and evaluate the organization and administration of South Africa's public service and government programs.

Although the SAPS's various integrity strategies were included in Annual Reports, disseminated to provincial police offices, and linked to the performance agreements of all divisional, provincial, and station commanders, the SAPS admitted in 2010 that these approaches had not yet been implemented as envisaged (Newham and Faull, 2011). In 2014 and 2015, the SAPS reported that it was once again pursuing many of the initiatives introduced in the various strategies but presented them as if they were new. It also stated its intention to finalize an "Anti-Corruption Strategy," which had first been reported on in 2010 as already implemented. By 2016, the SAPS stated that the implementation of this strategy had been delayed because of the changes in management (SAPS, 2014, p. 226, 2015, p. 24, 2016, p. 246). In a further contradiction, in 2016, the SAPS confessed that although "top management has on numerous occasions expressed its zero tolerance approach to criminality by its members, including offenses like corruption and fraud . . . [t]he organization has . . . for a number of years, not had an approved Anti-Corruption Strategy in place" (SAPS, 2016, p. 21). All these misleading or false reports on strategy implementations raise serious questions about the veracity of SAPS.

In 2015, the SAPS reported it had established an "Integrity Management Capability" to meet its obligations to the national government (SAPS, 2015, p. 290). It claimed this would ensure ethical management, good governance, and "fundamental moral duties," and stressed the importance of ethical leadership and accountability to internal and external stakeholders (SAPS, 2015, p. 290). It also mentioned an intention to develop a performance indicator related to the timely reporting and addressing of corruption-related activities. The indicator, first implemented in 2016, is "[p]ercentage of reported incidences of corruption by members" and the target is the instituting of steps in response to all reported corruption. In most years, the SAPS reports that it attained 90–100% of this goal.

While these integrity- and corruption-focused strategies were being developed, and again in response to incentives put in place by the national government, the SAPS introduced a Risk Management Strategy (RMS) in 2005 and Risk Management Framework (RMF) in 2007. These, it reported, pursued an "enterprise-wide risk management approach" to ensure its organizational objectives were achieved. This required that risk-management capacities be developed in all SAPS divisions and provinces and the establishment of a strategic risk register. Once identified, control mechanisms were expected to be developed to address risks (SAPS, 2008, p. 67). In 2008, the SAPS's National Risk Committee was established, together with the Risk Management National Instruction (SAPS, 2009, p. 52). By 2012, the organization was reporting that a number of key risks had been addressed through these and related interventions, including the theft of SAPS firearms, (ab)use of sick leave, and vehicle collisions (SAPS, 2011, p. 40). Senior managers in the SAPS are required to declare their financial interests each year (SAPS, 2013, p. 160), and vetting and lifestyle audits are applicable among those working in the areas vulnerable to corruption (e.g., procurement) (SAPS, 2015, p. 290). The SAPS also boasts regulating procedures in terms of remunerative work by employees outside of the SAPS.

Again, to comply with national regulations, the SAPS in 2016 began establishing an Integrity Management Service (IMS). The IMS would pursue a range of tasks, including—among other tasks reported on in previous years—the appointment of ethics officers, conducting annual ethics and corruption risk assessments, implement awareness campaigns, and the establishment of a whistleblowing policy (SAPS, 2017, p. 263). In 2018, the SAPS reported it had established the IMS "with a number of key issues that are in the process of being addressed," many of which are very similar to those outlined in the 1999 Service Integrity Framework (SAPS, 2018, p. 263), and announced the "launch" of yet another new "Anti-Corruption Strategy" (Sitole, 2018). Finally, in 2019, the SAPS reported that it had developed and implemented an "Ethics and Anti-Corruption Strategy" (SAPS, 2019, pp. 10, 22, 269). The above review suggests that the SAPS is good at complying with its obligations to establish committees, structures, and policies, and reporting its ostensible progress in this regard. However, the circular nature of the reporting, in which goals presented in 1999 are restated in 2019, suggests that moving from compliance to implementation remains a challenge for the organization. Such "paper performances" are reminiscent of activities observed by South African ethnographers, with frontline officers submitting reports of activities carried out in order to meet their reporting obligations, even when the work has not been done (e.g., Hornberger and Vigneswaran, 2009; Faull, 2018).

Shifts in reporting of disciplinary processes and data

With the closure of the ACU in 2002, the SAPS once again changed the way it reported its disciplinary data. In some respects, reporting became more detailed and informative, but shifts in style complicate the understanding and comparison of the data over many years. For example, from 2002, the SAPS began reporting on the number of officers suspended each year, and how many of these were "corruption-related." However, it interpreted this phrase quite broadly, as demonstrated in Tables 5.2 and 5.3, which list the types of alleged misconduct and criminality for which officers were suspended, but in a manner that is not simple to interpret (Table 5.2). The report notes that 872 officers were suspended, but lists reasons for only 657 of them (Table 5.3). It also notes that criminal cases were opened against 1,332 officers (Tables 5.2 and 5.4), but that only 641 departmental hearings were conducted (Table 5.4). It makes sense that there would be more departmental hearings than criminal cases, because most misconduct is not criminal. It is, however, less comprehensible that there appear to be three times as many criminal cases as departmental hearings. The only implication here could be that a substantial proportion of officers against whom criminal cases were opened were not subjected to internal disciplinary procedures.

Table 5.2 ACU corruption-related reports, arrests, and convictions

Year	*Reports received*	*Arrests/charges*	*Convictions*
1996	2,300	249	30
1997	3,108	429	78
1998	3,779	475	128
1999	4,618	844	147
2000	6,974	1,048	193
2001–2002	2,370	1,332	576

Source: SAPS (2001, 2002, 2003, 2004)

Table 5.3 Police officers suspended, 2002–2003—number and reason

Reason for suspension	*Number of officials suspended*	*Number & type of criminal charges*
Corruption	188	1251
Theft	109	696 ("Fraud/Theft")
Assisting Escapees	96	90
Murder	61	—
Fraud	57	696 ("Fraud/Theft")
AWOL	40	—
Attempted Murder	39	—
Defeating ends of Justice	34	293
Armed Robbery	33	—
Bribery/Extortion	—	40
TOTAL	657	2,370

Source: SAPS (2003)

Table 5.4 Outcomes of departmental hearings and criminal cases, 2002–2003

Outcome	*Criminal cases*	*Departmental hearings*
Guilty	252	143
Pending	704	250
Withdrawn	100	35
Not Guilty	39	164
Warrants Issued	3	—
Prosecution Declined	225	—
Prosecutor Decision Pending	9	—
Member left SAPS voluntarily	—	26
Dismissed from Service	—	18
Suspended	—	5
TOTAL	1,332	641

Regarding dismissals, in 2002–2003, when the SAPS employed 132,310 personnel, 102,737 of whom were police officers, only 18 officers were dismissed for misconduct (Table 5.4; SAPS, 2003). This equates to a dismissal rate of 18 per 100,000 in 2003.

Since 2011–2012, the SAPS has started reporting its disciplinary data in a slightly less nebulous manner (e.g., Tables 5.5–5.7) and in increasingly more detail, albeit with some lingering inconsistencies. For example, under a section titled "Fraud and Corruption," from 2011 to 2018, the SAPS kept constant the categories of charges classified as such (Table 5.5), while in other tables it presents the types of offenses for which disciplinary hearings were held (Table 5.6) and their outcomes (Table 5.7). The SAPS has even begun providing detailed tables listing the rank, unit, nature of misconduct committed, as well as the outcome of cases, per offense (e.g., SAPS, 2019, pp. 270–277).

Most notable about Table 5.5 is the sudden reduction in charges between 2016 and 2018. Although the organization shrunk by approximately 10,000 employees between 2011 and 2018, this hardly accounts for the sudden decline. A similar drop is seen in the SAPS's general discipline data (Table 5.6), which, like the corruption and fraud data since 2011, is quite detailed. This suggests that something changed in the way the SAPS approached disciplinary matters during this period.

Table 5.6 shows that 30–40% of hearings were related to common law and statutory offenses. These include theft, assault, robbery, and murder—crimes that far surpass mere "misconduct"—though the specific offense is not listed and would probably result in a further criminal investigation. Almost all other offenses each year relate to misconduct likely to be noticed only by SAPS personnel: failure to comply with or contravene an Act, Regulation, or legal obligation (a) (±15%); failure to carry out a lawful order or routine instruction without just or reasonable cause (i), (±15%); and absence from work without reason or permission (j), (±10–20%). In an average year, these four categories constituted more than two-thirds of disciplinary hearings, suggesting that some SAPS managers utilized the organization's regulations to initiate proceedings against subordinates. It might, of course, also suggest that relatively few disciplinary hearings emanate from complaints by the public.

Table 5.8 presents the outcomes of disciplinary hearings from 2011–2019. On average, around 50% of hearings resulted in a withdrawal, "not guilty" finding, or suspended dismissal. Charges might be withdrawn for a number of reasons, some of them benign and some less so. For example, if a whistleblower or witness feels intimidated by the accused or their allies, they may withdraw a charge. However, it is also possible that charges are maliciously and falsely lodged against officers, and that the exposure of this fact results in withdrawal. Malicious accusations may also explain some of the "not guilty" findings, though it is just as likely that guilt was simply not proven on a balance of probability. Suspended dismissal is a peculiar outcome, implying a finding of guilt worthy of dismissal, but which is then downgraded to an effective final written warning.

Table 5.5 Fraud and corruption-related departmental charges, 2011–2018

Offense	*2011–2012*	*2012–2013*	*2013–2014*	*2014–2015*	*2015–2016*	*2016–2017*	*2017–2018*
Corruption	383	319	332	328	469	115	113
Fraud	257	153	207	172	104	28	1
Aiding an escapee	352	349	421	340	407	152	109
Defeating the ends of justice	209	149	169	199	134	35	66
Extortion	67	46	34	38	13	10	11
Bribery	18	10	9	21	7	5	15
TOTAL	1,050	892	1,172	1,098	1,134	345	315

Source: SAPS (2012, 2013, 2014, 2015, 2016, 2017, 2018)

Table 5.6 SAPS disciplinary hearings 2011–2019, charge and number

Type of offense	*2011–2012*		*2012–2013*		*2013–2014*		*2014–2015*		*2015–2016*		*2016–2017*		*2017–2018*		*2018–2019*	
	Total	*%*	*Total*	*%*	*Total*	*%*	*Total*	*%*	*Total*	*%*	*Total*	*%*	*Total*	*%*	*Total*	*%*
Failed to comply with or contravened an Act, Regulation or legal obligation.	425	14.7	476	15.1	491	16.5	353	14.2	566	12.7	597	14.3	399	16.5	147	10.8
Conducted himself or herself in an improper, disgraceful and unacceptable manner while on duty.	98	3.4	65	2.1	75	2.5	66	2.7	130	2.9	126	3.0	315	13.0		
Contravened any prescribed Code of Conduct for the SAPS or the Public Service	47	1.6	75	2.4	65	2.2	101	4.1	97	2.2	138	3.3	189	7.8		
Was absent from work without reason or permission.	390	13.4	502	15.9	569	19.1	369	14.9	469	10.6	377	9.0	184	7.6		
Possessed or used the property of the State, another employee or a visitor without permission.	89	3.1	71	2.2	96	3.2	67	2.7	97	2.2	100	2.4	69	2.9		
Intentionally or negligently damaged and or caused loss of State property.	110	3.8	137	4.3	136	4.6	100	4.0	164	3.7	183	4.4	102	4.2		
Endangered the lives of others by disregarding safety rules and regulations.	20	0.7	15	0.5	13	0.4	12	0.5	24	0.5	23	0.6	63	2.6		
Prejudiced the administration, discipline or efficiency of a department, office or institution of the State.	109	3.8	115	3.6	122	4.1	92	3.7	189	4.3	205	4.9	62	2.6		
TOTAL	**1,288**	**44.5**	**1456**	**46.1**	**1567**	**52.6**	**1160**	**46.8**	**1736**	**39.1**	**1749**	**41.9**	**1,383**	**57.2**		

Table 5.7 Numbers and reasons for SAPS disciplinary hearings, 2011–2016

	Reason for disciplinary hearing	*2015–2016*		*2014–2015*		*2013–2014*		*2012–2013*		*2011–2012*	
		Total	*%*	*Total*	*%*	*Total*	*%*	*Total*	*%*	*Total*	*%*
a	Failed to comply with or contravened an Act, Regulation or legal obligation.	566	12.7	353	14.2	491	16.5	476	15.1	425	14.7
b	Willfully or negligently misused the finances of the State.	20	0.5	11	0.4	4	0.1	14	0.4	12	0.4
c	Possessed or used the property of the State, another employee or a visitor without permission.	97	2.2	67	2.7	96	3.2	71	2.2	89	3.1
d	Intentionally or negligently damaged and or caused loss of State property.	164	3.7	100	4	136	4.6	137	4.3	110	3.8
e	Endangered the lives of others by disregarding safety rules and regulations.	24	0.5	12	0.5	13	0.4	15	0.5	20	0.7
f	Prejudiced the administration, discipline or efficiency of a department, office or institution of the State.	189	4.3	92	3.7	122	4.1	115	3.6	109	3.8
g	Misused his or her position in the SAPS to promote or to prejudice the interest of any political party.	1	0.02	0	0	1	0	3	0.1	4	0.1
h	Accepted any compensation in cash or otherwise from a member of the public or another employee, for performing her or his duties without written approval from the employer.	3	0.07	0	0	3	0.1	1	0	1	0
i	Failed to carry out a lawful order or routine instruction without just or reasonable cause.	671	15.1	360	14.5	315	10.6	355	11.2	276	9.5
j	Was absent from work without reason or permission.	469	10.6	369	14.9	569	19.1	502	15.9	390	13.4
k	Committed an act of sexual harassment.	17	0.4	7	0.3	8	0.3	8	0.3	10	0.3
l	Unfairly discriminated against others on the basis of race, gender, disability, sexuality or other grounds prohibited by the Constitution.	2	0.05	2	0.1	0	0	3	0.1	12	0.4
m	Performed work for compensation in a private capacity for another person or organization either during or outside working hours, without written approval by the employer.	11	0.25	10	0.4	3	0.1	3	0.1	3	0.1
n	Slept on duty without permission/authorization.	7	0.02	4	0.2	12	0.4	13	0.4	7	0.2

(*Continued*)

Table 5.7 (Continued)

	Reason for disciplinary hearing	2015–2016		2014–2015		2013–2014		2012–2013		2011–2012	
		Total	%	Total	%	Total	%	Total	%	Total	%
p	Conducted himself or herself in an improper, disgraceful and unacceptable manner while on duty.	130	2.9	66	2.7	75	2.5	65	2.1	98	3.4
q	Contravened any prescribed code of conduct for the SAPS or the Public Service, whichever may be applicable to him or her.	97	2.2	101	4.1	65	2.2	75	2.4	47	1.6
r	Incited other employees to unlawful conduct or conduct in conflict with accepted procedure.	2	0.05	0	0	0	0	0	0	0	0
s	Displayed disrespect toward others in the workplace or demonstrated abusive or insolent behavior	61	1.4	32	1.3	53	1.8	72	2.3	49	1.7
t	Intimidated or victimized other employees.	8	0.2	3	0.1	2	0.1	3	0.1	2	0.1
u	Prevented other employees from belonging to any trade union.	0	0	0	0	0	0	0	0	0	0
v	Ran any money-lending scheme for employees during working hours or from the premises of SAPS.	0	0	1	0	0	0	0	0	0	0
w	Gave a false statement or evidence in the execution of his or her duties.	2	0.05	4	0.2	7	0.2	5	0.2	6	0.2
x	Falsified records or any other documentation.	17	0.4	10	0.4	12	0.4	13	0.4	13	0.4
y	Participated in any unlawful labour or industrial action.	0	0	57	2.3	27	0.9	0	0	2	0.1
z	Committed a common law or statutory offense.	1,798	40.5	747	30.1	846	28.4	1,059	33.6	1,098	37.9
	TOTAL	**4,443**	**100**	**2,482**	**100**	**2,980**	**100**	**3,156**	**100**	**2,900**	**100**

Table 5.8 Outcomes of SAPS disciplinary hearings, 2011–2018

Outcome of Disciplinary Hearings	*2011–2012*		*2012–2013*		*2013–2014*		*2014–2015*		*2015–2016*		*2016–2017*		*2017–2018*		*2018–2019*	
	Number	*%*	*Number*	*%*	*Number*	*%*	*Number*	*%*	*Number*	*%*	*Number*	*%*	*Number*	*%*	*Number*	*%*
Correctional counseling	136	2.5	150	2.6	178	2.9	151	2.9	96	2.2	113	2.5	90	4.6	71	3.7
Demotion	0	0	0	0	0	0	0	0	0	0	0	0	0	0	0	0
Dismissal	532	9.6	496	8.5	537	8.9	409	7.9	361	8.1	323	7.2	220	11.2	234	12.3
Final written warning	337	6.1	378	6.4	524	8.7	419	8.1	313	7.1	339	7.5	341	17.4	260	13.8
Fine	978	17.7	1112	19	995	16.4	807	15.5	629	14.2	512	11.4	8	0.4	0	0
Suspended action	130	2.3	78	1.3	60	1	38	0.7	0	0	0	0	0	0	0	0
Suspended dismissal	792	14.3	791	13.5	664	11	638	12.3	539	12.2	478	10.6	0	0	0	0
Case withdrawn	864	15.6	507	8.7	827	13.7	756	14.5	622	14	539	12	258	13.2	199	10.6
Not guilty	1,185	21.4	1,641	28	1,467	24.2	1,266	24.4	1,378	31.1	1,741	38.7	629	32.1	688	36.5
Suspended without payment	27	0.5	51	0.9	88	1.5	103	2	55	1.2	40	0.9	198	10.1	232	12.1
Verbal warning	80	1,4	126	2,1	72	1,2	51	1	40	0,9	36	0,8	12	0,6	9	0,5
Written warning	479	8,6	531	9,1	641	10,6	560	10,8	398	9	375	8,3	204	10,4	195	10,4
TOTAL	5,540	100	5,861	100	6,053	100	5,198	100	4,431	100	4,496	100	1,960	100	1,888	100

Source: SAPS Annual Reports, 2010–2019

The Directorates for Special Operations (DSO—Scorpions) and Priority Crime Investigations (DPCI—Hawks)

First announced by President Thabo Mbeki in 1999, the Directorate for Special Operations (DSO), better known as the Scorpions, is perhaps the most famous and respected of South Africa's post-apartheid law enforcement agencies. Between 2001 and 2009, it led investigations into crime syndicates and criminal politicians, including a future president, Jacob Zuma, and investigated former SAPS National Commissioner, Jackie Selebi, who was subsequently convicted for corruption.

The DSO was not part of the SAPS. It was located instead in the National Prosecuting Authority (NPA) through the NPA Amendment Act 61 of 2000. Mandated with investigating serious organized and priority crimes, it did so using a troika model which combined prosecutors, investigators, and analysts in investigation teams. It boasted a prosecution rate of over 90% (though the SAPS accused the DSO of pursuing only winnable cases) and strategically leveraged media coverage of its dramatic raids to create an image of itself as South Africa's solution to organized crime and state corruption.

The DSO's success was, however, also the author of its doom (Minnaar, 2008). The country's leading political party, the African National Congress (ANC), accused the DSO of targeting ANC members and resolved at its December 2007 conference to shut the directorate down (Bruce, 2008). It first sought to do so through a commission of inquiry, hoping that an independent judge would close the DSO. The Minister for Justice and Constitutional Development had declared that the relationship between the DSO and the SAPS had irretrievably broken down, and questions were raised about the directorate's constitutionality given that the Constitution (1996, Section 199) allows for only a "single police service" (Khampempe, 2008). Judge Khampepe was asked to clarify the location, mandate, and operation of the DSO in relation to those of other government departments, with the ANC hoping she would rule it unconstitutional. Although Khampepe agreed that the SAPS-DSO relationship was fraught, she ultimately did not support its closure or merger with the SAPS (Khampempe, 2008). Nevertheless, the ANC used the report to support its conference resolution and, in 2008, the DSO was closed down by the government, insidiously opening a path to the looting of state coffers and the abuse of the criminal justice system over the next decade.

The DSO was replaced by the Directorate for Priority Crime Investigation (DPCI), better known as the Hawks, which was located *within* the SAPS. The SAPS and NPA Acts were amended in 2008 and came into effect in 2009. However, the Scorpions' closure was robustly resisted by civil society, with businessman Hugh Glenister challenging the ANC's decision in the constitutional court—apparently in service of democracy (Mooki, 2011). The court ordered the revised SAPS Act amended to ensure the operational independence of the Hawks. This further delayed the new unit's formation. Although the SAPS

Act requires that the DPCI has its own division within the SAPS's structure, this had yet to be established by 2019. Instead, it had been located within the Detective Services (Lebeya, 2019), potentially hindering its independence.

The function of the DPCI is to prevent, combat, and investigate national priority offenses and corruption. The SAPS has also positioned it as its premier anti-corruption entity, establishing within it an Anti-Corruption Task Team (ACTT) and Anti-Corruption Component (SAPS, 2014, p. 177). The ACTT was established to investigate senior officers and major corruption within and outside the SAPS. In 2019, the SAPS reported that the "ACTT was revitalized in June 2018 after a period of inactivity" (SAPS, 2019, p. 217), suggesting that it may have been subdued during President Zuma's damaging state capture project (see Chapter 3). Since 2012–2013, the DPCI has also dealt with senior managers' disclosures of financial interests, lifestyle audits in key areas, integrity testing for drug and alcohol abuse by officers, the management of an internal corruption reporting email address, and related communication (SAPS, 2013, p. 159).

The DPCI has also suffered its own corruption and abuse scandals. For instance, in 2018, two Hawks's officers were arrested after allegedly conspiring to bribe a SAPS detective (Maromo, 2018), while in 2019, a court convicted a Hawks Captain of fraud and money laundering after he claimed subsistence and incidental costs for work-related travel that he did not undertake (SA News, 2019). To mitigate such abuse, the Hawks have established an internal Integrity Management Unit (IMU) to investigate allegations against its own officers (SAPS, 2015, p. 292). The SAPS Amendment Act (No. 57 of 2008) requires that the DPCI be overseen by a retired judge, providing another avenue through which complaints can be lodged and addressed. Although the DPCI has had some success in tackling corruption and abuse in the SAPS, it simply had too many accusations against and arrests made of its own for it to be likened to the Scorpions. In addition to the kinds of lower-level incidents described earlier, some of the most senior Hawks' officers have been accused of gross abuse. In 2018, its head, Major General Zinhle Mnonopi, was suspended after former finance minister Mcebisi Jonas testified she had attempted to force him to sign a statement to stop a high-level bribery investigation into President Zuma's allies (Sehloho, 2018; Smit, 2019). In 2019, Colonel Kobus Roelofse testified that one of the first actions undertaken by Berning Ntlemeza, a former Hawks head and Zuma ally, in 2014, was to threaten the Anti-Corruption Task Team (ACTT) officers with suspension if they challenged his decisions (Umraw, 2019). Not surprisingly, it was under Ntlemeza's leadership that the Hawks's integrity was most damaged (e.g., Pauw, 2018; Basson and du Toit, 2018; Chapter 3).

Internal control and the theory of police integrity

Internal control is a key element of the police integrity theory (see, e.g., Klockars et al., 2000; 2004, 2006; Klockars and Kutnjak Ivković, 2004, p. 1.4). The second dimension of the theory explicitly focuses on various internal control mechanisms and their link to police integrity. A police agency is viewed as being of high integrity if it effectively and consistently uses both reactive and proactive internal control mechanisms (Kutnjak Ivković, 2015a). While internal control mechanisms range from education in ethics and integrity testing to proactive and reactive investigations, one critical component is the effective and consistent investigation of police misconduct and the disciplining of officers who violate official rules.

To facilitate the empirical study of the second dimension of the theory, Klockars and colleagues (2000, 2004, 2006) incorporated questions about discipline severity into their police integrity questionnaire, exploring the kinds of discipline that officers participating in the survey think is appropriate and the discipline they expect their agencies to mete out. Yet answers accompanying these questions, unlike those for other questions in the survey, were adapted to each country's legal and social environment. In fact, these disciplinary choices are heavily dependent upon the country's and agency's legal rules, resulting in different disciplinary options in each country, and in decentralized police systems, in each police agency.

A cursory review of a sample of extant studies of police integrity yields a diversity of perceived internal disciplinary environments (e.g., Cheloukhine et al., 2015; Kang and Kutnjak Ivković, 2015; Khechumyan and Kutnjak Ivković, 2015; Kutnjak Ivković, 2015b; Kutnjak Ivković et al., 2015; Lobnikar and Meško, 2015; Phettthong and Kutnjak Ivković, 2015; Porter et al., 2015; Vallmuur, 2015). As Kutnjak Ivković and colleagues (2016, p. 340) state, "An example of expected discipline for the abuse of deadly force—the most serious type of force on the use of force continuum—is illustrative of its diversity." While in about one-half of these countries (five out of 10) the surveyed officers expected dismissal for officers who abused deadly force (Australia: Porter et al., 2015; Croatia: Kutnjak Ivković, 2015b; Estonia: Vallmuur, 2015; Slovenia: Lobnikar and Meško, 2015; the United States: Kutnjak Ivković et al., 2015), police officers surveyed in other countries expected much less severe discipline (Armenia: Khechumyan and Kutnjak Ivković, 2015; South Korea: Kang and Kutnjak Ivković, 2015; Thailand: Phettthong and Kutnjak Ivković, 2015), or no discipline at all (Russia: Cheloukhine et al., 2015) for the same offenses. Expectations of harsh discipline for an abuse of deadly force appear to be related to the country's overall standing on the Transparency International Corruption Perceptions Index (e.g., Kutnjak Ivković and Haberfeld, 2015). Accordingly, in a follow-up analysis in five countries, Kutnjak Ivković and colleagues (2016) found the perceived severity of the disciplinary system and the country's ranking on the CPI positively related.

The importance of societal expectations of integrity seems to transcend similarity of geographic location and political history, resulting in large differences in the perceived disciplinary environments. A study of police integrity in Bosnia and Herzegovina and the Czech Republic (Kutnjak Ivković and Shelley, 2008), two East-European countries in transition, provided clear support for this assertion. Police officers surveyed in Bosnia and Herzegovina, a country with more societal tolerance for corruption in general, expected less severe discipline for the same misconduct than the police officers surveyed in the Czech Republic did (Kutnjak Ivković and Shelley, 2008).

Because the specific disciplinary options vary across countries, a direct cross-country comparison of the respondents' answers is prone to considerable limitations. Yet studies reveal some common features across heterogeneous disciplinary systems. First, the ten-country study (Kutnjak Ivković and Haberfeld, 2015) implied that police officers from almost all countries expected at least some discipline for all examples of misconduct included in the questionnaire. In other words, police officers who participated in these studies did not expect that they would be completely ignored by their police agency if they engaged in such misconduct. Second, while there is some variation across the countries in how often officers expected dismissal, a common feature is that dismissal is expected rather infrequently and only for a small subset of scenarios (e.g., Kutnjak Ivković and Haberfeld, 2015). Third, examples of misbehavior that would result in more severe discipline, be it dismissal or a written reprimand, are relatively similar across the countries (e.g., Klockars et al., 2004; Kutnjak Ivković and Haberfeld, 2015). Based on a comparison of disciplinary questions from 14 countries, Klockars and colleagues (2004) noted that the country rankings of scenarios, based on the severity of expected discipline, are quite similar among such diverse countries as Canada, Croatia, Finland, Poland, South Africa, and the United States.

A handful of studies (e.g., Kutnjak Ivković and Sauerman, 2011; Matlala et al., 2016; Meyer et al., 2013; Reyneke-Cloete and Meyer, 2016; Newham, 2004) analyzed South African police officers' views about the disciplinary system; all of them used the police corruption questionnaire (see Chapter 1). Newham (2004) surveyed about 100 SAPS officers in Johannesburg and found a relation between expected discipline and perceptions of seriousness. Respondents thought that no discipline would result from the least serious scenarios, a written warning would probably be issued for the cases of intermediate seriousness, and one of the more serious disciplinary options—suspension without pay or even dismissal—would be applied only for the two most serious scenarios. Newham (2004, p. 245) also noted officers' lack of common understanding of the disciplinary environment:

> the spread of answers generally follows the respondents answers for both questions. Consequently, there are varying perceptions as to how the organization would respond to the cases presented. This reveals that the respondents have no common understanding of how the organization responds to misconduct.

In a 2005 study of 379 SAPS officers, Kutnjak Ivković and Sauerman (2015) found that the respondents—all of whom were supervisors—typically expected either no discipline or only minor discipline. Written reprimand was the most frequently selected discipline and dismissal was expected in only one scenario. Similar results were reported by Matlala and colleagues (2016), who surveyed 370 SAPS officers, municipal police, and traffic police from the Gauteng Province. They found that the written warning and dismissal were the most commonly expected disciplinary action in the majority of scenarios (Matlala et al., 2016). The most commonly selected disciplinary option was the same for expected and appropriate discipline in about three-quarters of the scenarios, leading Matlala and colleagues (2016) to conclude that the respondents perceived expected discipline as fair and, in a few scenarios in which disciplinary choices differed between appropriate and expected discipline, police officers evaluated the expected discipline as too harsh. Meyer and colleagues (Meyer et al., 2013; Reyneke-Cloete and Meyer, 2016) also surveyed 160 non-commissioned police from the same province and reported that reprimand, be it verbal or written, was the predominant discipline police officers expected, if any. Reyneke-Cloete and Meyer (2016) noted a high level of agreement between the respondents' views about the appropriate discipline and their estimates of the discipline their agency would mete out in response to such violations.

Survey findings

In the empirical part of the project on police integrity in South Africa, we surveyed 868 SAPS police officers. Our respondents received a copy of the police integrity questionnaire developed by Klockars and colleagues (2006) and revised by Kutnjak Ivković (2009). The questionnaire asked officers to evaluate 14 hypothetical scenarios describing a range of forms of police misconduct, including police corruption, use of excessive force, falsification of official documents, and the planting of evidence. After the respondents had read the description of each scenario, they were asked to assess whether the described behavior violated official rules, how serious they thought the behavior was, what they perceived the appropriate and expected discipline would be, and whether they would report such behavior (for further details, see Chapter 1).

The questions that measured the second dimension of the police integrity theory focused exclusively on internal disciplinary procedures and excluded criminal procedures. They tapped into the respondents' estimates of the discipline which the SAPS would mete out for such behavior ("If an officer in your agency engaged in this behavior and was discovered doing so, what, if any, discipline do YOU think *WOULD* follow?") and their views about the appropriate discipline for such behavior ("If an officer in your agency engaged in this behavior and was discovered doing so, what, if any, discipline do YOU think *SHOULD* follow?"). The respondents could have selected one answer from the six disciplinary options: 1 = "none" [no discipline], 2 = "verbal reprimand," 3 = "written reprimand," 4 = "reassignment to a different position," 5 = "demotion in rank," and 6 = "dismissal." As Table 5.9 reveals, all behaviors described in the questionnaire are violations of official rules and could result in discipline and/or conviction.

Table 5.9 Scenario descriptions and official normative punishment/discipline

Scenario description	*Statute articles (punishment)*	*Official SAPS rules (discipline)*
Scenario 1: A police officer is widely liked in the community. Local merchants and restaurant owners regularly show their appreciation for his attention by giving him gifts of food, cigarettes, and other items of small value.	**Section 26 of the Prevention and Combating of Corrupt Activities, Act 12 of 2004 determines:** (1) Any person who is convicted of an offence referred to in (a) Part 1, 2, 3 or 4, or section 18 of Chapter 2, is liable- (i) in the case of a sentence to be imposed by a High Court, to a fine or to imprisonment, including imprisonment for life; (ii) in the case of a sentence to be imposed by a regional court, to a fine or to imprisonment for a period not exceeding 18 years; or (iii) in the case of a sentence to be imposed by a magistrate's court, to a fine or to imprisonment for a period not exceeding five years; The Minimum Sentence Act 105 of 1997 provides in Schedule 2 in section 51 for minimum sentences if it was proved that the offence was committed by any law enforcement officer involving amounts of more than R10,000.00 or where the officer was a member of a group of persons, a syndicate or enterprise acting in the execution or furtherance of a common purpose or conspiracy.	The South African Police Service Act: Regulations: Discipline (Government Gazette 40389 published in Government Notice R1361 on 1 November 2016.) Contravening Regulation 5(4)(f) of the South African Police Service Discipline Regulations, 2016—Corruption - The institution of the expeditious procedure as provided for in regulations 5(4)(f) and 9(1) read with regulation 12 (1)(e) of the South African Police Service Discipline Regulations 2016—Dismissal.

(*Continued*)

Table 5.9 (Continued)

Scenario description	*Statute articles (punishment)*	*Official SAPS rules (discipline)*
Scenario 2: A police officer is aware that there is a felony warrant for a longtime friend of his. Although he sees his friend frequently over a period of more than a week and warns his friend of its existence, he does not arrest him.	Section 257 of the Criminal Procedure Act 1977 (Act 51 of 1977) determines that if the evidence in criminal proceedings does not prove the commission of the offence charged, but proves that the accused is guilty as an accessory after that offence or any other offence of which he may be convicted, the accused may be found guilty as an accessory after that offence or, as the case may be, such other offence, and shall, in the absence of any punishment expressly provided by law, be liable to the punishment at the discretion of the court: Provided that such punishment shall not exceed the punishment which may be imposed in respect of the offence with reference to which the accused is charged as an accessory.	Contravening Regulation 5(4)(h) of the South African Police Service Discipline Regulations 2016—Defeating the course of Justice—The institution of the expeditious procedure as provided for in regulation 9 read with regulation 12 (1)(e) of the South African Police Service Discipline Regulations 2016—Dismissal.
Scenario 3: A police officer discovers a burglary of a hardware store. The display cases are smashed and many items have obviously been taken. While searching the store, he takes an expensive pocketknife and slips it into his pocket. He reports that the knife has been stolen during the burglary.	First Charge: Theft. He stole the pocketknife. It was not his property. He put it in his pocket instead of handing it in as an evidence. This shows that he had the intention to deprive the owner permanently. Second Charge: Fraud: His report that the pocketknife was stolen is false. It is a misrepresentation that the pocketknife has been stolen. The misrepresentation is made not only to the police department but also to the shop owner and the insurance company. There is therefore prejudice to the department because it "beats the system" and brings it into discredit. There is also prejudice to the shop owner because the owner believes the pocketknife to be stolen and lists it as such. The insurance company believes the misrepresentation and pays the claim to the shop owner to their prejudice. The police officer's actions are clearly unlawful. He knew that taking the knife was illegal, and his actions were intentional.	Contravening Regulation 5(4)(s) of the South African Police Service Discipline Regulations 2016—Theft—The institution of the expeditious procedure as provided for in regulation 9 read with regulation 12 (1)(e) of the South African Police Service Discipline Regulations 2016—Dismissal. Contravening Regulation 5(4)(k) of the South African Police Service Discipline Regulations 2016—Fraud—The institution of the expeditious procedure as provided for in regulation 9 read with regulation 12 (1)(e)—Dismissal.

Scenario 4: An officer who was severely beaten by a person resisting arrest, has just returned to duty. On patrol, the officer approaches a person standing in a dimly lit alley. Suddenly, the person throws a gym bag at the officer and begins to run away. The officer fatally shoots the person, striking him in the back. It was later determined that the person was unarmed.	The scenario presents a number of possibilities. The first is that the officer may be guilty of murder if he had the intention to kill the deceased. There does not appear to be any grounds of justification in the scenario. If so, the killing is unlawful. There are also other aspects to consider such as the officer identifying him/herself and warning the deceased. This does not appear to have taken place. The second is the lesser charge (and usually the alternative charge) under the circumstances. This would be "Culpable Homicide." This is a competent verdict on a charge of murder where there is insufficient evidence to prove intention to murder. This offence is when the death of another person is caused negligently although still unlawfully. If the officer shot the deceased only as a warning and intent to kill was not present, the charge would be culpable homicide.	Contravening Regulation 5(4)(p) of the South African Police Service Discipline Regulations 2016—Murder—The institution of the expeditious procedure as provided for in regulation 9 read with regulation 12 (1)(e)—Dismissal. Should the officer be found guilty of another, lesser, Common Law offence like Culpable homicide or another Statutory offence, he would face a disciplinary hearing for serious misconduct in terms of regulation 8 of the South African Police Service Discipline Regulations 2016.
Scenario 5: A police officer is scheduled to work during coming holidays. The supervisor offers to give him these days off, if he agrees to run some personal errands for the supervisor. Evaluate the supervisor's behavior.	The supervisor's behavior is one of malfeasance in the form of Fraud. It is an abuse of official power and it "beats the system "because the supervisor does not have the legal authority to give the official off without the proper administrative formalities having been followed. If the police officer's absence is not covered by an approved leave of absence, any injury suffered by that police officer would have severe negative repercussions on the administration of the department. It would mean that the department would have to pay for any medical expenses arising from that injury while believing that the officer was on duty.	Although this is an evaluation of the supervisor's behavior, the officer may also face sanction in terms of Regulation 7 of the South African Police Service Discipline Regulations 2016—"Less serious misconduct." In this case Regulation 7 provides for: 1. Corrective counseling; 2. Verbal Warning; 3. Written warning; or 4. Final written warning.

(Continued)

Table 5.9 (Continued)

Scenario description	*Statute articles (punishment)*	*Official SAPS rules (discipline)*
Scenario 6: In responding with her male partner to a fight in a bar, a young, female officer receives a black eye from one of the male combatants. The man is arrested, handcuffed, and as he is led into the cells, the male member of the team punches him very hard in the kidney area saying, "hurts, doesn't it."	The officer would be guilty of assault. The sentence is at the discretion of the magistrate. The sanction may vary from a fine to a suspended sentence to imprisonment. Depending on the severity of the crime and the surrounding circumstances (like aggravating circumstances) which include any previous convictions, the sentence in a Regional Court may vary from a fine not exceeding R300,000.00 or imprisonment up to 20 years. In a Magistrate's Court the sentence may vary from a fine not exceeding R100,000.00 to imprisonment up to three years.	If the assault is of a serious nature the male member could face dismissal in terms of the provisions of Regulation 5(4)(d) read with regulations 9(1) and 12(1)(e) of the South African Police Service Discipline Regulations 2016. If the misconduct is regarded as being "less serious", Regulation 7 provides for: 1. Corrective counselling; 2. Verbal Warning; 3. Written warning; or 4. Final written warning. If the misconduct is regarded as being serious, Regulation 8 of the South African Police Service Discipline Regulations 2016 provides for the holding of a disciplinary hearing. Regulation 12 of the South African Police Service Discipline Regulations 2016 provides as follows: (1) Any of the following sanctions may be imposed: (a) counseling (b) a warning which will be valid for six (6 months; (c) a final written warning which will be valid for six (6) months; (d) suspension without salary for a period of between one (1) month and no more than two (2) months; or (e) dismissal.

Scenario 7: A police officer stops a motorist for reckless driving. As the officer approaches the vehicle, the driver yells, "What the hell are you stopping me for?" The officer replies, "Because today is 'Arrest an Asshole Day.'"	The officer would be guilty of *Crimen Iniuriae*. The sentence is at the discretion of the magistrate. The sanction may vary from a fine, a suspended sentence to imprisonment. Depending on the severity of the crime and the surrounding circumstances (like aggravating circumstances) which include any previous convictions the sentence in a Regional Court may vary from a fine not exceeding R300,000.00 or imprisonment up to 20 years. In a Magistrate's Court the sentence may vary from a fine not exceeding R100,000.00 to imprisonment up to three years.	Conduct of an improper, disgraceful, and unacceptable manner; If the misconduct is regarded as being less serious, Regulation 7 of the South African Police Service Discipline Regulations 2016, provides for: 1. Corrective counseling; 2. Verbal Warning; 3. Written warning; or 4. Final written warning.
Scenario 8: At 2:00 A.M. a police officer, who is on duty, is driving his patrol car on a deserted road. He sees a vehicle that has been driven off the road and is stuck in a ditch. He approaches the vehicle and observes that the driver is not hurt but is obviously intoxicated. He also finds that the driver is a police officer. Instead of reporting this accident and offense, he transports the driver to his home.	The officer would be guilty of Defeating the course of justice. The sentence is at the discretion of the magistrate. The sanction may vary from a fine to a suspended sentence to imprisonment. Depending on the severity of the crime and the surrounding circumstances (like aggravating circumstances) which include any previous convictions, the sentence in a Regional Court may vary from a fine not exceeding R300,000.00 or imprisonment up to 20 years. In a Magistrate's Court the sentence may vary from a fine not exceeding R100,000.00 to imprisonment up to three years.	Contravening Regulation 5(4)(h) of the South African Police Service Discipline Regulations 2016—Defeating the course of Justice—The institution of the expeditious procedure as provided for in regulation 9(1) read with regulation 12 (1)(e) of the South African Police Service Discipline Regulations 2016—Dismissal.

(*Continued*)

Table 5.9 (Continued)

Scenario description	*Statute articles (punishment)*	*Official SAPS rules (discipline)*
Scenario 9: A police officer has a private arrangement with a local auto body shop to refer the owners of cars damaged in accidents to the shop. In exchange for each referral, he receives a payment of 5% of the repair bill from the shop owner.	The officer would be guilty of corrupt activities relating to contracts. Section 26 of the Prevention and Combating of Corrupt Activities Act 12 of 2004: (1) Any person who is convicted of an offence referred to in (a) Part 1, 2, 3 or 4, or section 18 of Chapter 2, is liable- (i) in the case of a sentence to be imposed by a High Court, to a fine or to imprisonment, including imprisonment for life; (ii) in the case of a sentence to be imposed by a regional court, to a fine or to imprisonment for a period not exceeding 18 years; or (iii) in the case of a sentence to be imposed by a magistrate's court, to a fine or to imprisonment for a period not exceeding five years.	Corruption—The institution of the expeditious procedure as provided for in regulation 5(4)(f) read with Regulations 9(1) and 12 (1)(e) of the South African Police Service Discipline Regulations 2016—Dismissal.
Scenario 10: A police officer arrests two drug dealers involved in a street fight. One has a large quantity of heroin on his person. In order to charge them both with serious offenses, the officer falsely reports that the heroin was found on both men.	The sentence is at the discretion of the magistrate. The sanction may vary from a fine to a suspended sentence to imprisonment. Depending on the severity of the crime and the surrounding circumstances (like aggravating circumstances), which include any previous convictions, the sentence in a Regional Court may vary from a fine not exceeding R300,000.00 or imprisonment up to 20 years. In a Magistrate's Court the sentence may vary from a fine not exceeding R100,000.00 to imprisonment up to three years.	Defeating the course of Justice—The institution of the expeditious procedure as provided for in regulation 5(4)(h) read with Regulations 9(1) and 12(1)(e) of the South African Police Service Discipline Regulations 2016—Dismissal.

Scenario 11: A police sergeant, without intervening, watches officers under his supervision repeatedly strike and kick a man arrested for child abuse. The man has previous child abuse arrests. Evaluate the Sergeant's behavior.	Assault. The sentence is at the discretion of the magistrate. The sanction may vary from a fine to a suspended sentence to imprisonment. Depending on the severity of the crime and the surrounding circumstances (like aggravating circumstances), which include any previous convictions, the sentence in a Regional Court may vary from a fine not exceeding R300,000.00 to imprisonment up to 20 years. In a Magistrate's Court the sentence may vary from a fine not exceeding R100,000.00 to imprisonment up to three years.	The sergeants' behavior is an act of misconduct which detrimentally affects the image of the Service or brings the Service into disrepute in terms of regulation 5(4)(x) of the South African Police Service Discipline Regulations 2016. This could lead to the institution of the expeditious procedure as provided for in regulation 9(1) read with regulations 5(4)(x) and regulations 9(1) and 12 (1)(e) of the South African Police Service Discipline Regulations 2016—Dismissal.
Scenario 12: In responding to a stabbing, a police officer arrests a suspect under circumstances that suggest that the attack was racially motivated. The police officer fails to write on the official report that the attack was racially motivated.	For the crime of defeating or obstructing the course of justice, the sentence is at the discretion of the magistrate. The sanction may vary from a fine to a suspended sentence to imprisonment. Depending on the severity of the crime and the surrounding circumstances (like aggravating circumstances), which include any previous convictions, the sentence in a Regional Court may vary from a fine not exceeding R300,000.00 to imprisonment up to 20 years. In a Magistrate's Court the sentence may vary from a fine not exceeding R100,000.00 to imprisonment up to three years.	Regulation 5(4)(h) of the South African Police Service Discipline Regulations 2016- An employee will be guilty of misconduct if he/she:- Defeats the course of justice in terms of regulation 5(4)(h) of the South African Police Service Discipline Regulations 2016. This could lead to the institution of the expeditious procedure as provided for in regulation 9(1) and 5(4)(h) read with Regulations 12 (1)(e) of the South African Police Service Discipline Regulations 2016—Dismissal.

(*Continued*)

Table 5.9 Continued

Scenario description	*Statute articles (punishment)*	*Official SAPS rules (discipline)*
Scenario 13: A police officer stops a motorist for not obeying a red traffic light. The officer agrees to accept a personal gift for half of the amount of the fine in exchange for not issuing a citation.	The officer is guilty of the crime of corrupt activities relating to public officers set out in section 4(1) (a) read with section 26 (1) of the Prevention and Combating of Corrupt Activities Act 12 of 2004:-. **26.** (1) Any person who is convicted of an offence referred to in (a) Part 1, 2, 3 or 4, or section 18 of Chapter 2, is liable- (i) in the case of a sentence to be imposed by a High Court, to a fine or to imprisonment, including imprisonment for life; (ii) in the case of a sentence to be imposed by a regional court, to a fine or to imprisonment for a period not exceeding 18 years; or (iii) in the case of a sentence to be imposed by a magistrate's court, to a fine or to imprisonment for a period not exceeding five years.	Contravening Regulation 5(4)(f) of the South African Police Service Discipline Regulations, 2016—Corruption: —The institution of the expeditious procedure as provided for in regulations 9(1) and 5(4)(f) and read with regulation 12 (1)(e) of the South African Police Service Discipline Regulations 2016—Dismissal.
Scenario 14: While on patrol in his patrol car, a police officer notices a group of juveniles writing graffiti on a building. The police officer decides not to intervene and instead drives on.	For the crime of defeating or obstructing the course of justice, the sentence is at the discretion of the magistrate. The sanction may vary from a fine to a suspended sentence to imprisonment. Depending on the severity of the crime and the surrounding circumstances (like aggravating circumstances), which include any previous convictions, the sentence in a Regional Court may vary from a fine not exceeding R300,000.00 to imprisonment up to 20 years.	Regulation 5(4)(h) of the South African Police Service Discipline Regulations 2016- An employee will be guilty of misconduct if he/she:- Defeats the course of justice. This could lead to the institution of the expeditious procedure as provided for in regulation 9(1) and 5(4)(h) read with Regulations 9(1) and 12 (1)(e) of the South African Police Service Discipline Regulations 2016—Dismissal.

	In a Magistrate's Court the sentence may vary from a fine not exceeding R100,000.00 to imprisonment up to 3 years.	If the misconduct is regarded as being less serious, Regulation 7 provides for: 1. Corrective counseling; 2. Verbal Warning; 3. Written warning; or 4. Final written warning. If the conduct is regarded as being serious misconduct, Regulation 8 of the South African Police Service Discipline Regulations 2016 makes provision for a disciplinary hearing to be held in which case any of the sanctions referred to above, including dismissal, could be implemented.

Expected discipline

With one exception, the respondents' views about the expected discipline are widely spread across all disciplinary options (Table 5.10), from no discipline to dismissal, with no disciplinary option selected by the majority of officers. The only scenario in which the majority of the respondents agreed on the same discipline—dismissal—was for Scenario 13 (Bribe from reckless motorist).[2] At the same time, this was the only scenario in which the majority of the respondents expected dismissal.

To streamline the analysis, we classified disciplinary options into three categories: no discipline or very mild discipline (verbal reprimand), some discipline less serious than dismissal, and dismissal. The good news is that the majority of the respondents expected that SAPS managers would discipline an officer who engaged in the misconduct described in almost all scenarios (13 out of 14 or 93%; Table 5.10); in only one scenario (Scenario 14: Failure to react to graffiti), the majority of the respondents did not expect *any* discipline or only a verbal reprimand. The bad news is that the expected discipline was not very serious in the majority of the scenarios, with dismissal rarely expected by most respondents (Table 5.10). In fact, the majority of the respondents expected dismissal for only one scenario in the questionnaire (Scenario 13: Bribe from reckless motorist), although some of the behaviors described include severe violations of the SAPS code of conduct, disciplinary regulations, and criminal law. In addition, a large minority of the respondents—close to one-half—expected dismissal in two additional scenarios (Scenario 3: Theft of knife; Scenario 4: Abuse of deadly force; Table 5.10).

The severity of the expected discipline seems to be related to the respondents' evaluations of misconduct seriousness (Spearman's rho = .864, $p < .001$) and estimates that the behavior violates official rules (Spearman's rho = .833, $p < .001$). In the scenarios perceived as the least serious or the least likely to be evaluated as rule-violating (e.g., Scenario 1: Acceptance of gratuities; Scenario 7: Verbal abuse; Scenario 8: Cover-up of DUI; Scenario 12: Failure to report hate crime; Scenario 14: Failure to react to graffiti), the largest percent of the respondents expected either no discipline or a verbal reprimand.[3] On the other hand, in the scenarios evaluated as the most serious or the most likely to be rule-violating (Scenario 2: Failure to arrest; Scenario 3: Theft of knife; Scenario 4: Abuse of deadly force; Scenario 13: Bribe from reckless motorist), the respondents were the most likely to perceive dismissal as the most likely disciplinary outcome.

Appropriate discipline

The respondents' views about appropriate discipline varied greatly across the scenarios. On the one hand, about one-half of the respondents thought that no discipline or only a verbal reprimand would be appropriate in connection with three scenarios (Scenario 1: Acceptance of gratuities; Scenario 7: Verbal abuse; Scenario 14: Failure to react to graffiti), evaluated among the least serious

Table 5.10 Expected discipline

Scenario	*Expected discipline*									
	None	*Verbal rep.*	*Written rep.*	*Reassign.*	*Demotion*	*Dismiss.*	*None + verbal*	*Some disc.*	*Dismiss.*	*Rank*
Scenario 1: Acceptance of gratuities	11.8%	29.7%	34.3%	6.0%	5.1%	13.1%	41.5%	45.4%	13.1%	5
Scenario 2: Failure to arrest	4.3%	14.1%	46.0%	10.0%	4.7%	21.0%	18.3%	60.7%	21.0%	8
Scenario 3: Theft of knife	1.4%	5.6%	30.1%	10.6%	3.3%	49.0%	7.0%	44.0%	49.0%	13
Scenario 4: Abuse of deadly force	2.9%	9.3%	21.3%	17.5%	7.9%	41.0%	12.2%	46.8%	41.0%	12
Scenario 5: Supervisor corruption	7.8%	19.2%	39.3%	10.8%	10.7%	12.1%	27.1%	60.8%	12.1%	4
Scenario 6: Striking arrestee	7.0%	15.7%	46.8%	13.6%	2.6%	14.3%	22.7%	63.0%	14.3%	7
Scenario 7: Verbal abuse	14.2%	33.3%	39.1%	5.1%	2.4%	6.0%	47.5%	46.5%	6.0%	1
Scenario 8: Cover-up of DUI	13.8%	22.6%	38.4%	6.1%	5.9%	13.2%	36.4%	50.3%	13.2%	6
Scenario 9: Kickback	13.0%	16.8%	31.7%	8.1%	5.4%	25.0%	29.8%	45.2%	25.0%	10
Scenario 10: False report on drugs	5.6%	9.8%	36.2%	10.6%	6.8%	31.0%	15.4%	53.6%	31.0%	11
Scenario 11: Supervisor fails to stop excessive force	3.7%	14.9%	41.4%	9.6%	7.0%	23.5%	18.5%	57.9%	23.5%	9
Scenario 12: Failure to report hate crime	5.3%	22.7%	41.4%	9.8%	10.3%	10.6%	28.0%	61.4%	10.6%	3
Scenario 13: Bribe from reckless motorist	2.5%	7.8%	24.8%	8.9%	5.4%	50.5%	10.4%	39.1%	50.5%	14
Scenario 14: Failure to react to graffiti	16.8%	34.0%	35.9%	4.3%	2.4%	6.6%	50.8%	42.6%	6.6%	2

(Table 5.11). On the other hand, about one-half or close to one-half of respondents thought that a SAPS officer who engaged in the behavior described in the scenario should be dismissed in connection with three scenarios (Scenario 3: Theft of knife; Scenario 4: Abuse of deadly force; Scenario 13: Bribe from reckless motorist). Viewed differently, in most scenarios, the respondents expected some discipline, more serious than a verbal reprimand and less serious than dismissal (Table 5.11).

As with expected discipline, the severity of the discipline which the respondents thought appropriate for each scenario seems to be related their evaluations of misconduct seriousness (Spearman's rho = .859, $p < .001$) and to their estimates that the behavior violates official rules (Spearman's rho = .833, $p < .001$). In the five scenarios perceived as the least serious and least likely to be evaluated as rule-violating (e.g., Scenario 1: Acceptance of gratuities; Scenario 7: Verbal abuse; Scenario 8: Cover-up of DUI; Scenario 12: Failure to report hate crime; Scenario 14: Failure to react to graffiti), the largest percent of respondents—between one-third and one-half—thought that no discipline at all or only the mildest type—verbal reprimand—would be appropriate. On the other hand, in three out of the four scenarios evaluated as the most serious and most likely to be a violation of the official rules (Scenario 3: Theft of knife; Scenario 4: Abuse of deadly force; Scenario 13: Bribe from reckless motorist), respondents were most likely to report that dismissal would be appropriate.

Discipline fairness

As Table 5.12 shows, the respondents' estimates of expected and appropriate discipline are strongly related (chi-squares indicate a strong correlation). Overall, when they thought the SAPS would use more severe discipline, the respondents also typically assumed that the more severe discipline would be appropriate (Spearman's rho = .987, $p < .001$).

To determine the officers' perceptions of the expected discipline, we compared their views about the expected discipline with their views about the appropriate discipline (Tables 5.12 and 5.13). If the respondents selected the same answer for both appropriate and expected discipline (i.e., the difference is zero), we assumed they evaluated the severity of the expected discipline as fair. If the respondents stated that the discipline the SAPS would mete out was harsher than the discipline they thought appropriate (i.e., the difference is positive), we assumed that they evaluated the severity of the expected discipline as too harsh. Finally, if they stated that the discipline the SAPS would apply was more lenient than the discipline they thought appropriate (i.e., the difference is negative), we assumed they evaluated the severity of the expected discipline as too lenient.

A direct comparison of the individual choices of the discipline the respondents expected the SAPS to mete out and the discipline they considered appropriate indicates that the majority of officers evaluated the severity of

Table 5.11 Appropriate discipline

Scenario	*Appropriate discipline*									
	None	*Verbal rep.*	*Written rep.*	*Reassign.*	*Demotion*	*Dismiss.*	*None + verbal*	*Some disc.*	*Dismissal*	*Rank*
Scenario 1: Acceptance of gratuities	8.9%	41.1%	24.4%	7.8%	2.5%	15.2%	50.0%	34.8%	15.2%	7
Scenario 2: Failure to arrest	2.9%	16.4%	45.9%	9.7%	4.4%	20.9%	19.2%	59.9%	20.9%	8
Scenario 3: Theft of knife	0.8%	7.5%	28.6%	10.4%	3.1%	49.7%	8.3%	42.1%	49.7%	13
Scenario 4: Abuse of deadly force	3.3%	10.4%	21.2%	15.9%	8.5%	40.7%	13.7%	45.6%	40.7%	12
Scenario 5: Supervisor corruption	3.8%	23.0%	40.8%	10.3%	11.9%	10.3%	26.8%	62.9%	10.3%	4
Scenario 6: Striking arrestee	5.1%	18.7%	48.7%	10.1%	2.8%	14.6%	23.8%	61.6%	14.6%	6
Scenario 7: Verbal abuse	14.5%	35.4%	36.4%	6.5%	1.4%	5.8%	49.9%	44.3%	5.8%	1
Scenario 8: Cover-up of DUI	14.1%	23.0%	34.9%	10.5%	5.6%	11.9%	37.1%	51.0%	11.9%	5
Scenario 9: Kickback	11.4%	14.9%	32.4%	7.5%	8.6%	25.2%	26.3%	48.5%	25.2%	10
Scenario 10: False report on drugs	4.6%	11.3%	36.1%	10.6%	6.3%	31.1%	15.9%	53.0%	31.1%	11
Scenario 11: Supervisor fails to stop excessive force	4.1%	16.1%	38.2%	10.9%	7.1%	23.4%	20.3%	56.3%	23.4%	9
Scenario 12: Failure to report hate crime	6.5%	24.0%	42.2%	7.6%	10.8%	9.0%	30.4%	60.6%	9.0%	3
Scenario 13: Bribe from reckless motorist	1.4%	7.6%	23.4%	8.9%	6.7%	52.1%	9.0%	38.9%	52.1%	14
Scenario 14: Failure to react to graffiti	16.4%	33.6%	35.8%	5.1%	3.2%	5.9%	50.0%	44.1%	5.9%	2

Table 5.12 Expected vs. appropriate discipline

Scenario	*Expected discipline*				*Appropriate discipline*				*Chi-square (Corr.)*
	None + verbal	*Some disc.*	*Dismissal*	*Rank*	*None + verbal*	*Some disc.*	*Dismissal*	*Rank*	
Scenario 1: Acceptance of gratuities	50.0%	34.8%	15.2%	7	41.5%	45.4%	13.1%	5	840.21*** (.712***)
Scenario 2: Failure to arrest	19.2%	59.9%	20.9%	8	18.3%	60.7%	21.0%	8	923.31*** (.756***)
Scenario 3: Theft of knife	8.3%	42.1%	49.7%	13	7.0%	44.0%	49.0%	13	852.57*** (.824***)
Scenario 4: Abuse of deadly force	13.7%	45.6%	40.7%	12	12.2%	46.8%	41.0%	12	1,292.50*** (.894***)
Scenario 5: Supervisor corruption	26.8%	62.9%	10.3%	4	27.1%	60.8%	12.1%	4	1,152.46*** (.831***)
Scenario 6: Striking arrestee	23.8%	61.6%	14.6%	6	22.7%	63.0%	14.3%	7	1,264.87*** (.862***)
Scenario 7: Verbal abuse	49.9%	44.3%	5.8%	1	47.5%	46.5%	6.0%	1	1,329.19*** (.848***)
Scenario 8: Cover-up of DUI	37.1%	51.0%	11.9%	5	36.4%	50.3%	13.2%	6	1,088.25*** (.818***)
Scenario 9: Kickback	26.3%	48.5%	25.2%	10	29.8%	45.2%	25.0%	10	1,104.59*** (.843***)
Scenario 10: False report on drugs	15.9%	53.0%	31.1%	11	15.4%	53.6%	31.0%	11	1,070.20*** (.834***)
Scenario 11: Supervisor fails to stop excessive force	20.3%	56.3%	23.4%	9	18.5%	57.9%	23.5%	9	1,174.90*** (.861***)
Scenario 12: Failure to report hate crime	30.4%	60.6%	9.0%	3	28.0%	61.4%	10.6%	3	1,178.41*** (.819***)
Scenario 13: Bribe from reckless motorist	9.0%	38.9%	52.1%	14	10.4%	39.1%	50.5%	14	1,058.34*** (.811***)
Scenario 14: Failure to react to graffiti	50.0%	44.1%	5.9%	2	50.8%	42.6%	6.6%	2	1,235.48*** (.850***)

Table 5.13 Discipline fairness

Scenario	*Too harsh*	*Fair*	*Too lenient*	*Rank fair*
Scenario 1: Acceptance of gratuities	17.7%	71.5%	10.7%	1
Scenario 2: Failure to arrest	11.8%	76.8%	11.4%	2
Scenario 3: Theft of knife	9.9%	83.2%	6.9%	8.5
Scenario 4: Abuse of deadly force	7.1%	86.2%	6.7%	13
Scenario 5: Supervisor corruption	9.8%	83.4%	6.8%	9
Scenario 6: Striking arrestee	9.8%	81.8%	8.4%	5
Scenario 7: Verbal abuse	9.5%	84.4%	6.1%	12
Scenario 8: Cover-up of DUI	11.4%	80.2%	8.4%	4
Scenario 9: Kickback	7.8%	80.1%	12.1%	3
Scenario 10: False report on drugs	8.5%	83.5%	7.9%	10
Scenario 11: Supervisor fails to stop excessive force	9.3%	82.8%	7.8%	7
Scenario 12: Failure to report hate crime	10.1%	83.2%	6.7%	8.5
Scenario 13: Bribe from reckless motorist	6.9%	82.5%	10.6%	6
Scenario 14: Failure to react to graffiti	7.6%	83.8%	8.6%	11

expected discipline as fair (Table 5.13). In particular, about three-quarters of the respondents or more evaluated the expected discipline severity as fair (i.e., they selected the same disciplinary option for the appropriate discipline and expected discipline). The percentage is somewhat smaller for two scenarios (Scenario 1: Acceptance of gratuities; Scenario 2: Failure to arrest) than it is for the others (Table 5.13).

We also explored the answers provided by the respondents who evaluated the expected discipline as being unfair (Table 5.13). Although some respondents evaluated the discipline as either too harsh or too lenient for each scenario, the percentages are neither large nor substantively important (i.e., reaching 20% or more).[4] The only scenario that potentially indicated substantively important views regarding expected discipline as unfair by a large minority of respondents is Scenario 1 (Acceptance of gratuities), in which about 18% of respondents evaluated the expected discipline as too harsh.

Conclusion

According to the second dimension of the police integrity theory (e.g., Klockars and Kutnjak Ivković, 2004; Klockars et al., 2000, 2004, 2006; Kutnjak Ivković, 2015a), the effective use of reactive and proactive internal control mechanisms is an indicator of an accountable police agency. Both measures, and internal control in general, have always occupied an important place in the SAPS's identity. This is because the organization was formed against the backdrop of a near century of undemocratic policing and abuse of force (see Chapter 2). The establishment of the Civilian Secretariat for Police Services (CSP) to oversee police policy, the creation of the Independent Complaints

Directorate (ICD) to investigate serious abuses by the police, and the requirement that every police station hosts a Community Policing Forum (CPF) hint at the kind of accountability envisaged by the country's first democratic government. However, whereas South Africans trust the police more than residents of many other African countries do (Chingwete, 2016), South Africans still trust them less than any other national government entity (Research, 2018).

In this chapter, we have traced the evolution of the SAPS's internal control mechanisms and apparent efforts to address abuses within the organization from the early 1990s until 2019. While officers have been put through disciplinary processes and were sanctioned throughout this period, most cases resulted in no sanction or were withdrawn. It is also likely, despite our empirical findings to the contrary, that most infringements of the SAPS rules go unaddressed. Certainly, the popular perception among South Africa's public is that many, even most, police abuse their power and cannot be trusted (see Chapter 1).

Because our respondents expected some discipline—not necessarily severe—for the rule-violating behavior described in almost all scenarios, our empirical data should be read as if respondents assumed the violations in each scenario had come to the attention of a motivated *supervisor*. However, even in such circumstances when they anticipate some discipline, the expected discipline was typically not very serious. Nearly regardless of the circumstances, the respondents typically expected only a verbal or written reprimand.

Our empirical results demonstrate that, with one exception, the respondents did not expect the SAPS to dismiss officers who engaged in even very serious forms of misconduct, including the abuse of deadly force and the acceptance of a kickback. Rather, in the best case, they expected a milder form of discipline for police officers who engaged in, and were investigated for, almost all forms of misconduct described in the questionnaire. While the respondents did not expect the SAPS to dismiss police officers who engaged in such violations, the majority of respondents would support dismissal for behaviors as serious as the abuse of deadly force and theft from a crime scene.

Concurringly, while the SAPS has since 1999 reported that it had various anti-corruption strategies and systems either in development or in place, the consistency of these reports over two decades combined with the fact that the organization almost always implies that strategies are either being planned or are about to be concluded, suggests that, in reality, little has been put in place. If this rings true—and the fact that so much high- and low-level corruption is consistently identified by the media and others suggests that it is (see Chapter 1)—those who end up being sanctioned by the system are the unlucky few who, for some reason or another, have been acted against in an inconsistent, procedurally unfair effort to demonstrate the illusion of internal control where in fact, there is little or none.

Notes

1 Although terminology more at place in criminal cases, the SDR nevertheless stipulates such jargon for SAPS disciplinary hearings (Regulation 13), an anomaly inherited from its predecessors, which were in turn strongly influenced by the South African Criminal Procedure Act (Newham, 2000).
2 The SAPS does not generally enforce traffic or road legislation, nor issue tickets, which may reflect on this scenario somewhat differently than it does for the agencies that do carry out such enforcement.
3 The percentage of the respondents who expected no discipline or only a verbal reprimand was also higher for Scenario 9: Kickback, which was not evaluated as one of the 5 least serious forms of misconduct.
4 Klockars and colleagues (2006) defined 20% as the benchmark of large and substantive differences.

References

Basson, A. and P. Du Toit (2018). *Enemy of the People: How Jacob Zuma Stole South Africa and How the People Fought Back*. Johannesburg and Cape Town: Jonathan Ball.

Brodeur, J.P. (2010). *The Policing Web*. Oxford: Oxford University Press.

Bruce, D. (2008). Without fear or favour: The Scorpions and the politics of justice. *South African Crime Quarterly*, 4: 11–15.

Burger, J. (2019). South Africa's Hawks are preparing to fly. *ISS Today*, 1 October. Available at: https://issafrica.org/iss-today/south-africas-hawks-are-preparing-to-fly

Burger, J. and S. Grobler (2017). *Why the SAPS Needs an Internal Anti-Corruption Unit*. Policy Brief No. 101. Pretoria: Institute for Security Studies.

Cawthra, G. (1993). *Policing South Africa*. London: Zed Books.

Chabalala, J. (2018). #StateCaptureInquiry: Ntlemeza Was Central to State Capture Project: Gordhan. *News24*, 20 November. Available at: www.news24.com/SouthAfrica/News/statecaptureinquiry-ntlemeza-was-central-to-state-capture-project-gordhan-20181120 (Accessed 2 September 2019).

Cheloukhine, S., S. Kutnjak Ivković, Q. Haq, and M.R. Haberfeld (2015). Police integrity in Russia. In S. Kutnjak Ivković and M. Haberfeld (eds.), *Police Integrity across the World*. New York: Springer.

Chingwete, A. (2016). *In South Africa, Citizens' Trust in President, Political Institutions Drops Sharply*, Afrobarometer Dispatch No. 90. Afrobarometer and Institute for Justice and Reconciliation.

Constitution of the Republic of South Africa (1996). Act No. 108 of 1996. Available at: www.gov.za/documents/constitution-republic-south-africa-1996

De Langa, I. (2017). Ntlemeza told to vacate office immediately after losing bid to remain Hawks boss. *The Citizen*, 12 April. Available at: https://citizen.co.za/news/south-africa/1485820/ntlemeza-told-to-vacate-office-immediately-after-losing-bid-to-remain-hawks-boss/ (Accessed 2 September 2019).

Department of Police (2016). *Discipline Regulations for the South African Police Service*, Government Gazette No. 40389. Available at: www.gov.za/sites/default/files/gcis_document/201611/40389rg10662gon1361.pdf (Accessed 3 July 2019).

Faull, A. (2007). *Corruption and the South African Police Service: A Review and Its Implications*. Research Paper. Pretoria: Institute for Security Studies.

Faull, A. (2018). *Police Work and Identity: A South African Ethnography*. Abingdon: Routledge.

Hornberger, J. and D. Vigneswaran (eds.) (2009). *Beyond "Good Cop"/"Bad Cop": Understanding Informality and Police Corruption in South Africa*. Johannesburg: University of the Witwatersrand.

Kang, W. and S. Kutnjak Ivković (2015). Police integrity in South Korea. In S. Kutnjak Ivković and M. Haberfeld (eds.), *Police Integrity across the World*. New York: Springer.

Khampempe, S. (2006). *Khampepe Commission of Inquiry into the Mandate and Location of the Directorate for Special Operations ("The DSO"): Final Report*. Available at: www.justice.gov.za/commissions/2008_khampempe.pdf

Khechumyan, A. and S. Kutnjak Ivković (2015). Police integrity in Armenia. In S. Kutnjak Ivković and M. Haberfeld (eds.), *Police Integrity across the World*. New York: Springer.

Klockars, C.B. and S. Kutnjak Ivković (2004). Measuring police integrity. In A.R. Piquero, J.R. Greene, and M.J. Hickman (eds.). *Police Integrity and Ethics*. Belmont, CA: Wadsworth Publishing, 1.3–1.20.

Klockars, C.B., S. Kutnjak Ivković, and M.R. Haberfeld (eds.) (2004). *The Contours of Police Integrity*. Newbury Park, CA: Sage Publication.

Klockars, C.B., S. Kutnjak Ivković, and M.R. Haberfeld (2006). *Enhancing Police Integrity*. New York: Springer.

Klockars, C.B., S. Kutnjak Ivković, W.E. Harver, and M.R. Haberfeld (2000). *The Measurement of Police Integrity*, National Institute of Justice, May. Available at: www.ncjrs.gov/pdffiles1/nij/181465.pdf

Kriegler, A. and M. Shaw (2016). *A Citizen's Guide to Crime Statistics in South Africa*. Johannesburg and Cape Town: Jonathan Ball.

Kutnjak Ivković, S. (2009). The Croatian police, police integrity, and transition toward democratic policing. *Policing: An International Journal of Police Strategies and Management*, 32(3): 459–488.

Kutnjak Ivković, S. (2015a). Studying police integrity. In S. Kutnjak Ivković and M. Haberfeld (eds.), *Police Integrity across the World*. New York: Springer.

Kutnjak Ivković, S. (2015b). Police integrity in Croatia. In S. Kutnjak Ivković and M. Haberfeld (eds.), *Police Integrity across the World*. New York: Springer.

Kutnjak Ivković, S. and M. Haberfeld (2015). A comparative perspective on police integrity. In S. Kutnjak Ivković and M. Haberfeld (eds.), *Police Integrity across the World*. New York: Springer.

Kutnjak Ivković, S., M. Haberfeld, W. Kang, R. Peacock, and A. Sauerman (2016). A multi-country comparative study of the perceived police disciplinary environments. *Policing: An International Journal of Police Strategies and Management*, 39(2): 338–353.

Kutnjak Ivković, S., M. Haberfeld, and R. Peacock (2015). Police integrity in the United States. In S. Kutnjak Ivković and M. Haberfeld (eds.), *Police Integrity across the World*. New York: Springer.

Kutnjak Ivković, S. and A. Sauerman (2011). Measuring the code of silence among the South African police: Findings from a SAPS supervisor survey. In C. Gould and G. Newham (eds.). *Toward a Coherent Strategy for Crime Reduction in South Africa Beyond 2010*. Pretoria, South Africa: Institute for Security Studies. 74–87.

Kutnjak Ivković, S. and A. Sauerman (2015). Threading the thin blue line: Transition toward Democratic policing and the integrity of the South African police service. *Policing and Society*, 25(1): 25–52.

Kutnjak Ivković, S. and T.O. Shelley (2008). The contours of police integrity across Eastern Europe: The case of Bosnia and Herzegovina and the Czech Republic. *International Criminal Justice Review*, 18(1): 59–82.

Lebeya, S. (2019). *Briefing to the Portfolio Committee on Police on the Annual Performance Plan: 2019/2020 for the Directorate for Priority Crime Investigation.* Presentation to Parliament, Cape Town.

Lobnikar, B. and G. Meško (2015). Police integrity in Slovenia. In S. Kutnjak Ivković and M. Haberfeld (eds.), *Police Integrity across the World.* New York: Springer.

Maromo, J. (2018). Senior Hawks officers, drug dealer arrested for bribery. *IOL,* 13 December. Available at: www.iol.co.za/news/south-africa/gauteng/senior-hawks-officers-drug-dealer-arrested-for-bribery-18496463

Matlala, R.L.G., D. Mistry, and A. Phala (2016). Measuring the integrity of law enforcement officers in Gauteng Province. *The International Journal of Social Sciences and Humanities Invention,* 3(11): 2969–2980.

Meyer, M.E., J. Steyn, and N. Gopal (2013). Exploring the public parameter of police integrity. *Policing: An International Journal of Police Strategies & Management,* 36(1): 140–156.

Minnaar, A. (2008). The Scorpions lose their sting: Challenges to incorporation of the DSO into the SAPS. *South African Crime Quarterly,* 24: 23–28.

Mooki, O. (2011). One man's battle for justice. *Cape Argus,* 18 March. Available at: www.iol.co.za/capeargus/one-mans-battle-for-justice-1043793 (Accessed 13 September 2019).

Newham, G. (2000). Transformation and the Internal Disciplinary System of the South African Police Service. Research report written for the Centre for the Study of Violence and Reconciliation, November 2000. Available at https://pdfs.semanticscholar.org/b2de/f896d31474d48cdb5a8c1b9e349ad9f3c96d.pdf

Newham, G. (2004). Out of step: Integrity and the South Africa police service. In C.B. Klockars, S. Kutnjak Ivković, and M.R. Haberfeld (eds.). *The Contours of Police Integrity.* Thousand Oaks, CA: Sage Publications, 232–247.

Newham, G. and A. Faull (2011). Protector or predator? Tackling police corruption in South Africa. Institute for Security Studies Monograph. Available at https://journals.co.za/content/ismono/2011/182/EJC48727

Newham, G. and L. Gomomo (2003). Bad Cops get a break: The closure of the SAPS anti-corruption unit. *South African Crime Quarterly,* 4: 5–9.

Omar, B. (2008). *SAPS' Costly Restructuring: A Review of Public Order Policing Capacity.* Pretoria: Institute for Security Studies.

Parliamentary Monitoring Group (1998). *National Crime Prevention Strategy: Briefing by SAPS,* Minutes of Briefing to the Parliamentary Portfolio Committee on Safety and Security, 17 February. Available at: https://pmg.org.za/committee-meeting/6272/ (Accessed 1 September 2019).

Pauw, J. (2018). *The President's Keepers: Those Keeping Zuma in Power and Out of Prison.* Johannesburg: Tafelberg.

Personal Communication (2019). Email Correspondence between Andrew Faull and Brigadier Craig Mitchell of SAPS Strategic Management (Head Office), 2 September.

Pharoah, R. (2008). *National Victims of Crime Survey: Overview of Key Data.* ISS Occasional Paper No. 175. Pretoria: Institute for Security Studies.

Phetthong, N. and S. Kutnjak Ivković (2015). Police integrity in Thailand. In S. Kutnjak Ivković and M. Haberfeld (eds.), *Police Integrity across the World.* New York: Springer.

Pitchford, J. (2016). *Blood on Their Hands: General Johan Booysen Reveals His Truth.* Johannesburg, South Africa: Pan MacMillan.

Planning Commission (2012). *National Development Plan,* November. Pretoria: The Planning Commission.

Plus 94 Research (2018). *Afrobarometer: South Africa Results*. Available at: www.afrobarometer.org/publications/south-africa-summary-results-2018

Porter, L.E., T. Prenzler, and K. Hine (2015). Police integrity in Australia. In S. Kutnjak Ivković and M. Haberfeld (eds.), *Police Integrity across the World*. New York: Springer.

Rauch, J. (2000). *Police Reform and South Africa's Transition*. Available at: www.csvr.org.za/docs/policing/policereformandsouth.pdf (Accessed 5 March 2019).

Rauch, J. (2004). *Police Transformation and the South African TRC*. Available at: www.csvr.org.za/docs/policing/policetransformationandTRC.pdf (Accessed 5 March 2019).

Research (2018). *Afrobarometer: South Africa Results*. Available at: www.afrobarometer.org/publications/south-africa-summary-results-2018

Reyneke-Cloete, S. and M.E. Meyer (2016). Viewing integrity in the South African police service among non-commissioned officers in Gauteng Province. *Acta Criminologica: Southern African Journal of Criminology*, 29(2): 87–106.

SA News (2019). *Hawks Investigator Sentenced for Fraud, Money Laundering*, 31 January. Available at: www.sanews.gov.za/south-africa/hawks-investigator-sentenced-fraud-money-laundering

Sapa (2015). Sibiya: Judge Calls Acting Head of Hawks "Dishonest". *Mail & Guardian*, 23 March. Available at: https://mg.co.za/article/2015-03-23-judge-harsh-on-acting-head-of-hawks (Accessed 25 February 2018).

Sehloho, M. (2018). Hawks suspend anti-corruption task team head. *Cape Talk*, 3 September. Available at: www.capetalk.co.za/articles/317757/hawks-suspend-anti-corruption-task-team-head (Accessed 15 September 2019).

Sitole, K. (2018). *Welcoming Speech: General Sitole Launch of Anti-Corruption Strategy, SAPS Academy Tshwane, Gauteng*, 29 June. Available at: www.gov.za/speeches/general-khehla-sitole-anti-corruption-strategy-launch-29-jun-2018-0000

Smit, S. (2019). Jonas stands by his state capture testimony. *Mail & Guardian*, 15 March. Available at: https://mg.co.za/article/2019-03-15-jonas-stands-by-his-state-capture-testimony (Accessed 15 September 2019).

South African Government (2016). SAPS's Discipline Regulations. Available at: https://www.gov.za/documents/south-african-police-service-act-regulations-discipline-1-nov-2016-0000

South African Police (1993). *Annual Report of the Commissioner of the SA Police, 1993*. Pretoria: South African Police Service. Available at: https://www.saps.gov.za/about/stratframework/annualreports_arch.php

South African Police Service (2000). *Annual Report, 1999/2000*. Pretoria: South African Police Service. Available at: https://www.saps.gov.za/about/stratframework/annualreports_arch.php

South African Police Service (2001). *Annual Report, 2000/2001*. Pretoria: South African Police Service. Available at: https://www.saps.gov.za/about/stratframework/annualreports_arch.php

South African Police Service (2002). *Annual Report, 2001/2002*. Pretoria: South African Police Service. Available at: https://www.saps.gov.za/about/stratframework/annualreports_arch.php

South African Police Service (2003). *Annual Report, 2002/2003*. Pretoria: South African Police Service. Available at: https://www.saps.gov.za/about/stratframework/annualreports_arch.php

South African Police Service (2004). *Annual Report, 2003/2004*. Pretoria: South African Police Service. Available at: https://www.saps.gov.za/about/stratframework/annualreports_arch.php

South African Police Service (2008). *Annual Report, 2007/2008*. Pretoria: South African Police Service. Available at: https://www.saps.gov.za/about/stratframework/annualreports_arch.php

South African Police Service (2009). *Annual Report, 2008/2009*. Pretoria: South African Police Service. Available at: https://www.saps.gov.za/about/stratframework/annualreports_arch.php

South African Police Service (2010). *Annual Report, 2009/2010*. Pretoria: South African Police Service. Available at: https://www.saps.gov.za/about/stratframework/annualreports_arch.php

South African Police Service (2011). *Annual Report, 2010/2011*. Pretoria: South African Police Service. Available at: https://www.saps.gov.za/about/stratframework/annualreports_arch.php

South African Police Service (2012). *Annual Report, 2011/2012*. Pretoria: South African Police Service. Available at: https://www.saps.gov.za/about/stratframework/annualreports_arch.php

South African Police Service (2013). *Annual Report, 2012/2013*. Pretoria: South African Police Service. Available at: https://www.saps.gov.za/about/stratframework/annualreports_arch.php

South African Police Service (2014). *Annual Report, 2013/2014*. Pretoria: South African Police Service. Available at: https://www.saps.gov.za/about/stratframework/annualreports_arch.php

South African Police Service (2015). *Annual Report, 2014/2015*. Pretoria: South African Police Service. Available at: https://www.saps.gov.za/about/stratframework/annualreports_arch.php

South African Police Service (2016). *Annual Report, 2015/2016*. Pretoria: South African Police Service. Available at: https://www.saps.gov.za/about/stratframework/annualreports_arch.php

South African Police Service (2017). *Annual Report, 2016/2017*. Pretoria: South African Police Service. Available at: https://www.saps.gov.za/about/stratframework/annualreports_arch.php

South African Police Service (2018). *Annual Report, 2017/2018*. Pretoria: South African Police Service. Available at: https://www.saps.gov.za/about/stratframework/annualreports_arch.php

South African Police Service (2019). *Annual Report, 2018/2019*. Pretoria: South African Police Service. Available at: https://www.saps.gov.za/about/stratframework/annualreports_arch.php

Statistics South Africa (1999). *Victims of Crime Survey 1998*. Pretoria: Statistics South Africa.

Umraw, A. (2019). Acting Hawks boss BerningNtlemeza "threatened" anti-corruption colleagues. *Times Live*, 20 September. Available at: www.timeslive.co.za/politics/2019-09-20-acting-hawks-boss-berning-ntlemeza-threatened-anti-corruption-colleagues/ (Accessed 20 September 2019).

Vallmüür, B. (2015). Police integrity in Estonia. In S. Kutnjak Ivković and M. Haberfeld (eds.), *Police Integrity across the World*. New York: Springer.

Chapter 6

Curtailing the code of silence

Introduction

Police organizations tend to generate distinct organizational cultures. They emerge from the milieu of tasks, responsibilities, and pressures that officers face, and the unique position they occupy between the state and the citizens in democratic polities. These cultures foster occupational personalities and shared values among officers.

One of the most common tenets of police organizational culture is the "code of silence" (e.g., Crank, 1998; Kleinig, 1996; Skolnick, 2002). At its simplest, the code is a shared understanding among officers that they will not report the misconduct of their fellow officers or cooperate with investigations against fellow colleagues (Kutnjak Ivković et al., 2018). Articulated by a South African police officer who served from 1991 to 2006 (Faull, 2010, p. 179): "What happens on the shift stays on the shift. What happens in the station stays in the station. It always stays inside with the police or in the station. It's got nothing to do with outside. Nothing, nothing, nothing."

Police culture, characterized by secrecy and resistance to public accountability, creates peer pressure that pushes officers to abide by informal organizational norms and tolerate misconduct in silence. When police misconduct is revealed, police organizations with strong codes of silence tend to blame "a few rotten apples" and deny organizational complicity, while in truth it is the barrel—the police organization—that makes the "apples" rot (Punch, 2009).

The apartheid SAP was an exceptionally secretive organization. Breaking through its code of silence was central to the vision of professional, democratic policing envisaged by the African National Congress (ANC) in the early 1990s (Haysom, 1993). However, as elsewhere in the world, the code has constrained efforts to reform policing in South Africa. For example, by early 2020, almost none of the apartheid police found guilty of human rights violations during the Truth and Reconciliation Commission (TRC) had been prosecuted, while corruption, abuse of force, and dishonesty remained common in street-level policing, despite repeated efforts (or claims) to address them (see Chapter 5). Under President Jacob Zuma's administration, political sympathizers were deployed to key positions within the SAPS to illegally spy on and intimidate political opponents and to squash criminal cases opened against Zuma and his allies (Steinberg, 2014; Pauw, 2018; Wiener, 2018; Chapter 3).

This chapter explores the code of silence in South Africa and considers it in the context of society at large. We elaborate on the influence of the mid-1990s transition, the quality of external and internal systems of control, and the "remilitarization" of policing on the extent of the code of silence. Based on the theory of police integrity and the related methodology, we present the results of our SAPS police officer survey and discuss what the data reveal about the contours of the code of silence.

The code of silence as a global phenomenon

Knowing something about the culture of a police organization helps in understanding the way officers make sense of their occupational world (O'Neil et al., 2007; Cockcroft, 2013). Policing scholars have long recognized that officers often feel disconnected and isolated from the broader society, and at the same time bound together with one another by their own shared experiences. Required to intrude upon the lives of others on behalf of the state, believing their work to be dangerous and to necessitate rule-breaking to enforce order (Brodeur, 2010), and thinking that the public is more likely to complain about them than to support them, the police develop informal rules to get the work done, and according to Manning, "cover their asses" (Manning, 2009).

Perhaps the best known of these mechanisms is the "code of silence," understood as a product of the occupation (Wilson, 1963) and the "working personality" it fosters in officers (Skolnick, 2011). Regularly scrutinized by the public, police officers find a measure of support and protection in the code of silence. It lets them know that their colleagues empathize with their plight and limits what their supervisors and the public know of their activities. Although the code of silence may improve morale for some officers, it also undermines police integrity. Honest, professional officers may be disinclined to report misconduct, while managers and those investigating complaints against the police may struggle to address wrongdoing (Wood, 1997). As one self-identified "passionate" SAPS officer told Faull (2010, loc4281):

> Four [officers] one night went to a [drug] house. They didn't find any drugs, but they took the guy's money—twelve thousand rand. They split it four ways. I heard about it; I knew about it, but when the internal investigators came and asked me if I would speak against them I said, "I know nothing". . . . Two weeks later they did it at another house; they stole fourteen or fifteen thousand. The guy opened a case and internal came to me and I said I knew nothing. But I knew.

Discretion and the use of force are also central to police culture and the code of silence (Waddington, 2005). The range of tasks that the police *may* be required to carry out during the course of a shift is immense. Although the constitutional mandate of the SAPS is crime- and safety-focused, officers can be called to assist in any emergency or any mundanity. As a result, the police are given tremendous discretionary autonomy to decide when and how to respond in each public interaction, including whether to detain someone or not. The decisions the police make in the course of their duties can be morally ambiguous or wrong. Force might be used when it should not be, a person might be arrested when they need not be, a violent spouse might be given a warning rather than be detained, a victim might be flirted with rather than empathized with, and a police vehicle might be driven recklessly for no justifiable reason. All of these behaviors breach the SAPS code of conduct, ethics, and standing orders. If officers believed that such actions would likely result in a swift sanction, they may be less inclined to pursue them. Conversely, when they do not expect their violations of the official rules to be reported by their colleagues and, hence, investigated by their supervisors, they are likely to be adhere to the code of silence themselves.

The consequences of a strong and entrenched code of silence are serious. To begin with, it is "perhaps the greatest barrier to effective corruption control," as the Mollen Commission (1994, Exhibit 6, p. 6) declared. Furthermore, a strong code of silence allows police misconduct not only to continue without any negative consequences but also to grow both in prevalence and severity. In addition, a strong code of silence separates the police from the community (e.g., Mollen Commission, 1994, p. 58), makes policing of police more challenging, and diminishes police accountability.

Although the code of silence manifests in all police agencies as a consequence of their semi-military organization (e.g., Klockars and Kutnjak Ivković, 2004), its contours—the strength of the code and the behaviors it covers—differ from agency to agency. Codes of silence also differ in terms of whom they benefit (Klockars et al., 2004, p. 8).

The code of silence and the theory of police integrity

Studying the code of silence by directly asking police officers to report fellow officers' misconduct is riddled with pitfalls (see, e.g., Klockars et al., 2000; Kutnjak Ivković et al., 2018). Klockars and Kutnjak Ivković (2004) developed the theory of police integrity and the accompanying methodology (see Chapter 1) to sidestep these challenges and allow the empirical measurement of the code of silence. The third dimension of their theory proposes that curtailing the code of silence is critical for police integrity because the code hinders effective control of the police. In police agencies of high integrity, the code is neither strong nor protects serious forms of misconduct (Kutnjak Ivković, 2015). On the other hand, in police agencies of low integrity, the code is strong and protects more serious forms of police misconduct.

In their first U.S. study, Klockars and colleagues (2000, 2006) found large and substantial differences in the code of silence across 30 police agencies. They used examples of two agencies on the opposite ends of the integrity spectrum to illustrate the differences in the integrity profiles (Klockars et al., 2004, p. 36):

> It is clear that in Agency 23 "The Code of Silence" is so strong that the officer who takes a kickback, a bribe, steals from a found wallet or a crime scene may do so without much worry that his police colleagues will expose his misconduct. By contrast, a police officer in St. Petersburg probably will not be reported for taking a free meal or a discount, accepting a holiday gift from a merchant, or for not reporting a police officer for driving under the influence, but every other offense described in the scenario runs a substantial chance of being revealed by a fellow police officer.

Subsequent studies confirmed that the code of silence differs not only across agencies within the same country (e.g., Hickman et al., 2016a, 2016b; Khechumyan and Kutnjak Ivković, 2019; Kutnjak Ivković and Sauerman, 2016; Lim and Sloan, 2016; Marche, 2009; McDevitt et al., 2011; Micucci and Gomme, 2005; Raines, 2010; Rothwell and Baldwin, 2007; Schafer and Martinelli, 2008) but also across police agencies from different countries (see e.g., Klockars et al., 2004; Kutnjak Ivković and Haberfeld, 2015), confirming that, indeed, the code of silence is a global phenomenon. In a co-edited book analyzing the contours of police integrity across 14 countries, Klockars and colleagues (2004) noted extensive differences in the code of silence across countries as diverse as Croatia, Japan, Pakistan, South Africa, Sweden, and the United States. Similarly, in a comparative overview of the results from ten countries, including Armenia, Estonia, Russia, South Africa, South Korea, Thailand, and the United States, Kutnjak Ivković and Haberfeld (2015) found extensive variations in the code of silence across these countries. Despite these differences, Kutnjak Ivković and Haberfeld (2015) also reported several common features of the code of silence across countries, notably that some forms of police misconduct (e.g., accepting free meals, verbally abusing citizens) would be protected by the code very strongly in most countries, while other forms of police misconduct (e.g., stealing from a crime scene, abusing deadly force) would be much less likely to be protected by the code.

Empirical research has consistently found that perceptions of misconduct seriousness and the extent of the code of silence are negatively related. Put differently, the more serious police officers evaluate police misconduct to be, the more likely they are to say that they would report it (e.g., Klockars et al., 1997, 2004, 2006; Kutnjak Ivković et al., 2013; Kutnjak Ivković et al., 2016; Kutnjak Ivković and Sauerman, 2013; Reyneke-Cloete and Meyer, 2016). Perceptions of seriousness are one of the strongest predictors of respondents' willingness to report (e.g., Hickman et al., 2016a, 2016b; Kutnjak Ivković et al., 2018; Long et al., 2013; Lim and Sloan, 2016; Porter and Prenzler, 2016; Westmarland, 2005).

Scholars have also linked the extent of the code of silence to the views about appropriate and expected discipline, not only of severity of discipline (e.g., Kutnjak Ivković et al., 2016; Reyneke-Cloete and Meyer, 2016) but also of perceptions of its fairness. Lim and Sloan (2016) found that "the strength of discipline" (i.e., disciplinary fairness) is related to the supervisors' willingness to report misconduct.

Although these organizational variables are important, by far the most significant predictor of whether the police officers would report misconduct is their perception of most police officers' willingness to report (Datzer et al., 2019; Kutnjak Ivković et al., 2016; Lim and Sloan, 2016; Long et al., 2013; Reyneke-Cloete and Meyer, 2016). The greater the officers' perception of the endorsement of reporting misconduct, the more likely they were to say that they would report misconduct themselves.

With the exception of the supervisory status, other demographic characteristics, such as gender and race, do not seem to be related to the code of silence (Hickman et al., 2016b; Huon et al., 1995; Klockars et al., 2006; Kremer, 2000; Kutnjak Ivković et al., 2016; Kutnjak Ivković and Klockars, 1998, 2000; Kutnjak Ivković and Shelley, 2008; Long et al., 2013; Lobnikar et al., 2016; Pagon and Lobnikar, 2000; Porter and Prenzler, 2016; Weisburd and Greenspan, 2000).

The code of silence in South Africa

The code of silence and transition to democracy

At the time of its creation in the mid-1990s, the SAPS, which was amalgamated from a number of different police agencies, including the SAP and homeland forces, consisted of "140,000 police officers who spoke different languages and had different rankings and uniforms, different firearms, and different levels of training, dedication, and commitment" (Newham, 2004, p. 235). The merge was further aggravated by the fact that about one-third of officers were functionally illiterate, two-thirds did not have drivers' licenses, about one-eighth (20,000) had criminal records (Mabuza, 2001), and because apartheid-era officers were allowed to remain employed (Newham, 2004). In addition, police training academies were desegregated only two years before the end of apartheid. Before this, black, white, colored, and Indian officers had been trained at separate facilities in different provinces. South African society, including its police, bore the brunt of a complex, hard, and unusual experience associated with desegregation (see Chapter 8). On the one hand, the police culture was so diverse that there probably was no common code of silence in the first years of the SAPS's existence. On the other hand, the spirit of national tolerance and acceptance of difference being promoted by the new government, as well as sensitivity to being labeled racist or prejudiced, may have prevented some officers from reporting misconduct. By all accounts, the informal standards of acceptable behavior were low, a large proportion of police officers could not read and may not have understood the official rules, torture had historically been used as a way to obtain confessions, there were no expectations of severe discipline for any violation of official rules, and many officers had engaged in behavior considered criminal under the former and/or new administration (Brogden and Shearing, 1993). All these circumstances were conducive to the development of a strong code of silence.

In 1996, the SAPS embarked on extensive recruitment efforts during which stringent screening and selection procedures were ostensibly neglected (Bruce, 2008, p. 8). By 2000, as many as 14,600 officers faced criminal charges "ranging from murder, rape, armed robbery, assault, theft and bribery to reckless driving" (Newham and Faull, 2011, p. 25). Clearly, the underlying idea that the strict adherence to the code of conduct and imposition of internal discipline would naturally establish standards and procedures that improved policing was "notably insufficient" (Vigneswaran and Hornberger, 2009, p. 6).

Based on ethnographic research, Hornberger (2011) has argued that SAPS officers were by 2010 still struggling to adapt to the middle-class, human-rights, and service-oriented paradigm of policing introduced after 1994. The result of some officers' inability to live up to reform expectations has been their retreat to the backstage of policing, where violence and abuse earn respect (Hornberger, 2011; Faull, 2017a, 2018; Perkins, 2018). Perkins describes the police she shadowed for much of a year in Cape Town as regularly deploying coercive violence without any fear of consequence (2018, pp. 195–207), suggesting the presence of attitudes and practices that can only be sustained by a solid code of silence. Similarly, Kinnes (2017) describes the lack of services provided to, and violence used against, suspected gang members by police. These various observations suggest that efforts to reform policing after apartheid may have unintentionally strengthened the code of silence, partly because police do not have the skills and/or resources to succeed in their work, and partly as a result of the impossibility of the challenges they face (Manning, 1979; Burger, 2006; Faull, 2018).

The code of silence, performance, and deceit

Since 2003, the SAPS has employed a sophisticated performance management system akin to the NYPD's CompStat, recognized as an example of good police practice (Davis, 2012). The system is intended to assess organizational efficiency by measuring accomplishments against targets. It is expected to help combat crime and guide corrective action at underperforming stations. As in other jurisdictions where police targets and metrics are common (Eterno et al., 2014; Sparrow, 2016), the SAPS's system, interpreted by station officers as emphasizing reported crime, has led to the dishonest manipulation of data, especially relating to reported crime and activities carried out, such as roadblocks held or people searched (Bruce, 2010, 2011; Faull, 2016, 2018).

When residents of Khayelitsha, Cape Town, in 2012 called for a commission of inquiry into a breakdown in the relationship between police and the community, stations serving the area were scoring fairly well in terms of the SAPS's performance metrics (O'Regan and Pikoli, 2014). But, as the inquiry revealed, policing at the three stations was dire. As such, a system intended to hold police accountable was, in fact, hiding incompetence.

Based on ethnographic fieldwork and volunteer police duties, Faull has described how street-level officers seek to quantify their activities, or fake the quantification thereof in the absence of actual activity, to satisfy their supervisors. He describes such practices as "data performance deceptions" (Faull, 2018, p. 80), with the falsified data being percolated up the SAPS hierarchy and presented to the public, effectively disseminating lies. Faull also describes "internal deceptions," referring to how officers deceive one another to minimize stress or risk to themselves. For example, in 2013–2014, 80% of SAPS's officers took an average of ten sick days—their full allocation. As a result, officers do not always believe that colleagues are really ill when they report being sick (Faull, 2018, p. 84). Furthermore, Faull (2018) describes "external deception" as the deceit of the public to protect officers' interests, such as telling a civilian that they must visit a different police station to report their crime so that an officer does not have to do the paperwork, although the crime can be reported at any station, including this specific station. The irony is that officers both mistrust and protect their colleagues (Faull, 2018, p. 88).

The code of silence and external control

The SAPS's failure (Chapter 5) and the state's failure (Chapter 3) to effectively sanction apartheid officers responsible for abuses may have weakened the organization's ability to deter subsequent violations. Abusive officers retained in the SAPS may have continued their practices, regardless of whether they violated official rules or not, passing them on to new recruits hired into the new organization. Although the 1996 Constitution required the establishment of various oversight bodies, these have not successfully curtailed the code of silence or ended endemic abuse (see Chapter 5). In fact, they might have had the opposite effect, providing a veneer of progress behind which violations continued.

The Independent Complaint Directorate (ICD; Chapter 3), established in 1997 and reformed in 2011, struggled to secure general cooperation from police for their investigators (Bruce et al., 2000). Understaffed and struggling to deal with its caseload, the ICD could only recommend to the SAPS, not compel, disciplinary action against officers. Any potential effect it had on the code of silence was likely diminished by this "toothlessness"—its inability to compel cooperation or action within the SAPS.

The Independent Police Investigative Directorate (IPID, see Chapter 3) was established out of the ICD in 2011 in an effort to address some of the latter's shortcomings. Established through its own Act (the IPID Act No. 1 of 2011) rather than the SAPS Act, as was the ICD, the new IPID was given a narrower, investigative—rather than complaints—mandate and more power to achieve it (e.g., police officers were required by law to cooperate with it and SAPS's managers had to initiate disciplinary proceedings based on IPID's recommendations). But, as we saw in Chapter 3, the IPID still handles relatively few cases every year given the size of the SAPS (about 5,000–7,000 complaints annually for a police agency of about 150,000 sworn officers; IPID Annual Report, 2017/2018) and struggles immensely with this load. In addition, only about 10% of investigations generally result in a conviction or a referral (with more than 90% of complaints declined, dismissed, or deemed to be unsubstantiated or unfounded; IPID Annual Report, 2017/2018; Chapter 3). In addition, in late 2019, investigative journalists accused the IPID of prematurely and arbitrarily closing cases against the police in order to meet its performance targets (Viewfinder, 2019), allegedly falling victim to the same "data performance deceptions" carried out within the SAPS in general (Faull, 2018).

The code of silence and internal control

As explained in Chapter 3, the SAPS established the Anti-Corruption Unit (ACU) in 1996, as part of early police reform efforts. Its tasks were to investigate corruption within the SAPS, raise awareness about corruption, identify dysfunctions within the organization, and maintain an information system in support of organizational integrity (Newham and Faull, 2011). Despite its relative success (e.g., Burger and Grobler, 2017; Newham and Faull, 2011), National Commissioner Jackie Selebi, who was himself subsequently convicted of corruption, closed the ACU in 2002. The ACU's mandate was transferred to the SAPS's Organized Crime Unit, and later to the Directorate for Priority Crime Investigations (DPCI, or "Hawks"), both of which have been implicated in corruption and abuse themselves (see Chapter 5). Similarly reflecting on the ACU's demise and what followed for SAPS's internal governance, Bruce (2008, p. 9) observes that "not only has the SAPS actively undermined its corruption control mechanisms but it has done so whilst management systems, which constitute the basic mechanism of control, have been undermined."

Deception and the related code of silence function not only at the street level but also at the level of high policing. Some of the SAPS's top managers have been found lying (e.g., O'Regan and Pikoli, 2014; Farlam, 2015). When residents of Khayelitsha in Cape Town sought to initiate a commission of inquiry into a breakdown in relations with police, the Minister of Police and senior police managers in the province fought vociferously to block the commission's work. As noted in the Commission's report (O'Regan and Pikoli, 2014), when residents first complained to the SAPS, the administration ignored them. When the SAPS Head Office eventually responded with an internal investigation, it was shallow, did not consult communities, and did not result in any change. When the Commission began its work, its requests to the SAPS were ignored, forcing it to subpoena police managers. Then, the Minister of Police sought to block the Commission through a court intervention, but failed. All these actions are indicative of an effort to protect the organization and its members from being exposed to public scrutiny. They form the very essence of the code of silence.

Another example of the code of silence functioning at the very highest levels in the SAPS was revealed at the Farlam Commission of Inquiry, which investigated the SAPS's killing of 34 striking mine workers and injuring of over a hundred more at Marikana in 2013 (Farlam, 2015). The commission found that the SAPS leadership, headed by National Commissioner Riyah Phiyega, conspired at a special retreat, held in the aftermath of the massacre, to tailor a version of events that would exonerate the organization and its officers of wrongdoing (Farlam, 2015). The commission described senior officers' lack of cooperation with it as producing a "feeling of absolute despair," reflecting that "[t]hese are the most senior people in the South African Police Service. They're asked some very important questions . . . [about] the greatest catastrophe since we achieved democracy, and the answers are evasive and they are non-responsive" (Farlam, 2015, p. 184). The Farlam report described sworn statements submitted by officers involved in the shooting as "all [singing] much of a tune and they don't offer you much of a clue . . . they're a frustrating bunch of statements" (Farlam, 2015, p. 517).

Additionally, South Africa's courts—the most respected of its criminal justice institutions—described former National Commissioner Jackie Selebi as having "complete contempt for the truth" (Joffee, 2010, p. 211) and former "Hawks" head Berning Ntlemeza as being "biased and dishonest . . . [lacking] integrity and honor [and making] false statements under oath" (Sapa, 2015). Though not necessarily a fair reflection of all top managers, these examples provide little reason to think that the SAPS's code of silence is likely to be challenged from above.

The code of silence and high policing

Another reason to doubt that senior managers will weaken the SAPS's code of silence emerges when one probes the state of high policing in the organization. Defined by Brodeur (1983), high policing refers to the clandestine intelligence work carried out to protect governments from internal threats. Brodeur did not necessarily see high policing as illegitimate. Instead, he found it an inevitable element of modern policing that would increase with technological developments, and one that should be kept in check through special accountability mechanisms because of its secretive nature.

At the time of Brodeur's initial writing (1983), the apartheid state was heavily invested in covert policing in defense of the state. According to Cawthra, the SAP's Security Branch was at the heart of, and guided, the country's policing project (Cawthra, 1993; also Steinberg, 2017). This was possible, in part, because regular "low" police service to the public was considered less important than "high" policing (see Chapter 2). The provision of regular police services to the majority black population was seldomly prioritized, freeing up significant resources for investment in high policing and white neighborhoods.

The abuse of high policing did not end in 1994, though it may have died down for much of the 1990s as policing was restructured and uniformed officers assumed the center of the new community-policing project (Steinberg, 2017). However, from the early 2000s the African National Congress (ANC) increasingly abused the state's criminal justice institutions, notably its police, to wage internal battles against political opponents (Steinberg, 2012) and to protect its industrialist allies intent on getting rich through government contracts and the looting of the state (Steinberg 2017; Myburgh, 2019). Examples include the restructuring of the SAPS by, and conviction of, Jackie Selebi, the closure of the Scorpions, and the timing and dropping of corruption charges against Jacob Zuma.

Under President Zuma, the SAPS lost almost all credibility as one National Commissioner after another was dismissed or suspended for misconduct (see Chapter 3), dubious officers were elevated to key positions through which they allegedly protected the president and his criminal supporters (Pauw, 2018), and Police Minister Nathi Nhleko embarrassed himself trying to convince the Parliament that a swimming pool paid for with taxpayers' money at the president's private residence was in fact a safety feature intended to put out fires (PMG, 2015). During this period, particular abuse occurred within the SAPS's Crime Intelligence (CI) division, where, as predicted by Brodeur (1983), technological developments gave its officers access to more information and surveillance capabilities than other police. Combined with schemes developed by CI officers to plunder funds intended for the payment of informers, together with a near absence of accountability in their work, CI became a sort of criminal syndicate itself.

In her book, *Ministry of Crime* (rather than of "police"), investigative journalist Mandy Wiener quotes a former SAPS General, Solomon Makgale, who put the abuse of high policing under Zuma in the starkest of terms (Wiener, 2018, loc 6608):

> The higher echelons of the criminal justice system are in a mess. Most [police] leaders are associated with certain politicians. . . . In exchange for money and favors, such as promotions and political protection, they are prepared to look the other way when there is crime committed. . . . It is about politicians who want certain investigations to disappear—their children, girlfriends, wives, business partners, family members—and investigations against political opponents to be intensified. Business people also do the same. They can pay cops to do dirty work for them, including commercial espionage. So it is the nexus of bent cops, corrupt politicians and shady business people that compromises police investigations into organized crime. Some junior officers see the mess at the top and decide that they should also cash in. Consequently, there is no political will to sort out the mess that SAPS Crime Intelligence and the Hawks have become.

The code of silence, remilitarization, and trust in the police

In 2010, the SAPS announced that it would revert to the military ranks of apartheid's SAP (Burger, 2010, p. 42; Faull, 2013, p. 22). This effort was one of the most visible steps in a gradual "remilitarizing" process taking place at the time (Lamb, 2018; Chapter 3). Shortly thereafter, in August 2012, SAPS officials, including members of the Tactical Response Team (TRT), shot and killed 34 striking mine workers and injured over 100 more at Marikana. The subsequent Commission of Inquiry found that police were ill-trained and incorrectly equipped for the crowd management task (Farlam, 2015). Accompanying this apparent, if unofficial, remilitarization of the SAPS has been an increase in civil claims (lawsuits) against the police and payouts by the department, as well as a steady flow of deaths, assaults, and torture at the hands of the police (see Chapter 3). This trend may indicate an increasingly entrenched culture of rule-breaking protected by a strong code of silence.

In 2017–2018, the SAPS initiated a management intervention intended to curb civil claims. The intervention sought to address their root causes, including unlawful arrest and detention, assault and shooting incidents, as well as political condonation of police violence (SAPS Annual Report, 2018). The latter point is particularly important because it suggests that remilitarization of the SAPS has been at least partly politically driven. Indeed, while the NDP and White Paper on Policing suggest an official government stance that rejects militarized policing, ambiguous "tough on crime" talk from leaders remains fairly common. This approach directly undermines a new strategic direction toward professionalism and accountability by contributing toward strengthening the code of silence and thereby threatening to further damage trust between the police and the public. Relatedly, in November 2019, the SAPS issued an internal instruction to all officers to "re-iterate certain principles that govern the arrest and detention of person and to, once again, explain how those principles should be applied in practice" (SAPS, 2019, p. 2). The inference was that officers did not sufficiently understand when they may arrest or detain people and that these powers were being abused. A more cynical interpretation would be that they understood the formal rules, but that the code of silence allowed them to ignore these rules.

By protecting abusive officers, the strength of the code of silence in the SAPS undermines trust in the organization. In a national sample of 1,300 South Africans surveyed every month since July 2015, trust in the police ranges from a high of 56% in late 2016 to a low of 40% in the third quarter of 2018. According to the survey, just 46% of respondents trusted the police in the third quarter of 2019 (Citizen Surveys, 2019). The same survey found that by late 2019, 72% of respondents believed that the level of corruption in the country as a whole was increasing.

In a 2018 survey of adult South Africans, 53% of the respondents felt public officials did not treat them with respect (Research, 2018). Fewer people trusted the police than any other institution, with 66% of the respondents saying they trusted the police "just a little" or "not at all." The same survey found that 65% of the respondents believed corruption had increased in the preceding year (Research, 2018). Highlighting the harms of South Africa's inequality, most felt that "a rich person" was more likely than "an ordinary person" to be able to pay a bribe to avoid taxes (80% v. 51%), going to court (83% v. 57%), or to register land that did not belong to them (83% v. 54%).

The survey also revealed clear correlations between trust in institutions and beliefs about institutional corruption (Research, 2018). The more corrupt South Africans believe officials to be, the less likely they are to trust them. Forty-nine percent of respondents believed that "all" or "most" police were involved in corruption, more than any other institution or actor (Research, 2018). Experiences of corruption are even more worrisome. Of the 25% of respondents who requested assistance from police in the year preceding the survey, 12% paid a bribe, gave a gift, or did a favor to get police assistance. Of the 27% of respondents who interacted with police at a checkpoint, during a traffic stop, investigation, or a related encounter, 26% gave a bribe, gift, or favor to avoid a problem (Research, 2018).

Extant research regarding trust in the public among South African police officers reveals the reverse side of this coin regarding the police-community relationship. In a study of the first-year police officers, Meyer and Steyn (2009) found that 62% of the respondents felt that they identified ("belonged") more closely with their police colleagues than with the people they policed. Ninety-three percent believed that the public watched them critically, even when off-duty. Further, Meyer and Steyn (2008) found that 73% of respondents believed that most people lack the proper level of respect for police officers, and almost one-half of respondents felt that police would never trust community members enough to work together effectively. Among the same group of officers, Steyn (2017) observed in a follow-up study ten years later that 80% of respondents felt even closer to their colleagues and more distant from the people they policed, than they had in the initial study. Ninety-eight percent in the second study believed that they were being watched critically by members of the public, even when off-duty. Steyn (2017) also observed that the proportion of respondents in the follow up study who believed that most people lack respect for police had dropped to 65%, and the proportion of respondents who believed that police officers would never trust the community enough to work together effectively had declined to about 45%. Steyn and Mkhize (2016) also compared the attitudes of SAPS officers with 10, 20, and 30 years of police service, respectively, and found few differences in the opinions, attitudes, and behavioral indicators measured. Steyn and Mkhize (2016, p. 24) concluded that the officers saw their work as:

> [T]aking place in a dangerous and uncertain environment, highly skilled and with moral purpose, and only suitable for unique individuals with

> characteristics such as toughness and suspiciousness. Groups outside of the police have very little understanding of police work, as reflected in unsatisfactory monetary compensation, cockeyed criticism and ill-considered prescriptions. These police officials isolate themselves from outsiders (friends, family members/important others, community, courts and top-ranking officials), in favor of their colleagues. They believe most people lie when answering questions asked by police officials, would thieve if they knew they would not be caught, are untrustworthy and dishonest, not perturbed by the cries for help of others, unlikely to praise police, and resistant to the opinions and advice of police officials.

In sync with South Africa's ethnographic record, the data suggest that distrust of the public is common within the SAPS.

The code of silence and the struggle for dignity

SAPS officers' perceptions that the public does not respect them are shaped by another core element of organizational and national culture bearing on the code of silence. Many SAPS officers feel disrespected not only because of their association with the SAPS, but because they also feel uncertain about the degree to which they have "succeeded," particularly male officers, in terms of the narrative of the new South Africa. To be a successful man in South Africa means earning good money, being virile, showing leadership, and being physically and mentally tough (Ratele, 2008). Most police are recruited from lower-middle to low/no-income households and areas with high rates of poverty and violence. Hardship, unemployment, violence, and crime have been part of many officers' lives long before they joined the SAPS.

As noted in the ethnographies of Steinberg (2008), Faull (2017b, 2018), and Perkins (2018), most SAPS officers struggle to realize their financial aspirations, despite the relatively good incomes earned in the SAPS. Central to this failure is the fact that many officers are the sole breadwinners in networks of precarious kin, so that their incomes must be shared widely. To lose the job through disciplinary action is to leave one's precarious network to poverty.

Thus, the code of silence is strengthened by an additional layer of peer empathy. Even if officers disapprove of their colleagues' actions or abuses, they may hesitate to report them if only for the officers' family to maintain their access to income and health insurance in a country where dismissed officers are unlikely to find comparably remunerated or secure work again.

The contours of the code of silence

Apart from the work of Steyn and colleagues discussed above, empirical research on the code of silence in South Africa is limited. There were only a handful of studies assessing the code of silence, and they primarily relied on the police corruption questionnaire that measured only one type of police misconduct—police corruption (see Chapter 1). Also, most studies do not include national samples, instead relying on local, typically very small samples (e.g., one police station, one province).

In 2004, Newham measured the contours of the code of silence among officers at a single station in (2004). Based on the 104 respondents (or about one-half of the police officers from this station), Newham found that the majority of officers would not report the misconduct described in most corruption scenarios (6 out of 11). In five scenarios, featuring examples of police corruption evaluated as the most serious in the questionnaire, most respondents, though at times barely more than 50%, said that they would report the violation (Newham, 2004, p. 246). With one exception, the majority of officers did not expect that their colleagues would report the corruption described in *any* of the scenarios (Newham, 2004), suggesting an immensely strong code of silence. Newham (2004, p. 247) concluded that "[i]f most police officers do not believe that reporting the misconduct of colleagues will result in adequate disciplinary action being taken, then they are less likely to report misconduct themselves."

In the 2005 study of 379 SAPS supervisors, Kutnjak Ivković and Sauerman (2011) used the same police corruption questionnaire (see Chapter 1) and also detected a strong code of silence. In fact, Kutnjak Ivković and Sauerman (2013, p. 44) noted that,

> [f]or 9 out of 11 scenarios, the mean values of their own reporting practices are clustered very closely around the mid-point of the scale (between 2.5 and 3.5), suggesting the existence of a strong code of silence and a relatively narrow variation in their expressed willingness to report.

Even among police supervisors—those in the Kutnjak Ivković and Sauerman (2011) study—the code of silence seemed to be strong. In addition, the respondents assumed that their fellow officers would be even less likely to report than they themselves were to report misconduct (Kutnjak Ivković and Sauerman, 2011, p. 84).

Matlala and colleagues (2016) also used the police corruption questionnaire to measure police integrity in Gauteng province. Based on a sample of SAPS officers, municipal police, and traffic police, they reported that officers would not be willing to report instances of police criminality as serious as a theft from a crime scene, accepting a bribe from a motorist, and stealing money from a found wallet. The differences between their respondents' own expressed willingness to report and the estimates of most police officers' willingness to report were mostly small. In another study from the same province, Meyer and colleagues (Meyer et al., 2013; Reyneke-Cloete and Meyer, 2016) utilized the police corruption questionnaire to survey 160 non-commissioned officers. Although the code of silence which they documented seemed to be weaker than that identified by others, fewer than 75% of their respondents in any scenario said that they would report the violations.

Survey findings

Our research team surveyed 868 SAPS officers across the country and asked them to evaluate 14 scenarios describing hypothetical cases of police misconduct, including examples of police corruption, use of excessive force, falsification of official documents, and planting of evidence. The questionnaire was designed by Klockars and colleagues (2006) and revised by Kutnjak Ivković (2009). It has been successfully used in over a dozen countries (see, e.g., Kutnjak Ivković, 2015). For details about the methodology, please see Chapter 1. Upon reading the description of these hypothetical cases of police misconduct, the respondents were asked a series of questions. In particular, they were asked to assess misconduct seriousness, recognize whether misconduct violated official rules, estimate appropriate and expected discipline within the SAPS, and indicate their express willingness to report.

In light of the focus of this chapter on the code of silence, we will discuss the results primarily as they relate to the respondents' willingness to report misconduct. The first question about the code of silence asked respondents whether they would report a police officer who engaged in such behavior ("Do you think you would report a fellow police officer who engaged in this behavior?"), while the follow-up question asked them to estimate whether the majority of police officers would do the same ("Do you think most police officers in your agency would report a fellow police officer who engaged in this behavior?"). The respondents were then instructed to select one answer from a 5-point scale ranging from 1 = "definitely not" to 5 = "definitely yes."[1]

Our results from these questions are presented in Figures 6.1 and 6.2. The bars in the two figures present the code of silence, measured as the percent of respondents who said that they would not report. The scenarios in the figures are not shown in the order in which they were included in the questionnaire. Rather, they have been reordered on the basis of the percent of police officers who would not report, ranging with the scenario with the highest percent of the respondents who said that they would not report (Scenario 7) to the scenario with the lowest percent of the respondents who said that they would not report (Scenario 13).

Figures 6.1 and 6.2 show that the code of silence does not create a flat prohibition on reporting. Instead, the variation in the height of the bars across scenarios suggests that the contours of the code of silence vary greatly across scenarios. On the one hand, there are about five scenarios, Scenario 7 (Verbal abuse), Scenario 8 (Cover-up of police DUI), Scenario 14 (Failure to react to graffiti), Scenario 1 (Acceptance of gratuities), and Scenario 9 (Kickback), in which about one-quarter of the respondents said that they would not report (Figure 6.1), and over one-third of the respondents expected that most police officers would not report either (Figure 6.2). On the other hand, there are a few scenarios, such

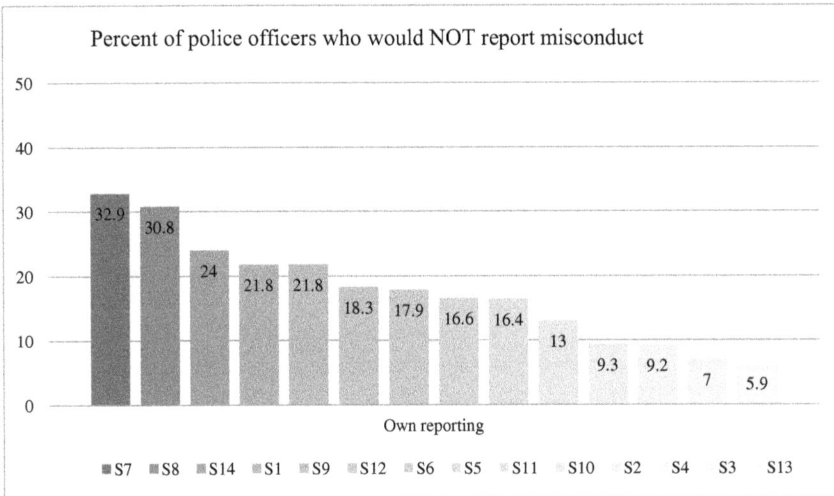

Figure 6.1 Own reporting by scenario

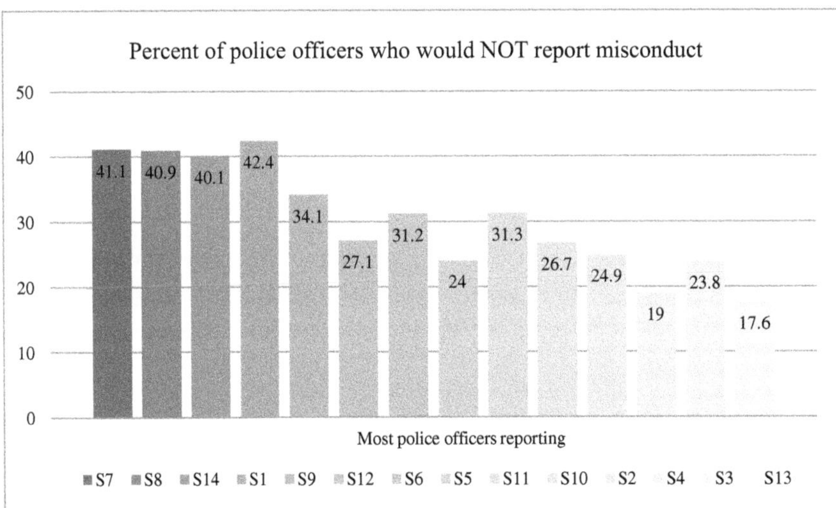

Figure 6.2 Most police officers' reporting by scenario

as Scenario 2 (Failure to arrest), Scenario 4 (Abuse of deadly force), Scenario 3 (Theft of knife), and Scenario 13 (Bribe from reckless motorist), in which one-tenth of the respondents said that they would not report (Figure 6.1), and a few scenarios, such as Scenario 3 (Theft of knife) and Scenario 13 (Bribe from reckless motorist), in which one-fifth of the respondents expected that most police officers would not report. These two opposite sides indicate where the code of silence is the strongest and the weakest, respectively.

Own willingness to report

A more detailed analysis of the code of silence, measured as the respondents' own expressed (un)willingness to report misconduct, is shown in Table 6.1 and Figure 6.1. While there were no scenarios in which the mean score fell between the end of the non-reporting side of scale (value of 1) and the midpoint of the scale (value of 3), there are three scenarios with means close to the midpoint (Scenario 7: Verbal abuse; Scenario 8: Cover-up of police DUI; Scenario 14: Failure to react to graffiti), all of which describe examples of police misconduct evaluated as the least serious in the questionnaire (see Chapter 4). The code of silence is the strongest for these scenarios.

Next is a group of scenarios (Scenario 1: Acceptance of gratuities; Scenario 5: Supervisor corruption; Scenario 6: Striking arrestee; Scenario 9: Kickback; Scenario 11: Supervisor fails to stop excessive force; and Scenario 12: Failure to report hate crime) for which the mean values are somewhere between 3.5 and 4, implying that the code of silence would not protect them as strongly as it would protect the scenarios with means closer to the midpoint of the scale. These middle-of-the-road scenarios constitute a heterogeneous group, diverse in terms of the types of police misconduct (e.g., police corruption, use of excessive force), severity of police misconduct within the same type (e.g., accepting gratuities v. accepting kickbacks), level of activity (e.g., hitting an arrestee v. omitting to report a relevant fact), and the status of the person who engages in such behavior (e.g., line officer v. supervisor).

Finally, there was a group of five scenarios (Scenario 2: Failure to arrest; Scenario 3: Theft of knife; Scenario 4: Abuse of deadly force; Scenario 10: False report on drugs; Scenario 13: Bribe from reckless motorist) with the means between 4 and 4.5, getting closer to the reporting side of the scale. These five scenarios have been evaluated as the most serious examples of police misconduct in the questionnaire (see Chapter 4) and incorporate severe abuses of police powers, such as an opportunistic theft, abuse of deadly force, and falsifying official reports, all of which, at the same time, violate norms of criminal law. The scenarios in this group appear least likely to be protected by the code of silence.

The respondents' expressed willingness to report misconduct and their evaluations of seriousness are strongly positively correlated (Spearman's rho = .969, $p < .001$). Put differently, when the respondents evaluate the misconduct as serious, they are less likely to say that they would adhere to the code of silence.

Most police officers' willingness to report

The second question about the code of silence asked the respondents to provide an estimate of how willing most SAPS officers would be to report the misconduct described in the scenarios. The results are shown in Table 6.1 and Figure 6.2.

Table 6.1 Own reporting and most police officers' reporting by scenario

Scenario	*Own reporting*			*Most police officers' reporting*			*t-test*	*Diff.*	*Cohen's d*
	Mean	*St. dev.*	*Rank*	*Mean*	*St. dev.*	*Rank*			
Scenario 1: Acceptance of gratuities	3.70	1.401	4	2.92	1.499	2	15.93***	.78	0.54 Medium
Scenario 2: Failure to arrest	4.21	1.112	11	3.46	1.351	10.5	17.12***	.75	0.61 Medium
Scenario 3: Theft of knife	4.39	1.031	13	3.61	1.361	12	17.57***	.78	0.65 Medium
Scenario 4: Abuse of deadly force	4.35	1.101	12	3.84	1.349	13	13.89***	.51	0.41 Small-medium
Scenario 5: Supervisor corruption	3.88	1.309	8	3.46	1.374	10.5	12.10***	.42	0.31 Small
Scenario 6: Striking arrestee	3.79	1.313	6	3.27	1.411	6	13.49***	.52	0.38 Small
Scenario 7: Verbal abuse	3.31	1.481	1	2.95	1.476	3	10.78***	.36	0.24 Small
Scenario 8: Cover-up of DUI	3.45	1.557	3	3.01	1.524	4	12.18***	.44	0.29 Small
Scenario 9: Kickback	3.75	1.437	5	3.26	1.487	5	12.73***	.49	0.33 Small
Scenario 10: False report on drugs	4.06	1.215	10	3.43	1.403	9	15.89***	.63	0.48 Small-medium
Scenario 11: Supervisor fails to stop excessive force	3.91	1.294	9	3.35	1.427	7	13.51***	.56	0.41 Small-medium
Scenario 12: Failure to report hate crime	3.81	1.260	7	3.37	1.333	8	11.71***	.44	0.34 Small
Scenario 13: Bribe from reckless motorist	4.46	.957	14	3.85	1.344	14	14.99***	.61	0.52 Medium
Scenario 14: Failure to react to graffiti	3.36	1.317	2	2.88	1.397	1	14.07***	.48	0.35 Small

Note: Cohen's *d* calculated at Becker, Effect size calculators, available at: www.uccs.edu/lbecker/

Four scenarios (Scenario 1: Acceptance of gratuities; Scenario 7: Verbal abuse; Scenario 8: Cover-up of police DUI; Scenario 14: Failure to react to graffiti) featured a mean score between the non-reporting side of scale (value of 1) and the midpoint of the scale (value of 3). Such mean values imply that the respondents thought that fellow officers would not be willing to report these examples of police misconduct, which were also evaluated as least serious by the respondents themselves (see Chapter 4).

An additional seven scenarios (Scenario 2: Failure to arrest; Scenario 5: Supervisor corruption; Scenario 6: Striking arrestee; Scenario 9: Kickback; Scenario 10: False report on drugs; Scenario 11: Supervisor fails to stop excessive force; Scenario 12: Failure to report hate crime) featured the mean values along the reporting side of the scale, though still near the mid-point (between 3 and 3.5). These scenarios include examples of serious corruption (e.g., a kickback), use of excessive force (e.g., hitting an arrestee), falsification of official reports (e.g., padding a report), and failure to execute an arrest warrant. Their seriousness notwithstanding, the respondents assumed that their colleagues would quite likely not report such misconduct.

Finally, only three scenarios (Scenario 3: Theft of knife; Scenario 4: Abuse of deadly force; Scenario 13: Bribe from reckless motorist) featured means closer to the reporting side of the scale (3.5–4), suggesting that the code of silence would not apply to them as forcefully as it would to the other scenarios. However, no scenarios featured means above 4, implying that respondents did not think that any of the misconduct examples in the questionnaire would definitely be reported by their fellow officers.

Our respondents' estimates of most officers' willingness to report were closely related to how seriously they perceived most officers would evaluate these forms of misconduct (Spearman's rho = .904; $p < .001$).

Comparing own and most officers' willingness to report

The next step in our analysis focuses on the comparison of the respondents' own expressed willingness to report and their estimates of most officers' willingness to report (Table 6.1). In all 14 scenarios, the respondents thought that they would be more likely than most officers to report the misconduct: In all scenarios, the mean values of their own reporting are significantly higher than the mean values of their estimates of most officers' reporting. To address the issue that these discrepancies might stem from self-serving bias, we will focus not only on the existence of the differences, but also on their magnitude.

Klockars and colleagues (2006, p. 26) developed a rule of thumb that, in the context of 5-point scales, the calculated differences in means of 0.5 or larger should be regarded as meaningful. An application of this rule to our data yields large and meaningful differences between the two means (i.e., own estimates v. most officers' estimates) in the majority of the scenarios (8 out of 14 scenarios; Table 6.1). In addition, there are five scenarios with differences getting close to the cut-off of 0.5 (Table 6.1). Another assessment of the magnitude of these differences relies on a calculation of Cohen's *d* (Table 6.1), a statistic based not only on the difference in means, but also on the standard deviation. In addition to the small differences in seven out of 14 scenarios, there were also small to medium-size differences in three scenarios and medium-size differences in four scenarios. In other words, Cohen's *d* implies that the differences were either of medium size or small-to-medium size in about one-half of the scenarios.

Own reporting and discipline fairness

To explore whether respondents' own expressed willingness to report is related to how they evaluate the fairness of expected discipline, we relied on the theoretical model developed by Kutnjak Ivković and Klockars (1998), that postulates that the relation between the code of silence and discipline fairness could take one of three forms (Figure 6.3). The first possibility, *the discipline indifference model*, assumes that there is no relation between the estimates of discipline fairness and adherence to the code of silence and that factors other than perceptions of discipline fairness influence the respondents' willingness to report misconduct. The second possibility, *the simple deterrence model*, assumes that the more severe the discipline, the less likely police are to adhere to the code of silence. The third model, *the simple justice model*, assumes that the more severe the discipline, the more likely officers are to adhere to the code of silence. However, the simple justice model only covers the comparison in views of those officers who evaluated discipline as fair with the officers who evaluated the discipline as too harsh (Kutnjak Ivković and Klockars, 1996).

Table 6.2 contains the results of our empirical analyses. In most scenarios, there were neither statistically significant nor large and meaningful differences

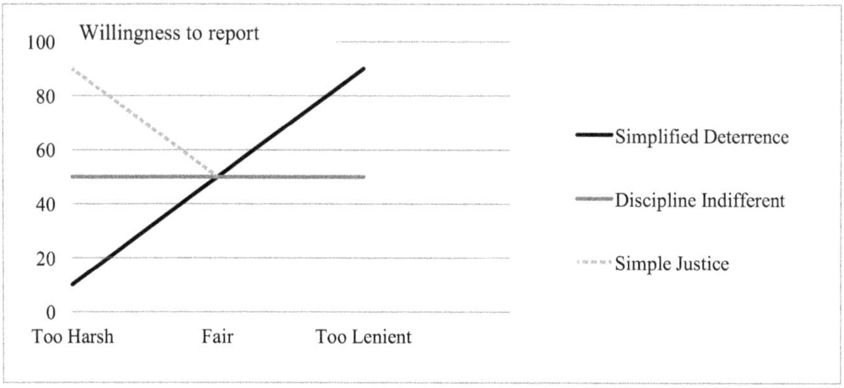

Figure 6.3 Relation between evaluations of discipline fairness and willingness to report misconduct

Table 6.2 Own reporting by perceptions of disciplinary fairness

Scenario	*Too harsh*		*Fair*		*Too lenient*		*F-test*	*Harsh v. fair*		*Fair v. lenient*	
	Mean	*St. dev.*	*Mean*	*St. dev.*	*Mean*	*St. dev.*		*Diff.*	*Cohen's d*	*Diff.*	*Cohen's d*
Scenario 1: Acceptance of gratuities	3.94	1.399	3.64	1.414	3.69	1.294	2.74	.30	.21 Small	−.05	−.04 Small
Scenario 2: Failure to arrest	4.27	1.195	4.24	1.073	3.98	1.262	2.49	.03	.03 Small	.26	.22 Small
Scenario 3: Theft of knife	3.88	1.418	4.44	.971	4.45	.891	11.76*** (H v. F)	−.56	−.46 Small-med.	−.01	−.01 Small
Scenario 4: Abuse of deadly force	4.02	1.324	4.41	1.078	4.00	1.009	6.95** (H v. F; F v. L)	−.39	−.32 Small	.41	.39 Small-med.
Scenario 5: Supervisor corruption	4.11	1.134	3.88	1.337	3.62	1.186	2.76	.23	.19 Small	.26	.21 Small
Scenario 6: Striking arrestee	3.59	1.474	3.83	1.282	3.59	1.416	2.08	−.24	−.17 Small	.24	.18 Small
Scenario 7: Verbal abuse	3.37	1.606	3.30	1.480	3.43	1.294	.28	.07	.05 Small	−.13	−.09 Small
Scenario 8: Cover-up of DUI	3.75	1.606	3.41	1.568	3.45	1.344	2.10	.34	.21 Small	−.04	−.03 Small
Scenario 9: Kickback	3.71	1.350	3.83	1.409	3.30	1.593	6.41** (F v. L)	−.12	−.09 Small	.53	.35 Small-med.
Scenario 10: False report on drugs	3.80	1.303	4.07	1.227	4.16	.949	2.00	−.27	−.21 Small	−.09	−.08 Small
Scenario 11: Supervisor fails to stop excessive force	3.46	1.566	3.96	1.279	4.03	.962	5.76** (H v. F)	−.50	−.35 Small-med.	−.07	−.06 Small
Scenario 12: Failure to report hate crime	3.39	1.317	3.87	1.228	3.66	1.446	6.34** (H v. F)	−.48	−.38 Small-med.	.21	.16 Small
Scenario 13: Bribe from reckless motorist	4.35	.988	4.54	.924	3.96	1.005	16.01*** (F v. L)	−.19	−.20 Small	.58	.60 Medium
Scenario 14: Failure to react to graffiti	3.62	1.250	3.30	1.343	3.69	1.026	4.49* (no pairs)	.32	.25 Small	−.39	−.33 Small

Note: Cohen's *d* calculated at Becker, Effect size calculators, available at: www.uccs.edu/lbecker/

in the respondents' expressed willingness to report across the spectrum of discipline fairness evaluation. Put differently, in eight out of 14 scenarios, the respondents' perceptions of disciplinary fairness were not related to their own willingness to report. Thus, most scenarios fit the discipline indifference model.

There are statistically significant differences in six scenarios (Scenario 3: Theft of knife; Scenario 4: Abuse of deadly force; Scenario 9: Kickback; Scenario 11: Supervisor fails to stop excessive force; Scenario 12: Failure to report hate crime; Scenario 13: Bribe from reckless motorist). The differences in these, measured through Cohen's *d*, ranged from small to medium-size to medium-size differences. Using the Klockars et al. (2006) rule of thumb, these differences were large and meaningful in most of these scenarios as well.

In four scenarios (Scenario 3: Theft of knife; Scenario 4: Abuse of deadly force; Scenario 11: Supervisor fails to stop excessive force; Scenario 12: Failure to report hate crime), the respondents who evaluated the expected discipline as too harsh were less likely to say that they would report than those who evaluated the discipline as fair. In these four scenarios, the simple justice model fits the reporting preferences of those who evaluated the expected discipline as too harsh and those who evaluated discipline as fair. In three scenarios (Scenario 4: Abuse of deadly force; Scenario 9: Kickback; Scenario 13: Bribe from reckless motorist), some of the most serious behaviors in the questionnaire, the simple deterrence model fits the data well because the respondents who evaluated the discipline as too lenient were less likely to say that they would report than those respondents who evaluated the discipline as fair.

Factors related to own willingness to report

The last set of analyses, shown in Table 6.3, explores factors that might be related to respondents' (un)willingness to report. We have included in the model two demographic variables (supervisory status, length of service), organizational variables (perceptions of seriousness, recognition of behavior as a violation of official rules, expected discipline, perceptions of disciplinary fairness), and one cultural variable (most officers' willingness to report).

The results show that *the* most important factor influencing respondents' own willingness to report in all 14 scenarios is their belief that most police officers would report as well (Table 6.3). Respondents who thought that most officers would report were more likely to say that they would report too.

Organizational variables mattered as well. Perceptions of misconduct seriousness were strongly and positively related to the respondents' own expressed willingness to report in the overwhelming majority of scenarios (12 out of 14 scenarios; Table 6.3). Similarly, severity of expected discipline was relevant; compared to the respondents who expected either no discipline or only a verbal reprimand, the respondents who expected some discipline other than dismissal had much higher odds of saying that they would report as well (12 out

Table 6.3 Logistic regression of own reporting[1]

	Scenario 1 (Gifts) Free Meals, Gifts from Merchants			*Scenario 2 (Warrant)*			*Scenario 3 (Theft)*			*Scenario 4 (Deady Force)*		
	B	s.e.	*Odds*	*B*	s.e.	*Odds*	*B*	s.e.	*Odds*	*B*	s.e.	*Odds*
Supervisory status[2]	−.086	.295	.917	.175	.378	1.192	2.768***	.570	15.927	−.557	.550	.573
Length of service[3]												
6–15 years	.932**	.317	2.539	−.202	.424	.817	−1.011	.543	.364	1.861**	.586	6.431
Over 15 years	1.360***	.374	3.896	−.165	.454	.848	−1.218*	.582	.296	1.774**	.642	5.897
Own seriousness[4]	1.475***	.255	4.372	1.461***	.379	4.310	2.248**	.668	9.470	1.445*	.723	4.241
Knowledge of rules[5]	1.388***	.260	4.005	.893*	.351	2.442	−.370	.637	.691	.342	.639	1.408
Expected discipline[6]												
Some discipline	1.390***	.330	4.015	2.375***	.387	10.749	1.993**	.635	7.337	1.625**	.590	5.080
Dismissal	1.858***	.562	6.411	1.304**	.429	3.685	2.445***	.664	11.536	3.433***	.675	30.980
Discipline fairness[7]												
Too harsh	−.897*	.375	.408	−.826	.446a	.438	−1.649***	.459	.192	−.742	.764	.476
Too lenient	.420	.345	1.522	.846	.451a	.2.331	2.304*	.947	10.011	3.800**	1.205	44.690
Most officers' reporting[8]	2.782***	.285	16.155	1.528***	.309	4.610	3.549***	.419	34.765	6.691***	.872	804.94
Constant	−2.957	.369		−1.681	.466		−2.294	.791		−4.525	.861	
Model Nagelkerke R Squared	.570			.420			.505			.747		

	Scenario 5 (Sup Corr) Free Meals, Gifts from Merchants			*Scenario 6 (Striking)*			*Scenario 7 (Verbal)*			*Scenario 8 (Cover up)*		
	B	s.e.	*Odds*	*B*	s.e.	*Odds*	*B*	s.e.	*Odds*	*B*	s.e.	*Odds*
Supervisory status[2]	.319	.355	1.375	.560	.348	1,751	1.104***	.316	3.018	.415	.275	1.515
Length of service[3]												
6–15 years	−1.136*	.479	.321	−.703	.428	.495	−.266	.456	.767	.439	.341	1.551
Over 15 years	−.718	.522	.488	−.548	.485	.578	−.808	.461	.446	.535	.365	1.707
Own seriousness[4]	1.367**	.400	3.925	1.854***	.345	6.383	2.136***	.366	8.465	.427	.314	1.532
Knowledge of rules[5]	.883*	.355	2.418	.894*	.386	2.444	.519	.352	1.680	1.261***	.303	3.530
Expected discipline[6]												
Some discipline	1.521***	.366	4.575	.707*	.348	2.028	1.364***	.296	3.911	1.198***	.277	3.314
Dismissal	.633	.504	1.883	1.542**	.579	4.674	−1.624*	.658	.197	−.177	.376	.837
Discipline fairness[7]												
Too harsh	.342	.552	1.408	−.993*	.455	.370	−1.242*	.543	.289	−.594	.380	.552
Too lenient	2.307***	.505	10.044	.788	.456	2.199	1.651***	.467	5.210	.640a	.359	1.896
Most officers' reporting[8]	4.087***	.327	59.587	4.147***	.388	63.232	4.906***	.375	135.16	3.510***	.271	33.438
Constant	−2.566	.547		−2.256	.460		−3.368	.499		−2.798	.365	
Model Nagelkerke R Squared	.657			.675			.761			.655		

(*Continued*)

Table 6.3 (Continued)

	Scenario 9 (Kickback) Free Meals, Gifts from Merchants			*Scenario 10 (Report)*			*Scenario 11 (Supervisor)*			*Scenario 12 (Hate Crime)*		
	B	s.e.	*Odds*	*B*	s.e.	*Odds*	*B*	s.e.	*Odds*	*B*	s.e.	*Odds*
Supervisory status[2]	−.194	.279	.824	.178	.387	1.195	.582a	.329	1.789	1.629***	.324	5.101
Length of service[3]												
6–15 years	.211	.379	1.235	−.376	.462	.687	−.595	.426	.552	−.443	.384	.642
Over 15 years	.198	.381	1.219	.273	.518	1.314	−.575	.467	.563	−.731a	.409	.481
Own seriousness[4]	.741*	.377	2.098	1.393**	.527	4.027	.528	.397	1.695	1.474***	.333	4.366
Knowledge of rules[5]	1.296***	.345	3.653	.760	.565	2.137	.479	.421	1.614	−.190	.317	.827
Expected discipline[6]												
Some discipline	.102	.303	1.107	1.728***	.427	5.630	1.907***	.344	6.732	1.292***	.275	3.641
Dismissal	.735a	.412	2.085	1.562**	.494	4.767	2.758***	.430	15.775	.655	.448	1.925
Discipline fairness[7]												
Too harsh	−.359	.502	.699	−.876a	.478	.416	−1.893***	.389	.151	−1.248**	.363	.287
Too lenient	.029	.331	1.030	2.035**	.645	7.652	2.341***	.625	10.396	−.814*	.405	.443
Most officers' reporting[8]	4.012***	.317	55.236	4.428***	.471	83.803	3.238***	.312	25.479	3.052***	.262	21.167
Constant	−2.054	.404		−2.784	.533		−1.647	.452		−1.536	.393	.215
Model Nagelkerke R Squared	.612			.658			.594			.546		

	Scenario 13 (Bribe) Free Meals, Gifts from Merchants			*Scenario 14 (Graffiti)*		
	B	*s.e.*	*Odds*	*B*	*s.e.*	*Odds*
Supervisory status[2]	1.154*	.526	3.171	.750*	.337	2.117
Length of service[3]						
6–15 years	–.534	.605	.586	–.370	.457.691	
Over 15 years	.582	.675	1.790	.219	.485	1.244
Own seriousness[4]	1.528*	.699	4.608	.918*	.358	2.505
Knowledge of rules[5]	1.752**	.598	5.767	1.818***	.351	6.159
Expected discipline[6]						
Some discipline	–.013	.680	.987	.693*	.348	2.000
Dismissal	–.181	.717	.835	–2.163	.816	.115
Discipline fairness[7]						
Too harsh	.179	.838	1.196	–1.434*	.622	.238
Too lenient	1.182	.731	3.263	1.099*	.530	3.001
Most officers' reporting[8]	4.521***	.554	91.932	5.396***	.518	220.55
Constant	–2.505	.772		–2.093	.440	
Model Nagelkerke R Squared	.577		.722			

* $p < .05$; ** $p < .01$; *** $p < .001$

[1]Reference category = "not reporting"; [2]Reference category = "non-supervisor"
[3]Reference category = "0 to 5 years"; [4]Reference category = "not serious"
[5]Reference category = "not"; [6]Reference category = "no discipline/verbal reprimand"
[7]Reference category = "fair"; [8]Reference category = "would not report"

of 14 scenarios; Table 6.3). In addition, compared to the respondents who expected some discipline less severe than dismissal, respondents who expected dismissal had higher odds of saying that they would report in about one-half of the scenarios (seven out of 14 scenarios; Table 6.3). Fairness of discipline was also related to respondents' willingness to report in about one-half of the scenarios. Compared to respondents who evaluated the discipline as too harsh, those who evaluated the discipline as fair had higher odds of saying that they would report in about one-half of the scenarios (seven out of 14 scenarios; Table 6.3). In addition, compared to the respondents who evaluated the discipline as fair, those who evaluated the discipline as too lenient also had higher odds of saying that they would report in about one-half of the scenarios (seven out of 14 scenarios; Table 6.3).

Finally, demographic characteristics did not carry as much weight as the organizational and cultural variables. Supervisory status was relevant in about one-third of the scenarios (five out of 14 scenarios; Table 6.3). The respondents who were supervisors had higher odds of saying that they would report than the respondents who were line officers. Length of service was relevant in only two scenarios.

Conclusion

The apartheid SAP was a secretive, distrusted organization in which power lay in covert, high police structures with the primary objective being the protection of the state and a minority white population. Restructuring the service post-apartheid to undo this secrecy and practices saw a dismantling of the intelligence capacity and an emphasis on street-level, uniformed policing (see Chapter 3). While this shift temporarily ended the police's political work, it quickly resurfaced again in 2000 with the appointment of Jackie Selebi and intensified under President Zuma. There are indications that some of this damage may be undone under President Ramaphosa, but it is far too early to confidently predict any positive changes.

The results of our empirical research demonstrate the presence of a strong code of silence. The police officers included in our study believed that most officers adhere to the code of silence and would not report most of the behaviors in the questionnaire, including not only the less severe forms of misconduct, such as the acceptance of gratuities and verbal abuse, but also the more serious forms of misconduct, such as use of excessive force on arrestee, a kickback, falsification of official reports, and a failure to execute an arrest warrant. The code did not seem to provide cover for only three scenarios describing the most serious forms of police misconduct, which also violated the norms of criminal law: theft from a crime scene, accepting a bribe, and abusing deadly force.

Our results support earlier survey studies (e.g., Kutnjak Ivković and Sauerman, 2011, 2013; Matlala et al., 2016; Meyer et al., 2013; Newham, 2004;

Reyneke-Cloete and Meyer, 2016) and ethnographic studies on the code of silence in South Africa (e.g., Marks, 2005; Vigneswaran and Hornberger, 2009; Hornberger, 2011; Faull, 2018; Perkins, 2018), all of which suggest the presence of a strong code of silence. Such results, demonstrating that the code is present throughout the country, are by no means surprising. In a context in which external mechanisms of control are dysfunctional (see Chapter 3) and where internal mechanisms suffer from inherent challenges (see Chapter 5), it is expected that the code of silence should be strong. In addition, amid the country's immense unemployment, SAPS officers see the job as essential to their survival and that of their social networks (Faull, 2018). Many may be more concerned with retaining their job than with delivering a professional service. This requires putting their heads down and ignoring the abuse of their colleagues and managers, which makes enforcing compliance and punishing misconduct difficult for dedicated officers. This, in turn, strengthens the code of silence.

In the meantime, South Africans continue to distrust the SAPS and police officers distrust the public. This is partly a result of perceptions of widespread corruption in the organization, both at the street and management levels, as well as the realization that the SAPS cannot keep South Africans safe. In response, SAPS officers work together to deceive the public and their leaders into believing that they are carrying out common-sense tasks that make South Africa safe, and generate or forge data to back this up. However, the conspiracy simply bolsters the culture of silence, suspicion, and mistrust. In circumstances where the public does not trust the police, and the police do not trust the public, the opposite can transpire. Citizens may opt out of the state, and police service delivery may become inefficient and/or abusive as officers neglect to share information with colleagues or report transgressions. This appears to be the case in South Africa.

Note

1 Both variables have been dichotomized in some analyses. Answers "1" and "2" were coded as "would not report" and answers "3," "4," and "5" were coded as "would report."

References

Brodeur, J.P. (1983). High policing and low policing: Remarks about the policing of political activities. *Social Problems*, 30(5): 507–520.

Brodeur, J.P. (2010). *The Policing Web.* New York: Oxford University Press.

Brogden, M. and C.D. Shearing (1993). *Policing for a New South Africa.* London: Routledge.

Bruce, D. (2008). Benign neglect? The politics and practise of controlling police corruption in South Africa. Paper presented at the Workshop Crime, Poverty and Police Corruption Annual Bank Conference on Development Economics Cape Town, 9–11 June. Available at: http://siteresources.worldbank.org/INTABCDESOUAFR2008/Resources/David Bruce.pdf

Bruce, D. (2010). "The ones in the pile were the ones going down": The reliability of violent crime statistics. *South African Crime Quarterly*, 31: 9–17.
Bruce, D. (2011). Measuring outputs, neglecting outcomes: The Auditor General's role in SAPS performance assessment. *South African Crime Quarterly*, 38: 3–13.
Bruce, D., K. Savage, and J. de Waal (2000). A duty to answer questions? The police, the independent complaints directorate and the right to remain silent. *CSVR*. Available at: www.csvr.org.za/docs/policing/dutytoanswer.pdf
Burger, J. (2006). Crime combating in perspective: A strategic approach to policing and the prevention of crime in South Africa. *Acta Criminologica*, 19(2): 105–118.
Burger, J. (2010). On the record: Interview with the minister of police, Nathi Mthethwa. *South African Crime Quarterly*, 34: 41–45.
Burger, J. and S. Grobler (2017). *Why the SAPS Needs an Internal Anti-Corruption Unit. Institute for Security Studies Policy Brief.* Available at: https://issafrica.org/research/policy-brief/why-the-saps-needs-an-internal-anti-corruption-unit
Cawthra, G. (1993). *Policing South Africa*. London: Zed Books Ltd.
Citizen Surveys (2019). *South African Citizen Survey*. Available at: https://citizensurveys.net/sa-citizens-survey/
Cockcroft, T. (2013). *Police Culture*. New York: Taylor and Francis. Kindle Edition.
Constitution of the Republic of South Africa (1996). Act No. 108 of 1996. Available at: www.gov.za/documents/constitution-republic-south-africa-1996
Crank, J.P. (1998). *Understanding Police Culture*. Cincinnati, OH: Anderson Publishing.
Datzer, D., E. Mujanović, and S. Kutnjak Ivković (2019). Assessments of discipline fairness by police officers from Bosnia and Herzegovina. *International Journal of Comparative and Applied Criminal Justice*. DOI: 10.1080/01924036.2019.1651749
Davis, R.C. (2012). *Selected International Best Practices in Police Performance Measurement*. RAND Centre on Quality Policing. Available at: https://www.rand.org/pubs/technical_reports/TR1153.html
Eterno, J.A., A. Verma, and E.B. Silverman (2014). Police manipulations of crime reporting: Insiders' revelations. *Justice Quarterly*, 33(5): 811–835. DOI: 10.1080/07418825.2014.980838
Farlam, I.G. (2015). Marikana Commission of Inquiry: Report on Matters of Public, National and International Concern Arising Out of the Tragic Incidents at the Lonmin Mine in Marikana, in the North West province. Available at: www.sahrc.org.za/home/21/files/marikana-report-1.pdf.
Faull, A. (2010). *Behind the Badge: The Untold Stories of South Africa's Police Service Members*. Cape Town: Penguin Random House South Africa. Kindle Edition.
Faull, A. (2013). Towards a new professionalism for the South African police service. *South African Review of Sociology*, 44(2): 18–35.
Faull, A. (2016). Measured governance? Policing and performance management in South Africa. *Public Administration and Development*, (36): 157–168.
Faull, A. (2017a). Fighting for respect: Violence, masculinity and legitimacy in the South African Police Service. In J. Beek, M. Gopfert, O. Owen, and J. Steinberg (eds.), *Police in Africa: The Street Level View*. London: Hurst.
Faull, A. (2017b). Police culture and personal identity in South Africa. *Policing: A Journal of Policy and Practice*, 11(3): 332–345. https://doi.org/10.1093/police/pax016
Faull, A. (2018). *Police Work and Identity: A South African Ethnography*. Abingdon: Routledge.
Haysom, F. (1993). *Policing the Transition: Transforming the Police, African National Congress*. Available at: www.anc.org.za/content/policing-transition-transforming-police

Hickman, M.J., A.R. Piquero, Z.A. Powell, and J. Greene (2016a). Expanding the measurement of police integrity. *Policing: An International Journal of Police Strategies & Management*, 39(2): 246–267.

Hickman, M.J., Z.A. Powell, A.R. Piquero, and J. Greene (2016b). Exploring the viability of an attitudes toward ethical behavior scale in understanding police integrity outcomes. *Policing: An International Journal of Police Strategies & Management*, 39(2): 319–337.

Hornberger, J. (2011). *Policing and Human Rights: The Meaning of Violence and Justice in the Everyday Policing of Johannesburg*. Abingdon, U.K.: Routledge.

Hornberger, J. (2017). The belly of the police. In J. Beek, M. Göpfert, O. Owen, and J. Steinberg (eds.), *Police in Africa: The Street Level View*. Cape Town: Oxford University Press, pp. 199–212.

Huon, G., B. Hesketh, M. Frank, K. McConkey, and G. McGrath (1995). *Perceptions of Ethical Dilemmas: Ethics and Policing-Study 1*. Payneham, Australia: National Police Research Unit.

Joffee, J. (2010). *Judgement: State vs Jackie Selebie, Case No 25/09*. Available at: www.saflii.org/za/cases/ZAGPJHC/2010/53.pdf

Khechumyan, A. and S. Kutnjak Ivković (2019). Exploring differences in police integrity within a centralized police system. In S. Kutnjak Ivković and M. Haberfeld (eds.), *Exploring Police Integrity*. New York: Springer.

Kinnes, I. (2017). Contested Governance: Police and Gang Interactions. PhD Dissertation. Cape Town: University of Cape Town.

Kleinig, J. (1996). *The Ethics of Policing*. New York: Cambridge University Press.

Klockars, C.B. and S. Kutnjak Ivković (2004). Measuring police integrity. In M.J. Hickman, A.R. Piquero, and J.R. Greene (eds.), *Police Integrity and Ethics*. Belmont, CA: Wadsworth Publishing.

Klockars, C.B., S. Kutnjak Ivković, and M.R. Haberfeld (2004). The contours of police integrity. In C.B. Klockars, S. Kutnjak Ivković, and M.R. Haberfeld (eds.), *The Contours of Police Integrity*. Thousand Oaks, CA: Sage Publications, pp. 1–18.

Klockars, C.B., S. Kutnjak Ivković, and M.R. Haberfeld (2006). *Enhancing Police Integrity*. Dordrecht, The Netherlands: Springer.

Klockars, C.B., S. Kutnjak Ivković, W.E. Harver, and M.R. Haberfeld (1997). *The Measurement of Police Integrity*. Final Report Submitted to the U.S. Department of Justice, Office of Justice Programs, National Institute of Justice.

Klockars, C.B., S. Kutnjak Ivković, W.E. Harver, and M.R. Haberfeld (2000). *The Measurement of Police Integrity*. Research in Brief. U.S. Department of Justice, Office of Justice Programs, National Institute of Justice. Washington, DC: Government Printing Office.

Kremer, F. (2000). Comparing supervisor and line officer opinions about the code of silence: The case of Hungary. In M. Pagon (ed.), *Policing in Central and Eastern Europe: Ethics, Integrity, and Human Rights*. Ljubljana, Slovenia: College of Police and Security Studies, pp. 211–219.

Kutnjak Ivković, S. (2009). The Croatian police, police integrity, and transition toward democratic policing. *Policing: An International Journal of Police Strategies and Management*, 32(3): 459–488.

Kutnjak Ivković, S. (2015). Studying police integrity. In S. Kutnjak Ivković and M. Haberfeld (eds.), *Police Integrity across the World*. New York: Springer.

Kutnjak Ivković, S. and M.R. Haberfeld (eds.) (2015). *Police Integrity across the World*. New York: Springer.

Kutnjak Ivković, S., M. Haberfeld, and R. Peacock (2013). Rainless West: The integrity survey's role in agency accountability. *Police Quarterly*, 16(2): 148–176.

Kutnjak Ivković, S., M. Haberfeld, and R. Peacock (2018). Decoding the code of silence. *Criminal Justice Policy Review*, 29(2): 172–189.

Kutnjak Ivković, S. and C.B. Klockars (1996). Police Perceptions of Disciplinary Fairness and the Formation of the Code of Silence. *The 1996 Annual Meeting of the Academy of Criminal Justice Sciences*, Las Vegas, Nevada, March.

Kutnjak Ivković, S. and C.B. Klockars (1998). The code of silence and the Croatian police. In M. Pagon (ed.), *Policing in Central and Eastern Europe: Organisational, Managerial, and Human Resource Aspects*. Ljubljana, Slovenia: College of Police and Security Studies, pp. 329–347.

Kutnjak Ivković, S., R. Peacock, and M. Haberfeld (2016). Does discipline fairness matter for the police code of silence? Answers from the U.S. supervisors and line officers. *Policing: An International Journal of Police Strategies and Management*, 39(2): 354–369.

Kutnjak Ivković, S. and A. Sauerman (2011). Measuring the code of silence among the South African police: Findings from a SAPS supervisor survey. In C. Gould and G. Newham (eds.). *Toward a Coherent Strategy for Crime Reduction in South Africa Beyond 2010*. Pretoria, South Africa: Institute for Security Studies. 74–87.

Kutnjak Ivković, S. and A. Sauerman (2013). Curtailing the code of silence among the South African Police. *Policing: An International Journal of Police Strategies and Management*, 36(1): 175–198.

Kutnjak Ivković, S. and A. Sauerman (2016). Police integrity in South Africa: A tale of three police agency types. *Policing: An International Journal of Police Strategies & Management*, 39(2): 268–283.

Kutnjak Ivković, S. and T.O. Shelley (2008). The police code of silence and different paths toward democratic policing. *Policing and Society*, 18(4): 445–473.

Lamb, G. (2018). Police militarisation and the "war on crime" in South Africa. *Journal of Southern African Studies*, 44(5): 933–949. DOI: 10.1080/03057070.2018.1503831

Lim, H. and J.J. Sloan (2016). Police officer integrity: A partial replication and extension. *Policing: An International Journal of Police Strategies & Management*, 39(2): 284–301.

Lobnikar, B., K. Prislan, B. Čuvan, and G. Meško (2016). The code of silence and female police officers in Slovenia: Gender differences in willingness to report police misconduct. *Policing: An International Journal of Police Strategies & Management*, 39(2): 387–400.

Long, M., J.E. Cross, T.O. Shelley, and S. Kutnjak Ivković (2013). The normative order of reporting police misconduct: Examining the roles of offense seriousness, legitimacy, and fairness. *Social Psychology Quarterly*, 76(3): 242–267.

Mabuza, E. (2001). Good economy can help fight crime. *Business Day,* 12

Manning, P. (1979). *Police Work: The Social Organization of Policing*. Cambridge, MA: MIT Press.

Manning, P. (2009). Bad cops. *Criminology and Public Policy*, 8(4): 787–794.

Marche, G.E. (2009). Integrity, culture, and scale: An empirical test of the big bad police agency. *Crime, Law, and Social Change*, 51: 463–486.

Marks, M. (2005). *Transforming Robocops: Changing Police in South Africa*. Scottsville, South Africa: University of Scottsville and KwaZulu-Natal Press.

Matlala, R.L.G., D. Mistry, and A. Phala (2016). Measuring the integrity of law enforcement officers in Gauteng Province. *The International Journal of Social Sciences and Humanities Invention*, 3(11): 2969–2980.

McDevitt, J., C. Posick, R. Zschoche, D.P. Rosenbaum, M. Buslik, and L. Fridell (2011). *Police Integrity, Responsibility, and Discipline*. National Institute of Justice. Available at: www.nationalpoliceresearch.org/storage/updated-papers/Police%20Integrity%20Responsibility%20and%20Discipline.pdf

Meyer, M.E. and J. Steyn (2008). Developing cynicism among male and female recruits in the South African Police Service. *Acta Criminologica*, 2: 1–20.

Meyer, M.E. and J. Steyn (2009). Nurturing isolation in the South African Police Service: A comparison of male and female recruits. *Policing: An International Journal of Police Strategies and Management*, 32(1): 108–127.

Meyer, M.E., J. Steyn, and N. Gopal (2013). Exploring the public parameter of police integrity. *Policing: An International Journal of Police Strategies & Management*, 36(1): 140–156.

Micucci, A.J. and I.M. Gomme (2005). American police and subcultural support for the use of excessive force. *Journal of Criminal Justice*, 33: 487–500.

[Mollen Commission] New York City Commission to Investigate Allegations of Police Corruption and the Anti-Corruption Procedures of the Police Department (1994). *Commission Report*. New York: Author.

Myburgh, P.L. (2019). *Gangster State*. Cape Town: Penguin Random House South Africa. Kindle Edition.

Newham, G. (2004). Out of step: Integrity and the South Africa police service. In C.B. Klockars, S. Kutnjak Ivković, and M.R. Haberfeld (eds.). *The Contours of Police Integrity*. Thousand Oaks, CA: Sage Publications, 232–247.

Newham, G. and A. Faull (2011). *Protector or Predator? Tackling Police Corruption in South Africa*. ISS Monograph 182. Pretoria: Institute for Security Studies.

O'Neil, M.M. and A.M. Singh (eds.) (2007). *Police Occupational Culture: New Debates and Directions*. Amsterdam: Elsevier.

O'Regan, J.C. and A.V. Pikoli (2014). *Towards a Safer Khayelithsa: Report of the Commission of Inquiry into Allegations of Police Inefficiency and a Breakdown in Relations Between SAPS and the Community of Khayelitsha*. Cape Town: Khayelitsha Commission of Inquiry. Available at: https://equaleducation.org.za/2014/09/09/the-oregan-pikoli-commission-of-inquiry-into-policing-in-khayelitsha-week-four/

Pagon, M. and B. Lobnikar (2000). Comparing supervisor and line officer opinions about the code of silence: The case of Slovenia. In M. Pagon (ed.), *Policing in Central and Eastern Europe: Ethics, Integrity, and Human Rights*. Ljubljana, Slovenia: College of Police and Security Studies, pp. 197–209.

Pauw, J. (2017). *The President's Keepers: Those Keeping Zuma in Power and out of Prison*. Cape Town: NB Publishers.

Pauw, J. (2018). *The President's Keeper: Those Keeping Zuma in Power and Out of Prison*. Cape Town: Tafelberg.

Perkins, G. (2018). *Danger and Death: Organisational and Occupational Responses to the Murder of Police in South Africa: A Case Study*. PhD Thesis. Cape Town: University of Cape Town.

Plus 94 Research (2018). *Afrobarometer: South Africa Results*. Available at: www.afrobarometer.org/publications/south-africa-summary-results-2018

PMG (Parliamentary Monitoring Group) (2015). Nkandla private residence security upgrades: Minister of Police briefing. Parliamentary meeting minutes, Ad Hoc Committee on Police Minister's Report on Nkandla, 20 July.

Porter, L.E. and T. Prenzler (2016). The code of silence and ethical perceptions: Exploring police officer unwillingness to report misconduct. *Policing: An International Journal of Police Strategies & Management*, 39(2): 370–386.

Punch, M. (2009). Police Corruption: Deviance, Accountability and Reform in Policing. London: Willan Publishing.

Raines, J. (2010). *Ethics in Policing: Misconduct and Integrity*. Sudbury, MA: Jones & Bartlett Learning.

Ratele, K. (2008). Masculinity and male mortality in South Africa. *African Safety Promotion: A Journal of Injury and Violence Prevention*, 6(2): 19–41.
Republic of South Africa (2011). *Independent Police Investigative Directorate Act, 1 of 2011.*
Reyneke-Cloete, S. and M.E. Meyer (2016). Viewing integrity in the South African Police Service among non-commissioned officers in Gauteng Province. *Acta Criminologica: Southern African Journal of Criminology*, 29(2): 87–106.
Rothwell, G.R. and J.N. Baldwin (2007). Whistle-blowing and the code of silence in police agencies policy and structural predictors. *Crime & Delinquency*, 53(4): 605–632.
Sapa (2015). Sibiya: Judge calls acting head of Hawks "dishonest". *Mail & Guardian*, 23 March. Available at: https://mg.co.za/article/2015-03-23-judge-harsh-on-acting-head-of-hawks (Accessed 25 February 2018).
SAPS (2019). *Instructions Relating to Arrest and Detention of Suspects*. Unpublished internal SAPS circular.
Schafer, J.A. and T.J. Martinelli (2008). First-line supervisor's perceptions of police integrity. *Policing: An International Journal of Police Strategies and Management*, 31(2): 306–323.
Skolnick, J.H. (2002). Corruption and the blue code of silence. *Police Practice and Research*, 31(1): 7–19.
Skolnick, J.H. (2011). *Justice without Trial: Law Enforcement in Democratic Society*. Classics of Law & Society. New Orleans, Louisiana: Quid Pro Books. Kindle Edition.
South African Police (1991). *Annual Report of the Commissioner of the South African Police 1991*. Pretoria: South African Police Service. Available at: https://www.saps.gov.za/about/stratframework/annualreports_arch.php
Sparrow, M. (2016). *Handcuffed: What Holds Policing Back, and the Key to Reform*. Washington, DC: Brookings Institution Press.
Steinberg, J. (2008). *Thin Blue: The Unwritten Rules of Policing South Africa*. Jeppestown: Jonathan Ball Publishers.
Steinberg, J. (2012). Establishing police authority and civilian compliance in post-apartheid Johannesburg: An argument from the work of Egon Bittner. *Policing and Society*, 22(4): 481–495.
Steinberg, J. (2014). Policing, state power, and the transition from apartheid to democracy: A new perspective. *African Affairs*, 113(451): 173–191.
Steinberg, J. (2017). Policing during and after apartheid. In J. Beek, M. Göpfert, O. Owen, and J. Steinberg (eds.), *Police in Africa: The Street Level View*. Cape Town: Oxford University Press, pp. 199–212.
Steyn, J. (2017). Lilac? Ten years of blue nurtured nature. *Police Practice and Research*. http://dx.doi.org/10.1080/15614263.2017.1292745
Steyn, J. and S. Mkhize (2016). Darker shades of blue: A comparison of three decades of South African Police Service culture. *South African Crime Quarterly*, 57: 15–26.
Viewfinder (2019). Viewfinder: IPID's cover-up of police brutality in SA. *Daily Maverick*, 7 October. Available at: www.dailymaverick.co.za/article/2019-10-07-viewfinder-ipids-cover-up-of-police-brutality-in-sa/
Vigneswaran, D. and J. Hornberger (eds.) (2009). *Beyond "Good Cop"/"Bad Cop": Understanding Informality and Police Corruption in South Africa*. Johannesburg: Forced Migration Studies Programme, University of Witwatersrand.
Waddington, P.A.J. (2005). Police (canteen) subculture: An appreciation. In T. Newburn (ed.), *Policing: Key Readings*. London: Routledge, pp. 364–385.

Weisburd, D. and R. Greenspan (2000). Police attitudes toward abuse of authority: Findings from a national study. *Research in Brief.* National Institute of Justice. Available at: www.ncjrs.gov/pdffiles1/nij/181312.pdf

Westmarland, L. (2005). Police ethics and integrity: Breaking the blue code of silence. *Policing and Society*, 15(2): 145–165.

Wiener, M. (2018). *Ministry of Crime: An Underworld Explored.* Johannesburg, South Africa: Pan Macmillan. Kindle Edition.

Wilson, J.Q. (1963). The police and their problems: A theory. *Public Policy*, 12: 189–216.

Wood, J.R.T. (1997). Final Report of the Royal Commission into the New South Wales Police Service, Volume 1: Corruption, Sydney: Government of the State of New South Wales. Available at: http:// www.pic.nsw.gov.au/files/reports/VOLUME1.pdf (Accessed 21 July 2011).

Chapter 7

Apartheid aftershock

Race, police, and police integrity

Introduction

Because of the country's past, race has been central to South Africa's state-making and nation-building projects, and, by extension, to its policing. For several reasons, today, more than 25 years after the end of apartheid—a system that classified people as "white," "Indian," "colored," or "native" ("black" or "African")—poverty and inequality remain highly racialized.

Until the early 1990s, race was primarily used to assign people to different life trajectories (Posel, 2001). Police were central to the apartheid system, which reserved 87% of the land, the best public services, and the best job opportunities for white people, who made up less than 20% of the total population. Apartheid laws required black people to carry passes confirming their right to be in white cities and neighborhoods. Police would frequently stop people to examine their passes and arrest and remove them if they did not have the necessary authority to be in the area (Cawthra, 1993; Shaw and Shearing, 1998). The pass laws were despised by black communities and occasionally resulted in public protests, with some of these demonstrations of dissent more violent than others. For example, in 1960 the police killed 69 people and wounded 180 others as they protested against these laws in the Sharpeville township. Essentially, most of apartheid policing focused on ensuring that people used only the doorways, benches, train carriages, beaches, and neighborhoods corresponding with their racial categorization. At its extreme, rogue operatives within the apartheid police established specialized units to torture and murder known opponents of the apartheid regime, initiated various secret operations to steer or undermine political resistance in black communities (TRC, 1998a, p. 529; Rauch, 2004), and armed and helped stoked conflict between affiliates of the liberation movements the Inkatha Freedom Party (IFP) and the African National Congress (ANC) (Brogden, 1994). With a call by the ANC in 1985 to make the country ungovernable, political violence between these groups escalated and, together with brutal attempts by the state to quell a perceived communist onslaught, a staggering death toll followed—in KwaZulu Natal province alone between 18,000 to 20,000 murders were documented (TRC, 1998b, pp. 160–163).

In this conflict for the very soul of the country, all sides felt justified in their actions, however gruesome the outcomes. Already in 1986, the ANC secretary-general commented that "collaborators with the enemy had to be eliminated" and "[w]hatever the people decide to use to eliminate these enemy elements is their decision" (Jeffery, 2019, Loc 1555). The "enemy"—the apartheid state—had by then established and enforced policies in line with the statements of the country's former Minister of Native Affairs and Prime Minister, Hendrik Verwoerd:

> [I]f mixed development is to be the policy of the future of South Africa, it will lead to the most terrific clash of interests imaginable. The endeavours and desires of the Bantu and . . . Europeans will be antagonistic.
>
> (1950)

> The Bantu must be guided to serve his own community in all respects. There is no place for him in the European community above the level of certain forms of labour. Within his own community, however, all doors are open.
>
> (1954)

> I say it unequivocally that the people of South Africa cannot accept the consequence of a multi-racial state unless the Whites, the Coloureds and the Indians are prepared to commit race suicide.
>
> (1962)

In contrast, the post-1994 era saw attempts to undo the harm done by decades of racial segregation and celebrate the country's diversity, in part by declaring that all residents were entitled to equal rights, including police services. Vice-President Thabo Mbeki's 1996, "I am an African" speech, delivered to mark the passing of the new constitution, is the antithesis of Verwoedian sentiments. In it, Mbeki defined himself—and all South Africans—as products of a shared land and of each other:

> I am an African. I owe my being to the hills and the valleys, the mountains and the glades . . . to the Khoi and the San. . . . I am formed of the migrants who left Europe. . . . Whatever their own actions, they remain still part of me. . . . In my veins courses the blood of the Malay slaves who came from the East. . . . I am the grandchild of the warrior men and women that Hintsa and Sekhukhune led, the patriots that Cetshwayo and Mphephu took to battle, the soldiers Moshoeshoe and Ngungunyane taught never to dishonour the cause of freedom. . . . I am the grandchild who lays fresh flowers on the Boer graves. . . . I come of those who were transported from India and China. . . . Being part of all of these people, and in the knowledge that none dares contest that assertion, I shall claim that—I am an African.

The phrase most famous for describing this aspect of post-apartheid South Africa is "the rainbow nation"—a term coined by Archbishop Desmond Tutu in the early 1990s. This metaphor, like a rainbow, has however not lasted. By the turn of the century, critics had begun referring pejoratively to "rainbowism" as a discourse that highlighted color, but ignored the power inequality between people (Gqola, 2001). A striking example of this push-back was seen in the 2015–2016 protests on university campuses across the country. Students who were mostly black, but few of whom had experienced apartheid, sought to shut down universities, often clashing violently with police and private security officers in the process. The protests were motivated by two primary factors. First, a perception that universities were places either unwelcoming or discriminating against black students as the curricula remained primarily Western, and that university staff were mostly white. Second, university education was perceived as unaffordable to most young and black people (Commission of Enquiry into Higher Education and Training, 2017).

The protests were not just about education but also about an unresolved, racialized past (Langa, 2017). As noted by Badat (2015, p. 82), student anger was succinctly summarized on a placard reading, "Our parents were SOLD dreams in 1994. We are just here for the REFUND." Not even Nelson Mandela, South Africa's first black president, was spared by the protesting students, with some accusing him of selling out to whites during the transition so that they retained their relative wealth while black South Africans remained poor.

While contemporary South Africa is officially a non-racial democracy, data make clear that race still predicts all manner of outcomes in the country, including health (Statistics South Africa, 2016, pp. 137–138), wealth (Statistics South Africa, 2012; Gumede, 2015), experiences of crime (Statistics South Africa, 2018b), and the lived experience of police officers (Steinberg, 2008; Faull, 2018), including their likelihood of being murdered (Perkins, 2018). Race also remains central to the country's politics. Politicians and political parties appeal to or demonize supposed members of particular "races" one day, and preach universal inclusiveness the next. For example, in 2018 the then black Free State Premier, Ace Magashule, responded to allegations of corruption by claiming the moral high ground, "I as premier, believe in radical economic transformation. I will not be bullied into not supporting black business because of all these allegations of corruption against me" (Myburgh, 2018, Loc 3591). Such rhetoric and posturing remains possible, in part, because in its pursuit of racial justice, South Africa continues to embrace racial labels, logic, and language throughout its bureaucracies and economy (Mare, 2001). When a victim of crime reports a case to the police, for example, they are requested to declare their own racial categorization. The same is true of almost all government forms. This could likely promote the notion that apartheid's race categories are indeed real, or at least confirmed by majority black rule.

Defining race in South Africa

The twentieth century saw philosophers postulate that truth was vested in national powers (Fanon, 1966; Foucault, 1991), that the law was underscored by social biases (van Doren, 1989), and that race was a social construct without any biological basis (Appiah, 1985, 1992). Apparent in Verwoerd's words above, such proposals did not sit well among South Africa's political elite. Rather, the National Party (NP) sought to define all residents of South Africa using one of three (later four) race categories and to link laws, service, governance, and opportunities to these groups. Ostensibly, this was in order to let each "race" thrive, but only once white South Africans had done so first.

The Population Registration Act (No. 30 of 1950) established a racial classification system. Based on physical appearance, social acceptance (perceived race), and descent, people were classified as "White," "Native," or "Colored." "Indian" was added at a later date. Race was clearly indicated in the identity numbers issued to citizens at birth. Ironically, South Africans could challenge their racial classification and have it changed, exposing its superficiality (Erasmus and Ellison, 2008). The ambiguity and common sense aspect of the system is apparent in the Act's definition of a "White" person as someone "who in appearance obviously is, or who is generally accepted as a white person, but does not include a person who, although in appearance obviously a white person, is generally accepted as a coloured person" (Population Registration Act, 1950).

The Population Registration Act was repealed in 1991, but the categories it had created live on. In part, this has been encouraged by the post-apartheid state (Republic of South Africa, 1998, 2003; Netschitenzhe, 2015). Apartheid distributed goods and opportunities based on the categories, and so the crafters of South Africa's post-apartheid constitution felt that redress should be based on such elements. "To promote the achievement of equality," the country's Bill of Rights allows that "legislative and other measures designed to protect or advance persons, or categories of persons, disadvantaged by unfair discrimination may be taken" (Republic of South Africa, 1996, pp. 5–6). Although this amounted yet again to state sanctioned or "legal" discrimination, the rational appears that color-blindness at the dawn of democracy would risk ignoring and perpetuating racial stratification. This occurred most clearly in Brazil, where an officially raceless democracy has disproportionately maintained the privilege of light-skinned citizens (dos Santos, 2006).

Race and its correlates in contemporary South Africa

A primary reason South Africa remains racialized is that resources and opportunities are scarce and access to them delineated along apartheid-era race lines. Racialized inequality is highly visible, generating social fractures, discontent, anger, and fear. These, in turn, feed racism, xenophobia, violence, and crime.

Table 7.1 shows the percentage of South Africans from each population group living below the upper- and lower-bound poverty lines in 2015. The lower-bound poverty line is defined as having access to less than R647 (US$43) per person per month, and the upper-bound as R992 (US$66). Although these figures represent marked improvements compared to the early 1990s, they also demonstrate that poverty remains racialized and that most black South Africans, representing almost 80% of the country's population, remain poor.

Household income correlates with numerous other variables that highlight the racial stratification of South African society and maintain the politicization of race. Chief among these is unemployment, which South Africans have long named the most important issue for government to address, three times more vital than crime or housing (Afrobarometer, 2013, 2018). In 2018, 30.4% of blacks were unemployed compared to 7.6% of whites (Table 7.2). Black citizens are less likely to complete school or post-school qualifications (Mlachila and Moeletsi, 2019; StatsSA, 2018a), less likely to have access to basic services (StatsSA, 2018a), and more likely to be victims of violent crimes (StatsSA, 2018b) than their white compatriots.

South African prisons show a similar race delamination. Colored South Africans, who make up less than 9% of the national population, account for 18% of the prison population. In contrast, whites and Indians, who represent 8% and 2.5% of the national population, respectively, account for only 1.6% and 0.6% of the prison population, respectively. Black South Africans account for 79% of the national and prison populations (Makou et al., 2017).

These differences produce attitudes of mistrust among groups, which, in turn, makes addressing those differences more difficult. In a series of surveys between 2003 and 2017, most South Africans indicated they would like South Africa to be a unified nation and for their children to see themselves as South African (IJR, 2017). In 2017, 77% agreed that people "should realize we are

Table 7.1 Percentage of population living in poverty by race, 2015

Poverty Line	*Black/African*	*Colored*	*Indian/Asian*	*White*
R647/$43 p/m	47.1%	23.3%	1.2%	0.4%
R992/$66 p/m	64.2%	41.3%	5.9%	1%

Source: Statistics South Africa (2017, p. 19)

Table 7.2 Unemployment by race, 1994 and 2018

Year	*Black/African*	*Colored*	*Indian/Asian*	*White*
1994	43.0	24.0	17.0	7.0
2018	30.4	21.6	12.4	7.6

Source: Statistics South Africa 2018

South Africans first" rather than other groups (IJR, 2017, p. 14), but most said their language and race formed their primary identity. Asked what they thought caused division in society, respondents highlighted inequality (31%) and race (21%) as the biggest factors (IJR, 2017, p. 30). Seventy-seven percent believed that inequality had remained the same or gotten worse since 1994 (IJR, 2017, p. 30). Income inequality, particularly wealth amid widespread poverty, correlates with all types of negative outcomes, including increased violence and obesity, and decreased trust, health, and educational attainment (Wilkinson and Picket, 2010; Elgar and Aitken, 2011; Seekings and Nattrass, 2005). Inequality is particularly destructive when correlated with clearly visible markers such as race or sex, especially when these generate social shame (Wilkinson and Picket, 2010).

Racial inequality remains a major obstacle to South Africa addressing its numerous challenges. Overcoming negativity about diversity is difficult in part because many South Africans do not seem to interact with members of other race groups in power-neutral settings (e.g., in general, only the elite attend schools or work in offices in which white and black occupy positions of comparable status). When surveyed, roughly one-half of all respondents reported interaction with other race groups at home or in social settings, on public transport or in public spaces, though many of these interactions may be superficial (IJR, 2017). Most interaction takes place at work, where power inequality is often stark and racialized (IJR, 2017).

Race and policing

The police have immense powers; their actions can alter the course of an individual's life. Yet police officers use their discretion when deciding how to wield these powers. An established body of research shows that, when people believe that the police will treat them and others fairly, they are more likely to cooperate with the police, obey the law in their absence, and identify with the authority that the police represent (Tyler, 2004; Hough et al., 2010; Bottoms and Tankebe, 2012; Stanko et al., 2012; Bradford et al., 2014). Such policing may indicate police integrity. However, the police may engage with people inconsistently, often based on the age, sex, class, nationality, sexuality, ethnicity, race, or other characteristics of the person with whom they interact.

Race is one of the most persistent themes in police literature. Decades of research show that in multiethnic societie racism has historically been a trait common to police officers and the police organizational culture (Reiner, 2010). As a result, minorities have historically been neglected, mistreated, or criminalized by the police. One response has been to change the demographic makeup of police organizations. In the majority of (historically) white countries such as the United States and the United Kingdom, police agencies intentionally recruited minorities to try to address this issue. Race-based tensions, however, still persists. In 2012, the United States saw the birth of the Black

Lives Matter movement, most prominently in response to the deaths of Trayvon Martin in 2012, as well as Michael Brown and Eric Garner in 2014. In the United Kingdom, riots were sparked in 2011 following the police shooting of Mark Duggen in London. Subsequent investigations found that protesters' anger was fueled by a perceived "lack of respect" and discrimination in the "everyday treatment" by the police (Roberts, 2011, p. 4).

Numerous South African communities, politicians, and civil society organizations have argued that policing (and other public services) is not delivered equally to all people and that race plays a part in this inequality (e.g., Equality Court of South Africa, 2018). Whereas it is true that, behind a rhetoric of human rights and rehabilitation, the county's criminal justice system remains black-focused and punitive (Super, 2013), it is likely that the uneven distribution of police resources and services in South Africa has more to do with class and income than with race. Specifically, the poorest South Africans are black and colored. Following the Khayelitsha Commission of Inquiry (O'Regan & Pikoli 2014), discussed in Chapter 3, the Social Justice Coalition took the SAPS to court to compel it to correct what it argued was a discriminatory (irrational and unjust) allocation of police resources in the City of Cape Town. The crux of the argument was that the police-to-population ratio in the city's most violent police precincts was ten times lower than in the city's more affluent areas. In 2018, the Equality Court ruled in the SJC's favor, finding that police resource allocation in the Western Cape unfairly discriminated against black and poor people. Although the case was specific to the Western Cape, the same resource allocation system is used across the country, and similar patterns of discriminatory allocation of police resources are, therefore, likely to be present elsewhere.

It has also been observed that the police interact with the public and public areas differently, depending on the prevailing affluence in a particular area. For instance, in his ethnography exploring policing in a poor township and a more affluent area in Cape Town, Faull (2018) found that the police were generally more law abiding in affluent (historically white) areas and less so in more impoverished (historically black) areas. Although most people were treated with relative respect in both spaces, those suspected of wrongdoing in poor areas were treated more harshly than those in wealthy areas (Faull, 2018).

The SAPS and its officers have also routinely been accused of xenophobia, usually in relation to corruption and abuse of asylum seekers from sub-Saharan Africa (Masuku, 2006). Various ethnographies of South African policing offer examples of such prejudice (Hornberger, 2011, p. 173; Altbeker, 2005, p. 26; Steinberg, 2011; Faull, 2018, p. 169), with other research suggesting that a disproportionate amount of the SAPS's time and budget is spent policing foreign nationals (e.g., Vigneswaran and Duponchel, 2009). Steinberg (2011) has even suggested that, through their practice of targeting immigrants, the police have helped construct immigrants as a social threat in the South-African mind and contributed to the nationwide xenophobic violence in 2008. However,

South Africans in general hold xenophobic sentiments (Research, 2018), and the SAPS is routinely called upon to protect foreign Africans from South African assailants. In 2008, when foreign nationals were attacked and chased from townships across the country by South Africans who blamed them for a range of social ills, they found shelter at police stations (Wolby et al., 2008; Steinberg, 2008a, 2011, 2014). In early 2019, xenophobic attacks continued to flare up (Fabricius, 2019).

Unlike the case in multicultural first-world democracies, most police officers in South Africa are black, as are most of the country's citizens. Therefore, any perceived discrimination cannot be explained with a logic of white government/police discriminating against minority citizens. Although seldom mentioned, it is notable that South African minorities—other than poor colored men and Africans from elsewhere on the continent—are more likely to be treated with deference than suspicion by the police (Faull, 2018). Because South Africa's minorities are generally more financially secure than black citizens, they are also able to access private security and therefore are less reliant on the public police. More financial power also means minorities have more access to legal practitioners who could pursue legal action against the police for potential abuses of their rights.

Race and the SAPS

Although the SAPS appears demographically balanced in 2018 (Table 7.3), it is not a race-neutral organization. Since 1994, race politics have continued to complicate management and morale in the SAPS as the organization has sought to correct its racially skewed composition. Between 2002 and 2012, for example, the period during which the SAPS grew by 65%, between 94% and 98% of recruits were black (Bruce, 2013, p. 20).

In 2001, the SAPS Conditions of Service Regulations provided for recruitment that ensured the organization was "broadly representative of the South African people, including representation according to race, gender and disability" (Republic of South Africa, 2001, p. 10), a sentiment repeated in the 2008

Table 7.3 SAPS and national demographics, 2018

	African	*Colored*	*Indian*	*White*	*Total*
Number of SAPS officers	115,757	15,760	3,723	14,643	149,833
Percentage of SAPS officers	77.3%	10.5%	2.5%	9.8%	100%
Number of civilian staff	32,700	4,751	1,034	3,985	42,470
Percentage of civilian staff	77%	11%	2.4%	9.4%	100%
Percentage of all staff	77.1%	10.7%	2.5%	9.7%	100%
Percentage of national population	80.90%	8.80%	2.50%	7.80%	100%

Source: SAPS (2018)

regulations. Also in 2001, the SAPS produced a *Promotion of Employment Equity and Elimination of Unfair Discrimination* policy. The policy was framed by the notion that an organization without discrimination would be more efficient and effective, and thus able to provide better services to the public.

Like the average South African, it is highly probable that most SAPS officers do not routinely interact with different race groups outside of work. In two 2016 projects exploring attitudes toward their colleagues and transformation efforts in the SAPS, the authors found that some officers utilized stereotypical judgments of their colleagues or even mistrusted them based on race (Newham et al., 2006; Faull, 2008). Many officers felt they had been treated unfairly as a result of their race or language (a proxy for race), and saw initiatives such as diversity training as ineffective. Most officers in these two studies (Newham et al., 2006; Faull, 2008) were unhappy with the way promotions were managed, a matter that has remained contentious ever since (Burger, 2019; SAPU, 2017).

These tensions have also played out in South African courts. For example, between 2006 and 2014, a white police officer, Lt-Col. Renate Barnard, and the trade union Solidarity, fought a succession of court battles against the SAPS affirmative action policies. Barnard alleged she had been denied promotion twice because she was white. She and Solidarity won a series of judgments until 2014, when the Constitutional Court ruled in favor of the SAPS. However, in 2016 the Johannesburg Labor Court found that the SAPS employment equity plan was unlawful because it promoted a system of quotas (DOJ, 2016) and the Constitutional Court ruled that the Department of Correctional Services must consider both regional and national demographics when setting numerical targets for staff (Constitutional Court, 2016). The latter compelled the SAPS to promote officers previously denied advancement on the grounds of race (van der Bijl, 2017). It is foreseeable that the obvious damage to the trust relationship between such discriminated members and the organization will likely impact the integrity levels of the SAPS in the future.

Having at some point stopped recruiting white officers, the SAPS has more recently struggled to recruit any. In 2017, only 12 of 3,499 trainees were white, resulting in the Minister of Police appealing for more white people to apply to the posts (Mlangeni, 2017). In 2018, the SAPS executive management comprised just one colored, Indian, and white officer, and 28 black officers, while 65% of 877 senior managers were black (SAPS, 2018).

Racism within the SAPS

Expressions of blatant racism by some police officers have continued after 1994. Between 1996 and 1999, there were at least six investigations into allegations of racism within the SAPS (Newham et al., 2006). One of these, the Independent Committee of Inquiry into Racism in the South African Police Service (Zulu et al., 1999), found that racism was institutionalized in the structure, practices, and procedures of the SAPS. Another investigation was conducted by the Human Rights Commission into complaints of racism against black officers by their fellow officers and the white farming community in Vryburg in the North West province (SAHRC, 1999). The Commission found that discrimination, "racial attitudes," and racism prevailed at the station and that these accusations were denied by a white management that appeared ignorant of racial challenges. These and other investigations resulted in renewed pressure for better employment equity and affirmative action policies in the Service.

In 2001, four white SAPS officers were imprisoned after filming themselves using racist language to encourage their police dogs to attack three black Mozambican men (McGreal, 2001). In 2017, the SAPS dismissed an employee for calling for "black apartheid" and "[white] genocide" on Facebook (Dagane vs. SSSBC and Others, 2018). The officer appealed the dismissal in the Labor Court but the decision was ultimately confirmed. In 2019, the SAPS dismissed its Head of Basic Police Development for commenting that there were too many white trainers at the SAPS Oudtshoorn Training Academy, and for implying that they discriminated against black students (Evans, 2019).

Race and discipline in the SAPS

In another indication of a racialized government bureaucracy, the SAPS presents the outcome of disciplinary hearings by race. A simplified version of the data for 2017–2018 is presented in Tables 7.4 and Table 7.5.

Table 7.4 shows that compared to the racial distribution of SAPS's officers, in 2017–2018, colored and Indian employees were more likely to be disciplined than African or white employees. White employees were significantly less likely than others to be formally charged.

Table 7.4 SAPS disciplinary hearings by race, 2017–2018—as a percentage of all employees

	African	*Colored*	*Indian*	*White*
Percent of all SAPS staff	77.13%	10.66%	2.49%	9.72%
Percent of disciplinary cases	76.93%	14.89%	3.87%	4.28%
Not guilty/withdrawn	650 (43%)	144 (49%)	52 (68%)	41 (49%)

Table 7.5 SAPS disciplinary hearings by race, 2017–2018—outcomes

	African	*Colored*	*Indian*	*White*
Disciplinary cases	1508 (100%)	292 (100%)	76 (100%)	84 (100%)
Not guilty/withdrawn	650 (43%)	144 (49%)	52 (68%)	41 (49%)
Guilty/sanctioned	858 (57%)	148 (51%)	24 (32%)	43 (51%)

Source: South African Police Service Annual Report (2017/2018, p. 280)

Table 7.5 suggests that African (black) officers were slightly more likely than others to be sanctioned in a disciplinary hearing, while Indian employees were least likely to be sanctioned.

Race and police integrity

The theory of police integrity (e.g., Klockars et al., 2000, 2004, 2006; Klockars and Kutnjak Ivković, 2004; Kutnjak Ivković, 2015) stresses that establishing and maintaining high levels of police integrity within the police organization are primarily organizational and administrative responsibilities. To that end, the explanations of how police integrity in a police agency is achieved are grounded in three organizational dimensions (emphasizing the role of the police agency in it) and one societal dimension (emphasizing the role of the society at large in shaping the organizational level of police integrity; see Chapter 1 and Chapters 4 to 6) of the theory.

Klockars and colleagues defined police integrity as "the normative inclination among *police* to resist temptations to abuse the rights and privileges of their occupation" (emphasis added, 2006, p. 1). Based on this definition, the police may include not only police agencies but also individual police officers. In fact, Klockars and colleagues (2006, p. 4) noted that "[i]ts deliberate choice (instead of 'police officer' or 'police agency') reflects that police integrity is a characteristic of individual police officers, groups of police officers, or entire police organizations." Such an approach opens avenues to the exploration of whether police integrity differs across groups of police officers, be they defined by their race, gender, rank, or police assignment.

The methodology that Klockars and colleagues developed to measure police integrity empirically (e.g., Klockars et al., 2000, 2004, 2006) relies on a questionnaire with hypothetical scenarios. To minimize the possibility that the respondents could be identified, particularly in smaller police agencies, Klockars and colleagues have purposely limited the number of demographic questions included in the questionnaire (e.g., Kutnjak Ivković, 2015). Hence, there was no question in the original police integrity questionnaire asking the respondents about race.[1] Consequently, we could not find any studies that explored the effect of race on police integrity. Instead, we provide an exploration of the extant literature concerning the effect race has on *any* police attitudes and behavior.

The way police interact with people, particularly in public space, generates social meaning (Loader, 2006). They signal who belongs and what is acceptable in a political community. The record on police organizational culture makes clear that societal prejudices have historically traveled into police organizations, taking center stage in their cultures (Cockcroft, 2013; Reiner, 2010; Skolnick, 2011). As a result, policing often does, or is perceived to, unfold in different ways, depending on the people involved. For instance, Zimring shows that, in a sample of police calls in the United States, African-Americans and minorities are significantly overrepresented, but more so in calls requiring criminal law intervention or arrest than those that do not request such services (Zimring, 2015, p. 54).

Despite this history, Loftus (2009, p. 193) and Skolnick (2008) indicate that the police they had in recent years observed in the U.K. and in the U.S.A. were aware that racist attitudes were no longer acceptable. Loftus believes that the focus on diversity in the British police has significantly reduced overt expressions of racism and that an emerging culture is developing that challenges old cultural values. Nevertheless, she warns that it would be premature to overstate the power of these counter-narratives. Sklansky suggests that the emergent diversity in many American police agencies has opened up spaces for a debate challenging cultural homogeneity and that such trends are positive (Sklansky, 2007, p. 36). While the SAPS has always been culturally and racially diverse, it has never been a purely meritocratic organization, and social tension persists among employees (see also Chapters 2 and 10).

When Marks ended her ethnography with the Public Order Unit in KwaZulu-Natal in 2001, she believed that, despite efforts at reform, it remained plagued by class, race, and gender divisions (Marks, 2005). Hornberger (2011) found that the human rights sensitization workshops she attended in the SAPS focused almost exclusively on issues of race. One session was canceled because the facilitator did not see the point in continuing when the only people attending were black. Similarly, Faull's (2008) mini-audit of diversity training in the SAPS found that this diversity training reinforced, rather than challenged, apartheid race labels, while in the organizational environment language replaced race as the source of internal solidarities and divisions.

On his ride-alongs with the police, Altbeker (2005) was told that crime in a historically white suburb was committed by outsiders. Altbeker reports that policing in the area involved regular and random stopping and searching of black people (2005, pp. 115–126). Steinberg (2011) found the same with the policing of poor (usually African) foreign nationals, while Faull (2018) found the same of poor black men in townships in general (the latter being stopped and searched by black police). Diphoorn (2015) again observed old race-based tropes and actions from the private security officers she shadowed, many of whom were former police.

A number of officers whom Faull interviewed in 2009 reported the most important thing they had learned from serving in the police was that the stereotypes with which they had been raised were false (Faull, 2010). Yet, as the following quote from a white constable illustrates, even in the act of challenging old narratives, stereotypes are often reinforced (Faull, 2010, p. 252):

> You can't just say "Okay, he's black; he's bad." Some of the black colleagues can really work; they're a lot like myself. One or two guys, if you wrote our personalities down on a piece of paper without our names, you might say we are family, we are brothers.

This muddled reinforcement of decades-old racial essentialism suggests that SAPS officers not only signal who and what is acceptable, but they also carry cultural meaning across the boundary that separates the police institution from the broader society. As they transition from the lives of (generally unemployed) civilians to those of working police, and then daily between their work and home lives *as* police, officers draw on community and national narratives to amend and update both their personal identities and the organizational culture of their institution.

Survey findings

We surveyed 868 SAPS officers across all provinces and territories. Respondents were requested to assess 14 hypothetical scenarios covering a range of forms of police misconduct. The questionnaire was designed by Klockars and colleagues (2006), and revised by Kutnjak Ivković (2009). For details about the methodology, please see Chapter 1.

After the respondents had read each of the hypothetical scenarios, they were asked to answer a series of seven questions. One of the first questions asked the respondents to assess whether the behavior described in the questionnaire violates the official rules ("Would this behavior be regarded as a violation of official policy in your agency?"). Possible answers ranged on a 5-point scale from 1 = "definitely not" to 5 = "definitely yes." Two questions asked the respondents to assess misconduct seriousness ("How serious do YOU consider this behavior to be?" "How serious do most police officers in your agency consider this behavior to be?"). The respondents could have selected an answer on a 5-point scale from 1 = "not at all serious" to 5 = "very serious." Then, two questions inquired about the appropriate discipline and expected discipline ("If an officer in your agency engaged in this behavior and was discovered doing so, what, if any, discipline do you think should follow?" "If an officer in your agency engaged in this behavior and was discovered doing so, what, if any, discipline do you think would follow?"). The respondents could have selected one of the six disciplinary options, ranging from 1 = "no discipline" to 6 = "dismissal." Finally, two questions inquired about the code of silence ("Do you think you would report a fellow police officer who engaged in this behavior?" "Do you think most police officers in your agency would report a fellow police officer who engaged in this behavior?"). The respondents could have selected one answer from a 5-point scale ranging from 1 = "definitely not" to 5 = "definitely yes."

To facilitate the respondents' comfort with filling in the questionnaire, we purposely limited the number of demographic questions. One of the demographic questions we did include inquired about the respondents' race. Possible answers were "white," "black," and "other" (Colored, Indian). About 16% of the respondents described themselves as "white," 58% as "black," and 25% as "other."

Familiarity with the official rules

The respondents were asked to assess whether the behavior described in each of the scenarios in the questionnaire violated the SAPS official rules. In about one-half of the scenarios (7 out of 14 or 50%; Table 7.6), there were statistically significant differences in the mean evaluations of seriousness by race.[2] However, although statistically significant, these differences were neither large[3] nor substantive.[4]

We analyzed the pairs that yielded statistically significant differences (e.g., white and black respondents, black and other respondents). Blacks' and others' familiarity with the rules was very similar; there was no scenario in which there were statistically significant differences (Table 7.6). On the other hand, familiarity with the official rules was different across white respondents and black respondents, and across white respondents and other respondents. Specifically, in about one-half of the scenarios (7 out of 14, or 50%), the average familiarity with the official rules was different for white and other respondents, and in the overwhelming majority of these scenarios (6 out of 7 or 86%), other respondents seemed more likely to evaluate these behaviors as rule-violating than white respondents did. Differences in the familiarity with official rules between white and black respondents were somewhat less pronounced; they were statistically significant in about one-third of the scenarios (5 out of 14, or 35%) and, in all of these scenarios, black respondents seemed more likely to recognize behaviors as rule-violating.

The relative order of familiarity with the official rules—how familiar respondents are with each scenario compared to the other scenarios in the questionnaire—is shown in Table 7.6 (ranks). Black and other respondents were very similar in how likely they were to say that the scenario is a violation of official rules compared to other scenarios in the questionnaire (Spearman's rho = .920; $p < .001$). While still strongly correlated, the relative order of familiarity with official rules for white respondents and black respondents (Spearman's rho = .727; $p < .01$) and for white respondents and other respondents (Spearman's rho = .662; $p < .05$) was not as similar.

Perceptions of misconduct seriousness

Upon reading the description of police misconduct described in each scenario, the respondents were asked to assess its seriousness and to predict how serious other police officers would evaluate the same behavior to be. In about two-thirds of the scenarios (9 out of 14 or 64%; Table 7.7), respondents' race was related to their evaluations of seriousness.[5] However, while statistically significant, only three of these differences were large[6] and only one was substantively important.[7] The mean differences were large in only one scenario (Scenario 12), getting close to, or even above, the threshold of .50 that indicates substantive differences, as established by Klockars and colleagues (2006).

Table 7.6 Violation of official rules by race

Scenario	*White (1)*		*Black (2)*		*Other (3)*		*F-test (Tukey b's post-hoc)*	*1 v. 2*	*1 v. 3*	*2 v. 3*
	Mean	*Rank*	*Mean*	*Rank*	*Mean*	*Rank*				
Scenario 1: Acceptance of gratuities	3.97	3	4.27	5	4.42	8	6.34** (1 v. 2; 1 v. 3)	−.30	−.45	−.15
Scenario 2: Failure to arrest	4.38	11	4.63	12	4.61	11	4.62* (1 v. 2; 1 v. 3)	−.25	−.23	.02
Scenario 3: Theft of knife	4.68	14	4.77	14	4.75	13.5	.812	−.09	−.07	.02
Scenario 4: Abuse of deadly force	4.51	13	4.43	8	4.57	10	1.62	.08	−.06	−.14
Scenario 5: Supervisor corruption	4.35	10	4.42	7	4.22	5	2.75	−.07	.13	.20
Scenario 6: Striking arrestee	4.03	4	4.46	9	4.32	7	9.16*** (1 v. 2; 1 v. 3)	−.43	−.29	.14
Scenario 7: Verbal abuse	4.07	5	4.00	3	3.76	1	3.76* (1 v. 3)	.07	.31	.24
Scenario 8: Cover-up of DUI	4.24	7	3.94	2	4.00	3	2.63	.30	.24	−.06
Scenario 9: Kickback	4.23	6	4.37	6	4.28	6	1.22	−.14	−.05	.09
Scenario 10: False report on drugs	4.32	9	4.49	10	4.62	12	4.65* (1 v. 3)	−.17	−.30	−.13
Scenario 11: Supervisor fails to stop excessive force	4.46	12	4.51	11	4.49	9	.17	−.05	−.03	.02
Scenario 12: Failure to report hate crime	3.92	2	4.22	4	4.14	4	4.59* (1 v. 2; 1 v. 3)	−.30	−.22	.08
Scenario 13: Bribe from reckless motorist	4.31	8	4.75	13	4.75	13.5	20.13*** (1 v. 2; 1 v. 3)	−.44	−.44	.00
Scenario 14: Failure to react to graffiti	3.81	1	3.67	1	3.89	2	2.49	.14	−.08	−.22

Table 7.7 Own views about misconduct seriousness by race

Scenario	*White (1)*		*Black (2)*		*Other (3)*		*F-test (Tukey b's post-hoc)*	*1 v. 2*	*1 v. 3*	*2 v. 3*
	Mean	*Rank*	*Mean*	*Rank*	*Mean*	*Rank*				
Scenario 1: Acceptance of gratuities	3.65	1	3.94	2	4.05	4	4.12* (1 v. 2; 1 v. 3)	−.29	−.40	−.11
Scenario 2: Failure to arrest	4.43	12	4.60	11	4.70	12	4.10* (1 v. 3)	−.17	−.27	−.10
Scenario 3: Theft of knife	4.69	14	4.81	14	4.78	13	2.48	−.12	−.09	.03
Scenario 4: Abuse of deadly force	4.38	10	4.61	12	4.60	11	3.70* (1 v. 2; 1 v. 3)	−.23	−.22	.01
Scenario 5: Supervisor corruption	4.03	7	4.40	7	4.45	9	7.91*** (1 v. 2; 1 v. 3)	−.37	−.42	−.05
Scenario 6: Striking arrestee	3.94	5	4.42	8	4.22	7	12.06*** (1 v. 2; 1 v. 3)	−.48	−.28	.20
Scenario 7: Verbal abuse	3.75	3	4.05	4	3.57	1	12.65*** (1 v. 2; 2 v. 3)	−.30	.18	.48
Scenario 8: Cover-up of DUI	3.99	6	3.95	3	3.90	3	.16	.04	.09	.05
Scenario 9: Kickback	4.42	11	4.36	6	4.18	6	2.78	.06	.24	.18
Scenario 10: False report on drugs	4.45	13	4.52	9	4.57	10	.70	−.07	−.12	−.05
Scenario 11: Supervisor fails to stop excessive force	4.24	8	4.58	10	4.32	8	9.06*** (1 v. 2)	−.34	−.08	.26
Scenario 12: Failure to report hate crime	3.68	2	4.31	5	4.17	5	20.80*** (1 v. 2; 1 v. 3)	−.63	−.49	.14
Scenario 13: Bribe from reckless motorist	4.33	9	4.78	13	4.81	14	23.29*** (1 v. 2; 1 v. 3)	−.45	−.48	−.03
Scenario 14: Failure to react to graffiti	3.81	4	3.68	1	3.75	2	.65	.13	.06	−.07

With one exception (Scenario 7: Verbal abuse), evaluations of seriousness provided by black and other respondents were very similar, never resulting in a statistically significant difference (Table 7.7). On the other hand, estimates of misconduct seriousness provided by white respondents were significantly different from the estimates of seriousness provided by black respondents in eight scenarios and by other respondents in seven scenarios (Table 7.7). In all eight scenarios with statistically significant differences, black respondents evaluated misconduct as more serious than white respondents did (Table 7.7). Similarly, in all seven scenarios with statistically significant differences, other respondents evaluated misconduct as more serious than white respondents did (Table 7.7).

The relative order of seriousness evaluations—how serious respondents evaluated each scenario compared to the other scenarios in the questionnaire—is shown in Table 7.7 (ranks). Black and other respondents were very similar in regard to the seriousness with which they evaluated each scenario compared to the other scenarios in the questionnaire (Spearman's rho = .938; $p < .001$). While still strongly correlated, the relative order of seriousness for white respondents and black respondents (Spearman's rho = .771; $p < .01$) and for white respondents and other respondents (Spearman's rho = .771; $p < .01$) was not as similar.

The next question focuses on the respondents' estimates of how serious most police officers in their police agency would evaluate the same behavior. The opinions among the three groups of respondents were even more heterogeneous (Table 7.8); there were statistically significant differences in over three-quarters of the scenarios (11 out of 14 scenarios or 78%; Table 7.3).[8] In five scenarios (Table 7.8), the differences were on the larger side, above .50, and, hence, substantively important.[9] They were also considered to be of medium to large size, according to Cohen's *d* statistic.[10]

Most police officers' estimates of seriousness are by far the most divergent between white and black respondents (10 out of 14 scenarios or 71%; Table 7.8). Differences were statistically significant in all of these scenarios, with black respondents estimating that most police officers would evaluate misconduct as more serious than white respondents did. Although other respondents agreed with white and black respondents about how seriously most police officers would evaluate the behavior in the majority of the scenarios, their views were different in four scenarios from the views expressed by white respondents and in five scenarios from the views expressed by black respondents (Table 7.8). In such scenarios with statistically significant differences, other respondents estimated that most SAPS officers would evaluate the behavior as more serious than white respondents did. On the other hand, in the scenarios in which the estimates of other respondents' evaluations of most police officers' evaluations of seriousness were significantly different from the black police officers' views, other respondents assumed that most police officers would be less likely to evaluate them as serious than black officers did.

The relative order of most officers' seriousness evaluations—how serious the respondents perceived that most officers would evaluate each scenario

Table 7.8 Most police officers' views about misconduct seriousness by race

Scenario	*White (1)*		*Black (2)*		*Other (3)*		*F-test (Tukey b's post-hoc)*	*1 v. 2*	*1 v. 3*	*2 v. 3*
	Mean	*Rank*	*Mean*	*Rank*	*Mean*	*Rank*				
Scenario 1: Acceptance of gratuities	3.17	2	3.44	1	3.42	3	2.17	−.27	−.25	.02
Scenario 2: Failure to arrest	3.69	10	4.08	7	3.95	10	6.39** (1 v. 2; 1 v. 3)	−.39	−.26	.13
Scenario 3: Theft of knife	3.94	12	4.40	13	4.14	12	12.31*** (1 v. 2)	−.46	−.20	.26
Scenario 4: Abuse of deadly force	4.14	14	4.35	12	4.30	13.5	2.11	−.21	−.16	.05
Scenario 5: Supervisor corruption	3.37	5.5	4.01	5	3.92	9	13.60*** (1 v. 2; 1 v. 3)	−.64	−.55	.09
Scenario 6: Striking arrestee	3.37	5.5	4.03	6	3.52	5	20.51*** (1 v. 2; 2 v. 3	−.66	−.15	.51
Scenario 7: Verbal abuse	3.34	4	3.71	4	3.23	1	11.10*** (1 v. 2; 2 v. 3)	−.37	.11	.48
Scenario 8: Cover-up of DUI	3.46	7	3.68	3	3.46	4	2.39	−.22	.00	.22
Scenario 9: Kickback	3.86	11	4.10	9	3.65	6	10.42*** (2 v. 3)	−.24	.21	.45
Scenario 10: False report on drugs	3.96	13	4.24	11	4.05	11	4.26* (1 v. 2)	−.28	−.09	.19
Scenario 11: Supervisor fails to stop excessive force	3.65	8.5	4.18	10	3.82	7.5	12.51*** (1 v. 2; 2 v. 3)	−.53	−.17	.36
Scenario 12: Failure to report hate crime	3.20	3	4.09	8	3.82	7.5	30.76*** (1 v. 2; 1 v. 3; 2 v. 3)	−.89	.27	−.62
Scenario 13: Bribe from reckless motorist	3.65	8.5	4.51	14	4.30	13.5	37.91*** (1 v. 2; 1 v. 3; 2 v. 3)	−.86	−.65	.21
Scenario 14: Failure to react to graffiti	3.08	1	3.47	2	3.39	2	4.45* (1 v. 2; 1 v. 3)	−.39	−.31	.08

compared to the other scenarios in the questionnaire—is shown in Table 7.8 (ranks). Black and other respondents were very similar in how serious they assume most officers would evaluate each scenario compared to the other scenarios in the questionnaire (Spearman's rho = .870; $p < .001$). While still strongly correlated, the relative order of seriousness by most police officers for white respondents and black respondents (Spearman's rho = .784; $p < .01$) and for white respondents and other respondents (Spearman's rho = .762; $p < .01$) was not as similar.

Finally, we compared the respondents' own evaluations of seriousness with what they assumed would be most police officers' evaluations of seriousness (Table 7.9). Because of self-serving bias, it is not surprising that all three racial groups expected that most police officers would be significantly less likely to evaluate misconduct as serious than the respondents themselves would (all 14 scenarios resulted in statistically significant differences for all three groups of respondents; Table 7.9). On the other hand, the size of these differences was not the same across all three groups; while white and other respondents expected that most police officers would evaluate misconduct as *substantively* less serious than the respondents themselves would in about two-thirds of the scenarios (Table 7.9),[11] black respondents' views were much more closely aligned with how they expected most officers to evaluate misconduct, resulting in a discrepancy in only two scenarios.

Views about discipline

The respondents were also asked to determine the appropriate discipline for a police officer who engages in the behaviors described in the questionnaire and state what discipline they anticipate the SAPS would mete out for those behaviors. In about three-quarters of the scenarios (12 out of 14, or 86%, for appropriate discipline, Table 7.9;[12] 10 out of 14, or 71%, for expected discipline, Table 7.10),[13] race was related to the respondents' views about both appropriate and expected discipline.

We explored which of the pairs with differences were statistically significant and found that in all 12 scenarios with statistically significant differences for appropriate discipline, the percentage of white respondents who thought that more severe discipline is appropriate is larger than the corresponding percentage of black respondents (Table 7.10). Differences in the views about appropriate discipline are less pronounced between white and other respondents, as there were only four scenarios with statistically significant differences, and in all of those scenarios, a higher percentage of white respondents expected more severe discipline than other respondents did. Finally, there were only two scenarios in which the differences in the appropriate discipline between black and other respondents were statistically significant; in both of them, the percentage of other respondents who stated that the more severe discipline is

Table 7.9 Views about misconduct seriousness by race

Scenario	*White Own*		*White Most*		*t-test*	*Diff.*	*Black Own*		*Black Most*		*t-test*	*Diff.*
	Mean	*Rank*	*Mean*	*Rank*			*Mean*	*Rank*	*Mean*	*Rank*		
Scenario 1: Acceptance of gratuities	3.65	1	3.17	2	4.00***	.48	3.94	2	3.44	1	8.42***	.50
Scenario 2: Failure to arrest	4.43	12	3.69	10	7.86***	.74	4.60	11	4.08	7	9.89***	.52
Scenario 3: Theft of knife	4.69	14	3.94	12	7.25***	.75	4.81	14	4.40	13	9.65***	.41
Scenario 4: Abuse of deadly force	4.38	10	4.14	14	2.70**	.24	4.61	12	4.35	12	5.82***	.26
Scenario 5: Supervisor corruption	4.03	7	3.37	5.5	5.36***	.66	4.40	7	4.01	5	7.65***	.39
Scenario 6: Striking arrestee	3.94	5	3.37	5.5	5.36***	.57	4.42	8	4.03	6	8.55***	.39
Scenario 7: Verbal abuse	3.75	3	3.34	4	5.92***	.41	4.05	4	3.71	4	7.54***	.39
Scenario 8: Cover-up of DUI	3.99	6	3.46	7	5.60***	.53	3.95	3	3.68	3	6.66***	.34
Scenario 9: Kickback	4.42	11	3.86	11	6.97***	.56	4.36	6	4.10	9	6.35***	.26
Scenario 10: False report on drugs	4.45	13	3.96	13	5.46***	.49	4.52	9	4.24	11	7.22***	.28
Scenario 11: Supervisor fails to stop excessive force	4.24	8	3.65	8.5	5.05***	.59	4.58	10	4.18	10	9.18***	.39
Scenario 12: Failure to report hate crime	3.68	2	3.20	3	4.82***	.48	4.31	5	4.09	8	5.78***	.22
Scenario 13: Bribe from reckless motorist	4.33	9	3.65	8.5	6.96***	.68	4.78	13	4.51	14	7.03***	.26
Scenario 14: Failure to react to graffiti	3.81	4	3.08	1	7.52***	.73	3.68	1	3.47	2	6.11***	.22

Scenario	*Other Own*		*Other Most*		*t-test*	*Diff.*
	Mean	*Rank*	*Mean*	*Rank*		
Scenario 1: Acceptance of gratuities	4.05	4	3.43	3	7.22***	.62
Scenario 2: Failure to arrest	4.70	12	3.95	10	10.66***	.75
Scenario 3: Theft of knife	4.78	13	4.14	12	9.13***	.64
Scenario 4: Abuse of deadly force	4.60	11	4.30	13.5	5.10***	.30
Scenario 5: Supervisor corruption	4.45	9	3.92	9	6.27***	.53
Scenario 6: Striking arrestee	4.22	7	3.52	5	9.85***	.70
Scenario 7: Verbal abuse	3.57	1	3.23	1	5.89***	.34
Scenario 8: Cover-up of DUI	3.90	3	3.46	4	6.47***	.44
Scenario 9: Kickback	4.18	6	3.65	6	6.50***	.53
Scenario 10: False report on drugs	4.57	10	4.05	11	7.86***	.52
Scenario 11: Supervisor fails to stop excessive force	4.32	8	3.82	7.5	7.74***	.50
Scenario 12: Failure to report hate crime	4.17	5	3.82	7.5	5.63***	.35
Scenario 13: Bribe from reckless motorist	4.81	14	4.30	13.5	7.23***	.51
Scenario 14: Failure to react to graffiti	3.75	2	3.39	2	5.90***	.36

Table 7.10 Views about the appropriate discipline by race

Scenario	*White (1)*		*Black (2)*		*Other (3)*		*Chi-square/Phi (Post-hoc)*	*1 v. 2*	*1 v. 3*	*2 v. 3*
	% serious	*Rank*	*% serious*	*Rank*	*% serious*	*Rank*				
Scenario 1: Acceptance of gratuities	34.5%	5	21.0%	3	30.3%	5	14.14**/.128 (1 v. 2)	13.5%	3.2%	−.9.7%
Scenario 2: Failure to arrest	50.7%	8	29.3%	8	37.6%	8	23.25***/.164 (1 v. 2)	21.4%	13.1%	−8.3%
Scenario 3: Theft of knife	78.2%	12	56.8%	13	67.9%	12	24.54***/.168 (1 v. 2)	21.4%	10.3%	−11.1%
Scenario 4: Abuse of deadly force	90.8%	14	53.7%	12	74.7%	14	79.38***/.302 (1 v. 2; 1 v. 3; 2 v. 3)	37.1%	16.1%	−21.0%
Scenario 5: Supervisor corruption	43.0%	6	27.9%	6	35.7%	7	12.99**/.122 (1 v. 2)	15.1%	7.3%	−7.8%
Scenario 6: Striking arrestee	17.6%	1	28.8%	7	31.2%	6	8.90*/.101 (1 v. 2; 1 v. 3)	−11.2%	−13.6%	−2.4%
Scenario 7: Verbal abuse	23.2%	3	10.3%	1	14.9%	1	16.18***/.137 (1 v. 2)	12.9%	8.3%	−4.6%
Scenario 8: Cover-up of DUI	46.5%	7	23.8%	4	25.8%	3	29.08***/.183 (1 v. 2; 1 v. 3)	22.7%	20.7%	−2.0%
Scenario 9: Kickback	51.4%	9	38.2%	10	42.1%	10	8.02*/.096 (no pairs)*	13.2%	9.3%	−3.9%
Scenario 10: False report on drugs	68.3%	11	39.0%	11	55.7%	11	45.01***/.228 (1 v. 2; 2 v. 3)	29.3%	12.6%	−16.7%
Scenario 11: Supervisor fails to stop excessive force	62.7%	10	36.4%	9	39.4%	9	31.98***/.192 (1 v. 2; 1 v. 3)	26.3%	23.2%	−3.0%
Scenario 12: Failure to report hate crime	30.3%	4	26.3%	5	28.1%	4	.93/.033	4.0%	2.2%	−1.8%
Scenario 13: Bribe from reckless motorist	79.6%	13	63.2%	14	70.1%	13	14.48**/.129 (1 v. 2)	16.4%	9.5%	−6.9%
Scenario 14: Failure to react to graffiti	21.1%	2	11.5%	2	15.8%	2	9.15/.103	9.6%	5.3%	−4.3%

*Based on the results of bivariate analyses.

appropriate was higher than the corresponding percentage of black respondents (Table 7.10).

In just below one-half of all scenarios (6 out of 14 scenarios; Table 7.10), these differences were also substantively important, indicating that white respondents thought that more severe discipline was appropriate in these scenarios than black respondents did (Table 7.10). There were even fewer substantively important differences between white and other respondents' views about appropriate discipline (Table 7.10). Finally, there were no large and substantive differences in the views about the appropriate discipline between black and other respondents.

The relative order of appropriate discipline—how severely respondents evaluated the appropriate discipline to be for each scenario compared to the other scenarios in the questionnaire—is shown in Table 7.9 (ranks). The relative order for all three groups is very similar (black and other respondents: Spearman's rho = .969; $p < .001$; white and black respondents: Spearman's rho = .864; $p < .01$; white and other respondents: Spearman's rho = .895; $p < .01$).

In the majority of the scenarios (10 out of 14 or 71%), there were statistically significant differences in the respondents' views about the discipline they expected the SAPS to mete out as well (Table 7.11). Of the pairs with statistically significant differences, views about expected discipline are the most different between white and black respondents (8 out of 14 scenarios; Table 7.11), and in all these scenarios with statistically significant differences, the percentage of white respondents who expected more severe discipline was higher than the corresponding percentage of black respondents. Although expectations of discipline that the SAPS would mete out are more similar between white and other respondents, resulting in statistically significant differences in only about one-third of the scenarios, in most of these scenarios with statistically significant differences, a higher percentage of white respondents than other respondents expected more severe discipline (Table 7.11). Finally, the views about the expected discipline are most similar for black and other respondents, resulting in only one scenario with statistically significant differences (Scenario 4).

Although the views about anticipated discipline differed by race, the differences were not large and only rarely reached the level of substantive differences (Table 7.11). There was only one scenario (Scenario 4) in which a substantively smaller percentage of black respondents expected more severe discipline than either the white respondents or the other respondents did.

The relative order of expected discipline—how severe expected discipline is for each scenario compared to the other scenarios in the questionnaire—is shown in Table 7.10 (ranks). The relative order for all three groups is very similar (black and other respondents: Spearman's rho = .969; $p < .001$; white and black respondents: Spearman's rho = .864; $p < .01$; white and other respondents: Spearman's rho = .895; $p < .01$).

Table 7.11 Views about the expected discipline by race

Scenario	*White (1)*		*Black (2)*		*Other (3)*		*Chi-square/Phi (Post-hoc)*	*1 v. 2*	*1 v. 3*	*2 v. 3*
	% serious	*Rank*	*% serious*	*Rank*	*% serious*	*Rank*				
Scenario 1: Acceptance of gratuities	33.8%	4	20.2%	3	27.1%	4	12.60**/.120 (1 v. 2)	13.6%	6.7%	−6.9%
Scenario 2: Failure to arrest	47.2%	9	30.9%	7	39.4%	9.5	14.54**/.129 (1 v. 2)	16.3%	7.8%	−8.5%
Scenario 3: Theft of knife	67.6%	13	59.6%	13	67.4%	12	5.63/.081	8.0%	0.2%	−7.8%
Scenario 4: Abuse of deadly force	90.1%	14	55.2%	12	76.9%	14	75.08***/.294 (1 v. 2; 1 v. 3; 2 v. 3)	34.9%	13.2%	−21.7%
Scenario 5: Supervisor corruption	40.1%	5.5	31.1%	8	35.3%	7	4.43/.071	9.0%	4.8%	−4.2%
Scenario 6: Striking arrestee	28.9%	3	30.7%	6	31.2%	5	.24/.017	−1.8%	−2.3%	−0.5
Scenario 7: Verbal abuse	24.6%	2	10.9%	2	12.2%	1	18.39***/.146 (1 v. 2; 1 v. 3)	13.7%	12.4%	−1.3%
Scenario 8: Cover-up of DUI	41.5%	7	22.6%	4	20.8%	3	24.22***/.167 (1 v. 2; 1 v. 3)	18.9%	20.7%	1.8%
Scenario 9: Kickback	40.1%	5.5	37.6%	9	39.4%	9.5	.40/.021	2.5%	0.7%	−1.8%
Scenario 10: False report on drugs	60.6%	11.1	41.6%	11	56.1%	11	23.06***/.163 (1 v. 2)	19.0%	4.5%	−14.5%
Scenario 11: Supervisor fails to stop excessive force	51.4%	10	37.8%	10	38.0%	8	9.05*/.102 (no pairs)*	13.6%	13.4%	−0.2%
Scenario 12: Failure to report hate crime	43.7%	8	26.5%	5	31.7%	6	15.45***/.133 (1 v. 2)	17.2%	12.0%	−5.2%
Scenario 13: Bribe from reckless motorist	60.6%	11.5	62.9%	14	71.9%	13	6.87*/.089 (no pairs)*	−2.3%	−11.3%	−9.0%
Scenario 14: Failure to react to graffiti	22.5%	1	10.5%	1	13.6%	2	14.01**/.127 (1 v. 2)	12.0%	8.9%	−3.1%

*Based on the results of bivariate analyses.

Expected willingness to report

The last two questions asked the respondents to express their own willingness to report misconduct and estimate how willing most police officers would be to report it. In about one-half of the scenarios (about 7 out of 14 or 50%; Table 7.12),[14] race was related to the respondents' expressed willingness to report. In all but one scenario with differences in the expressed willingness to report between white and black respondents, black respondents were statistically more likely to say that they would report misconduct (Table 7.12). Similarly, in all scenarios (5 out of 14 or 36%) in which there were statistically significant differences between the respondents' expressed willingness to report between white and other respondents, other respondents seemed more likely to say that they would report. Finally, the expressed willingness to report was very similar between black and other respondents; in the only two scenarios with the statistically significant differences in the expressed willingness to report between black and other respondents, black respondents were statistically more likely to say that they would report. However, although these differences were statistically significant, none of them passed the threshold of substantive differences established by Klockars and colleagues (2006). They were also not considered to be large.[15]

The relative order of officers' own expressed willingness to report—how willing the respondents are to report behavior described in other scenario compared to the behaviors in other scenarios—is shown in Table 7.12 (ranks). Black and other respondents (Spearman's rho = .894; $p < .001$) and white and other respondents (Spearman's rho = .887; $p < .001$) were more similar in their relative evaluations than white and black respondents were (Spearman's rho = .787; $p < .01$).

The second question invited the respondents to assess how willing most police officers would be to report such behavior (Table 7.13). In three-quarters of the scenarios (about 11 out of 14 or 78%), race was related to how willing most police officers would be to report misconduct (Table 7.13).[16] These differences were not only statistically significant but also large in four scenarios (Scenario 2, Scenario 3, Scenario 9, and Scenario 11),[17] and of substantive importance in seven scenarios.[18]

The effect of race was particularly strong when the white respondents' estimates were compared to the black respondents' estimates (Table 7.13); in all of these scenarios (11 out of 14), black respondents were more likely to assume that most police officers would report such behavior than white respondents were. Similarly, with one exception (Scenario 9), black respondents were also more likely to assume that most police officers would report than other respondents were in six scenarios with statistically significant differences (Table 7.13). In addition, while the differences were less pronounced between white and other respondents, white respondents were less likely than other respondents to assume that most police officers would report in five scenarios with statistically significant differences (Table 7.13).

Table 7.12 Own willingness to report by race

Scenario	*White (1)*		*Black (2)*		*Other (3)*		*F-test*	*1 v. 2*	*1 v. 3*	*2 v. 3*
	Mean	*Rank*	*Mean*	*Rank*	*Mean*	*Rank*	*(Tukey b's post-hoc)*			
Scenario 1: Acceptance of gratuities	3.37	2	3.81	5	3.66	6	5.51** (1 v. 2)	−.44	−.29	.15
Scenario 2: Failure to arrest	3.85	9	4.33	11.5	4.16	11	10.96*** (1 v. 2; 1 v. 3)	−.48	−.31	.15
Scenario 3: Theft of knife	4.44	13.5	4.38	13	4.37	12	0.24	.06	.07	.01
Scenario 4: Abuse of deadly force	4.23	12	4.33	11.5	4.48	13	2.41	−.10	−.25	−.15
Scenario 5: Supervisor corruption	3.67	6	3.98	8	3.78	7	3.96* (1 v. 2)	−.31	−.11	.20
Scenario 6: Striking arrestee	3.82	8	3.90	7	3.53	4	5.98** (1 v. 3; 2 v. 3)	−.08	.29	.37
Scenario 7: Verbal abuse	3.21	1	3.47	2.5	3.01	1	8.07*** (2 v. 3)	−.26	.20	.46
Scenario 8: Cover-up of DUI	3.51	3	3.47	2.5	3.36	2	0.55	.04	.15	.11
Scenario 9: Kickback	3.55	4	3.89	6	3.57	5	5.52** (1 v. 2)	−.34	−.02	.32
Scenario 10: False report on drugs	3.86	10	4.10	10	4.09	10	2.25	−.24	−.23	.01
Scenario 11: Supervisor fails to stop excessive force	3.80	7	4.00	9	3.80	8	2.51	−.20	.00	.20
Scenario 12: Failure to report hate crime	3.98	11	3.76	4	3.81	9	1.73	.14	.17	−.05
Scenario 13: Bribe from reckless motorist	4.44	13.5	4.43	14	4.54	14	1.18	.01	.10	−.12
Scenario 14: Failure to react to graffiti	3.58	5	3.24	1	3.49	3	5.35** (1 v. 2)	.34	.09	−.25

Table 7.13 Most police officers' willingness to report by race

Scenario	*White (1)*		*Black (2)*		*Other (3)*		*F-test (Tukey b's post-hoc)*	*1 v. 2*	*1 v. 3*	*2 v. 3*
	Mean	*Rank*	*Mean*	*Rank*	*Mean*	*Rank*				
Scenario 1: Acceptance of gratuities	2.73	4	3.07	2	2.71	1	6.07** (1 v. 2)	−.34	.02	.32
Scenario 2: Failure to arrest	2.83	7	3.68	11	3.36	9	24.15*** (1 v. 2; 1 v. 3; 2 v. 3)	−.85	−.53	.32
Scenario 3: Theft of knife	3.07	11	3.80	12	3.52	11.5	16.95*** (1 v. 2; 1 v. 3; 2 v. 3)	−.73	−.45	.28
Scenario 4: Abuse of deadly force	3.73	14	3.89	13	3.80	13	.98	−.16	−.07	.09
Scenario 5: Supervisor corruption	3.09	12	3.67	10	3.21	7	15.25*** (1 v. 2; 2 v. 3)	−.58	−.12	.46
Scenario 6: Striking arrestee	2.96	8	3.51	8	2.91	5	18.44*** (1 v. 2; 2 v. 3)	−.55	.05	.60
Scenario 7: Verbal abuse	2.71	3	3.11	3	2.73	2	7.50** (1 v. 2; 2 v. 3)	−.40	−.02	.38
Scenario 8: Cover-up of DUI	2.75	6	3.17	4	2.79	3	7.52** (1 v. 2; 2 v. 3)	−.42	−.04	.38
Scenario 9: Kickback	2.70	2	3.09	6	3.48	6	17.69*** (1 v. 2; 1 v. 3; 2 v. 3)	−.39	−.78	−.39
Scenario 10: False report on drugs	3.03	9	3.50	7	3.52	11.5	7.02*** (1 v. 2; 1 v. 3)	−.47	−.49	−.02
Scenario 11: Supervisor fails to stop excessive force	2.74	5	3.52	9	3.34	8	17.43*** (1 v. 2; 1 v. 3)	−.78	−.60	.18
Scenario 12: Failure to report hate crime	3.06	10	3.41	5	3.45	10	4.70* (1 v. 2; 1 v. 3)	−.35	−.39	−.04
Scenario 13: Bribe from reckless motorist	3.63	13	3.91	14	3.84	14	2.30	−.28	−.21	.07
Scenario 14: Failure to react to graffiti	2.64	1	2.94	1	2.89	4	2.63	−.30	−.25	.05

The relative order of most officers' expressed willingness to report—how willing the respondents assumed that most officers would be to report behavior described in each scenario compared to the behaviors in other scenarios—is shown in Table 7.13 (ranks). The relative order of most officers' expressed willingness to report was very similar across the three racial groups (black and other respondents: Spearman's rho = .823; $p < .001$; white and other respondents: Spearman's rho = .781; $p < .001$; white and black respondents: Spearman's rho = .798; $p < .01$).

Finally, we compared the respondents' own expressed willingness to report with how likely they anticipated that most police officers would be to report (Table 7.14). Because of the self-serving bias, it is not surprising that the respondents from all three racial groups expected that they would be more likely to report misconduct than most police officers would be (14 out of 14 scenarios resulted in statistically significant differences for all three groups of respondents; Table 7.14). By contrast, the size of these differences varied greatly across the three groups; while white and other respondents respectively thought that most police officers would be substantively less likely to expect most police officers to report than the respondents themselves would be in about three-quarters of the scenarios (Table 7.14),[19] black respondents expected that most police officers would be less likely to report in only about one-third of all the scenarios.

Conclusion

As is the case with many countries across the globe, the issues of race, racism, and racialization are at the heart of South Africa's modern history and contemporary society. Often provoked and invoked by unscrupulous politicians, these societal elements are also ingrained in the makeup, politics, and functions of the SAPS. As Newham and colleagues (2006, p. 5) argue, when the country started on a democratic path, "one of the world's largest attempts at police transformation was launched. From a militant and racist organization serving the interests of a numerically small white population, attempts were made to create a new police agency that would reflect the demographic diversity of the country and serve the interest of all South Africans." Four years later, in 1999—the year of South Africa's second democratic election—an independent committee of inquiry reported that structural racism was (still) present within the SAPS (Newham et al., 2006). The policies that followed, including the SAPS pursuit of employment equity and affirmative action, should have changed the SAPS membership and worldview. Indeed, by 2005 the SAPS looked significantly different in terms of its race and gender profile.

Nevertheless, research conducted in 2006 in Johannesburg by the Centre for the Study of Violence and Reconciliation revealed that race remained a salient issue within the organization (Newham et al., 2006). One of the findings showed a substantial minority of police officers perceiving that most of their colleagues had problems working with officers of a different race. Faull's

Table 7.14 Willingness to report by race

Scenario	*White Own*		*White Most*		*t-test*	*Diff.*	*Black Own*		*Black Most*		*t-test*	*Diff.*
	Mean	*Rank*	*Mean*	*Rank*			*Mean*	*Rank*	*Mean*	*Rank*		
Scenario 1: Acceptance of gratuities	3.37	2	2.73	4	4.38***	.64	3.81	5	3.07	2	11.99***	.74
Scenario 2: Failure to arrest	3.85	9	2.83	7	8.88***	1.02	4.33	11.5	3.68	11	11.67***	.65
Scenario 3: Theft of knife	4.44	13.5	3.07	11	12.12***	1.37	4.38	13	3.80	12	11.09***	.58
Scenario 4: Abuse of deadly force	4.23	12	3.73	14	5.71***	.50	4.33	11.5	3.89	13	9.80***	.44
Scenario 5: Supervisor corruption	3.67	6	3.09	12	5.74***	.58	3.98	8	3.67	10	7.33***	.31
Scenario 6: Striking arrestee	3.82	8	2.96	8	7.69***	.86	3.90	7	3.51	8	8.00***	.39
Scenario 7: Verbal abuse	3.21	1	2.71	3	5.53***	.50	3.47	2.5	3.11	3	8.39***	.36
Scenario 8: Cover-up of DUI	3.51	3	2.75	6	8.30***	.81	3.47	2.5	3.17	4	6.52***	.30
Scenario 9: Kickback	3.55	4	2.70	2	7.65***	.85	3.89	6	3.48	6	8.52***	.41
Scenario 10: False report on drugs	3.86	10	3.03	9	8.62***	.83	4.10	10	3.50	7	11.43***	.60
Scenario 11: Supervisor fails to stop excessive force	3.80	7	2.74	5	8.51***	1.06	4.00	9	3.52	9	9.05***	.48
Scenario 12: Failure to report hate crime	3.98	11	3.06	10	7.51***	.92	3.76	4	3.41	5	7.91***	.35
Scenario 13: Bribe from reckless motorist	4.44	13.5	3.63	13	6.14***	.81	4.43	14	3.91	14	10.60***	.52
Scenario 14: Failure to react to graffiti	3.58	5	2.64	1	8.78***	.94	3.24	1	2.94	1	7.74***	.30

Scenario	*Other Own*		*Other Most*		*t-test*	*Diff.*
	Mean	*Rank*	*Mean*	*Rank*		
Scenario 1: Acceptance of gratuities	3.66	6	2.71	1	10.71***	.95
Scenario 2: Failure to arrest	4.16	11	3.36	9	9.18***	.80
Scenario 3: Theft of knife	4.37	12	3.52	11.5	8.80***	.85
Scenario 4: Abuse of deadly force	4.48	13	3.80	13	8.10***	.68
Scenario 5: Supervisor corruption	3.78	7	3.21	7	8.22***	.57
Scenario 6: Striking arrestee	3.53	4	2.91	5	8.48***	.62
Scenario 7: Verbal abuse	3.01	1	2.73	2	4.20***	.28
Scenario 8: Cover-up of DUI	3.36	2	2.79	3	7.65***	.57
Scenario 9: Kickback	3.57	5	3.09	6	6.11***	.48
Scenario 10: False report on drugs	4.09	10	3.52	11.5	7.39***	.57
Scenario 11: Supervisor fails to stop excessive force	3.80	8	3.34	8	6.15***	.46
Scenario 12: Failure to report hate crime	3.81	9	3.45	10	5.00***	.36
Scenario 13: Bribe from reckless motorist	4.54	14	3.84	14	9.11***	.70
Scenario 14: Failure to react to graffiti	3.49	3	2.89	4	8.93***	.60

(2008) assessment of the SAPS's diversity training and its effect at a police station revealed similar dynamics.

The results of our empirical research, involving a nationwide sample of SAPS officers and carried out several years after these earlier studies, provide further evidence that race continues to shape social and organizational dynamics within the SAPS. The most prominent differences by race in the integrity-related views that we found across all dimensions of police integrity always involved black and white respondents. Based on these findings, officers identifying as black and white—labels that have historically shaped very different life trajectories—still hold different views. While black officers tended to evaluate misconduct as more serious and state that they would be more willing to report it, white officers seemed to both expect and support more severe discipline. Such differing views are not surprising. Prior research on black and white perceptions of discrimination within the SAPS has indicated that perceptions across race are divergent; "almost all white officers feeling discriminated against when it comes to promotions, while black officers tend to see discrimination being exercised primarily through the allocation of work duties and resources" (Newham et al., 2006, p. 6).

On the other hand, the integrity-related views of black respondents and other (non-white) respondents tended to be quite similar, regardless of the police integrity measure used. These results are by no means shocking; both population groups were victims of varying forms of racism during the apartheid era (see Chapter 2), discriminated against by the white minority government, and were generally treated poorly as police officers (Brogden and Shearing, 1993; Shear, 1998). The apartheid ideology covered all aspects of police officers' lives, from the lack of power to arrest whites, different uniforms, and lower pay to limited career opportunities (Newham et al., 2006). Being "stuck in the same boat" for a long period of time could probably also have resulted in similar integrity-related views.

The most compelling finding from our empirical research is that the integrity-related views of white and other (non-white) respondents were more similar than the views of white and black respondents. In about one-half of the scenarios, there were some systematic differences, indicating the other (non-white) respondents, just like black respondents, were more likely to perceive misconduct as serious and evaluate it as rule-violating. On the other hand, just like white respondents were more likely than black respondents to expect and support more severe discipline, in the few cases with differences in the views between white and other (non-white) respondents about expected and appropriate discipline, white respondents expected more severe discipline than other (non-white) respondents did. Although the size of the differences between white and other (non-white) respondents is smaller than the size of the differences between white and black respondents, differences do exist.

Notwithstanding our findings that race is strongly correlated with SAPS officers' views on police integrity, the relation is far from straightforward.

Evidence suggests that racial dynamics in the SAPS are improving and that the wounds of apartheid are beginning to heal (e.g., Newham et al., 2006; Faull, 2008). Nevertheless, as police officers participating in research studies claim they had no difficulties working with colleagues of different races, the country remains stratified by race, with opportunity, health, and wealth differing vastly between (but increasingly also within) historical race categories. As a result, race myths persist, including narratives that the "black government" is corrupt in ways the "white government" was not. As we have outlined, corruption and nepotism have run throughout both the pre- and post-apartheid states and private sectors. As such, constantly blaming the previous administration for the current government's troubling record with corruption further contributes to the dearth of meaningful solutions to the country's inequality problems. Until these challenges are met with significant political will and focused, equitable strategies and allocated resources, the work of South Africa's police officers will continue to be not only arduous but certainly rooted in segregation by class, culture, and race.

Notes

1 The question about the respondents' race was added to the South African version of the questionnaire.

2 Analyzing racial differences in familiarity with the official rules while controlling for other demographic variables by means of linear regression preserved all the statistically significant differences in familiarity with the official rules previously detected in simpler, bivariate analyses. In addition, the linear regression analyses revealed one additional scenario in which familiarity with the official rules was significantly different at the .05 level for white and black respondents (Scenario 8). In Scenario 8, white respondents were more likely to say that the behavior violates the official rules.

3 Cohen's *d* indicated that the differences were small for all pairs, except for Scenario 13, in which the differences between 1 v. 2 and 1 v. 3 were of medium size.

4 We rely on the threshold of .50 to ascertain substantive differences, as established by Klockars and colleagues (2006).

5 Analyzing racial differences in perceptions of misconduct seriousness while controlling for other demographic variables by means of linear regression preserved all the statistically significant differences in perceptions of misconduct seriousness previously detected in simpler, bivariate analyses, with one exception (white v. other evaluations of seriousness in Scenario 4 were marginally significant at the .10 level). In addition, the linear regression analyses revealed one additional scenario in which familiarity with the official rules was significantly different at the .05 level for white and black respondents (Scenario 3). In Scenario 3, black respondents were more likely to say that this is a serious violation.

6 Cohen's *d* indicates that the differences were small for all pairs, except for Scenario 12 (1 v. 2) and Scenario 13 (1 v. 2; 1 v. 3), for which they were of medium size.

7 We rely on the threshold of .50 to ascertain substantive differences, as established by Klockars and colleagues (2006).

8 Analyzing racial differences in police officers' estimates of most police officers' evaluations of seriousness while controlling for other demographic variables by means of linear regression preserved all the statistically significant differences in estimates of most police officers' evaluations of seriousness previously detected in simpler, bivariate

analyses. In addition, the linear regression analyses revealed two additional scenarios in which perceptions of most police officers' seriousness were significantly different at the .05 level for white and black respondents (Scenarios 1 and 4), in which black respondents were more likely to say that others would evaluate behavior as serious than white respondents were.

9 We rely on the threshold of .50 to ascertain substantive differences, as established by Klockars and colleagues (2006).

10 Cohen's *d* indicates that the differences were of medium size in three scenarios (Scenario 5: 1 v. 2, Scenario 6: 1 v. 2, and Scenario 13: 1 v. 3) and were large in two scenarios (Scenario 12: 1 v. 2; Scenario 13: 1 v. 2).

11 We rely on the threshold of .50 to ascertain substantive differences, as established by Klockars and colleagues (2006).

12 Analyzing racial differences in police officers' views about appropriate discipline while controlling for other demographic variables by means of logistic regression preserved all the statistically significant differences in appropriate discipline previously detected in simpler, bivariate analyses. In addition, the logistic regressions added one more Scenario 13 (1 v. 3) that was statistically significant.

13 Analyzing racial differences in police officers' views about expected discipline while controlling for other demographic variables by means of logistic regression preserved all the statistically significant differences in expected discipline previously detected in simpler, bivariate analyses.

14 Analyzing racial differences in own willingness to report while controlling for other demographic variables by means of linear regression preserved almost all statistically significant differences in own willingness to report. In addition, in two other scenarios (Scenarios 4 and 10), previously statistically insignificant differences became statistically significant in linear regression.

15 Cohen's *d* indicated that the differences were small.

16 Analyzing racial differences in the estimates of most police officers' willingness to report while controlling for other demographic variables by means of linear regression preserved all the statistically significant differences in most police officers' willingness to report. In addition, the linear regression analyses revealed two additional scenarios (Scenarios 13 and 14) in which black respondents were significantly more likely to say that most police officers would report than white respondents assumed.

17 Cohen's *d* indicated larger differences in these scenarios.

18 They passed the threshold of substantive importance, as established by Klockars and colleagues (2006).

19 We rely on the threshold of .50 to ascertain substantive differences, as established by Klockars and colleagues (2006).

References

Afrobarometer (2013). *Summary of Results: Afrobarometer Round 5 Survey in South Africa.* Available at: http://afrobarometer.org/sites/default/files/publications/Summaryofresults/saf_r5_sor.pdf

Afrobarometer (2018). *Summary of Results: Afrobarometer Round 7 Survey in South Africa.* Available at: http://afrobarometer.org/sites/default/files/publications/Summary%20of%20results/saf_r7_sor_13112018.pdf

Altbeker, A. (2005). *The Dirty Work of Democracy: a year on the streets with the SAPS.* Johannesburg and Cape Town: Jonathan Ball.

Appiah, A. (1985). The uncompleted argument: Du Bois and the illusion of race. *Critical Inquiry*, 12(1): 21–37.

Appiah, A. (1992). *In my father's house: Africa and the philosophy of culture*. Oxford: Oxford University Press.

Badat, S. (2015). Deciphering the meanings and explaining the South African higher education student protests of 2015–16. *Pax Academica, African Journal of Academic Freedom*, 1&2: 71–106.

Bottoms, A. and J. Tankebe (2012). Beyond procedural justice: A dialogic approach to legitimacy in criminal justice. *Journal of Criminal Law and Criminology*, 102(1): 119–170.

Bradford, B., K. Murphy, and J. Jackson (2014). Officers as mirrors. *British Journal of Criminology*, 54(4): 527–550.

Brogden, M. (1994). Reforming police powers in South Africa. *Police Studies*, 17(1): 25–44.

Brogden, M. and C. Shearing (1993). Policing for a *New* South Africa. London: Routledge.

Bruce, D. (2013). New blood: Implications of en masse recruitment for the South African Police Service. *South African Crime Quarterly*, 43: 17–28.

Burger, J. (2019). Secretive promotions threaten South Africa's policing. *ISS Today*. Available at: https://issafrica.org/iss-today/secretive-promotions-threaten-south-africas-policing

Cawthra, G. (1993). *Policing South Africa: The SAP and the Transition from Apartheid*. London: Zed Books Ltd.

Cockcroft, T. (2013). *Police Culture*. New York: Taylor & Francis. Kindle Edition

Commission of Enquiry into Higher Education and Training (2017). *Report of the Commission of Enquiry into Higher Education and Training to the President of the Republic of South Africa*. Available at: www.thepresidency.gov.za/sites/default/files/Commission%20of%20 Inquiry%20into%20Higher%20Education%20Report.pdf

Con Court (2016). Solidarity and Others v Department of Correctional Services and Others (CCT 78/15) [2016] ZACC 18; (2016) 37 ILJ 1995 (CC); 2016 (5) SA 594 (CC); [2016] 10 BLLR 959 (CC); 2016 (10) BCLR 1349 (CC), 15 July. Available at: www.saflii.org/za/cases/ZACC/2016/18.pdf

Dagane v SSSBC and Others (JR2219/14) [2018] ZALCJHB 114; [2018] 7 BLLR 669 (LC); (2018) 39 ILJ 1592 (LC), 16 March.

Department of Justice (2016). *Solidarity vs the Minister of Safety and Security and Others*. Case no: J 879/12. Available at: www.justice.gov.za/labourcourt/jdgm-lbc/2016lab/2016-J%20879-12.pdf

Diphoorn, T. (2015). The "Bravo Mike Syndrome": Private security culture and racial profiling in South Africa, *Policing and Society*, 27(5): 525–540.

dos Santos, Sales Augusto (2006). Who is black in Brazil? A timely or a false question in Brazilian race relations in the era of affirmative action? *Latin American Perspectives*, 149(33: 4), July, 30–48. DOI: 10.1177/0094582X06290122

Elgar, F.J. and N. Aitken (2011). Income inequality, trust and homicide in 33 countries. *The European Journal of Public Health*, 21(2): 241–246. Available at: https://academic.oup.com/eurpub/article/21/2/241/498070

Equality Court of South Africa (2018). *Social Justice Coalition and Others v Minister of Police and Others*. Available at: www.saflii.org/za/cases/ZAWCHC/2018/181.pdf

Erasmus, Y. and G.T.H. Ellison (2008). What can we learn about the meaning of race from the classification of population groups during apartheid? *South African Journal of Science*, 104: 450–452.

Evans, G. (2019). *Skin Deep: Journeys in the Divisive Science of Race*. London: Oneworld.

Fabricius, P. (2019). Xenophobia? What xenophobia? We love foreigners! *ISS Today*. Available at: https://issafrica.org/iss-today/xenophobia-what-xenophobia-we-love-foreigners

Fanon, F. (1963). *The Wretched on the Earth*. New York: Grove Press.

Faull, A. (2008). *Policing Diversity: An Analysis of a Diversity Intervention and Its Impact on a South African Police Service Station*. Amsterdam: Rozenberg Publishers.
Faull, A. (2010). *Behind the Badge: The Untold Stories of South Africa's Police Service Members*. Cape Town: Penguin Random House South Africa. Kindle Edition.
Faull, A. (2018). *Police Work and Identity: A South African Ethnography*. Abingdon, U.K.: Routledge.
Foucault, M. (1991). *Discipline and Punish: The Birth of the Prison*. New York: Penguin Social Sciences.
Gqola, P. (2001). Defining people: Analyzing power, language and representation in metaphors of the New South Africa. *Transformation*, 47: 94–106.
Gumede, V. (2015). Inequality in democratic South Africa. In X. Mangcu (ed.), *The Colour of Our Future: Does Race Matter in Post-Apartheid South Africa?* Johannesburg: Wits University Press.
Hornberger, J. (2011). *Policing and Human Rights: The Meaning of Violence and Justice in the Everyday Policing of Johannesburg*. Abingdon: Routledge.
Hough, M., J. Jackson, B. Bradford, A. Myhill, and P. Quinton (2010). Procedural justice, trust, and institutional legitimacy. *Policing*, 4(3): 203–310.
Institute for Justice and Reconciliation (2017). *SA Reconciliation Barometer 2017*. Available at: http://www.ijr.org.za/home/wp-content/uploads/2017/12/IJR-Barometer-Report-2017-web.pdf
Jeffery, A. (2019). *People's War: New Light on the Struggle for South Africa*. Jeppestown, South Africa: Jonathan Ball Publishers. Kindle Edition.
Kutnjak Ivković, S. (2009). The Croatian police, police integrity, and transition toward democratic policing. *Policing: An International Journal of Police Strategies and Management*, 32(3): 459–488.
Kutnjak Ivković, S. (2015). Studying police integrity. In S. Kutnjak Ivković, and M. Haberfeld (eds.), *Police Integrity across the World*. New York: Springer.
Klockars, C.B., and S. Kutnjak Ivković (2004). Measuring police integrity. In M.J. Hickman, A.R. Piquero, and J.R. Greene (eds.), *Police Integrity and Ethics*. Belmont, CA: Wadsworth Publishing.
Klockars, C.B., S. Kutnjak Ivković, and M.R. Haberfeld (eds.). (2004). *The Contours of Police Integrity*. Newbury Park, CA: Sage Publication.
Klockars, C.B., S. Kutnjak Ivković, and M.R. Haberfeld. (2006). *Enhancing Police Integrity*. Dordrecht, The Netherlands: Springer.
Klockars, C.B., S. Kutnjak Ivković, W.E. Harver, and M.R. Haberfeld (2000). *The Measurement of Police Integrity*. Research in Brief. U.S. Department of Justice, Office of Justice Programs, National Institute of Justice: Washington, DC: Government Printing Office. Available at: https://www.ncjrs.gov/pdffiles1/nij/181465.pdf
Langa, M. (ed.) (2017). *#Hashtag: An Analysis of the #FeesMustFall Movement at South African Universities*. Johannesburg: Centre for the Study of Violence and Reconciliation. Available at: www.csvr.org.za/pdf/An-analysis-of-the-FeesMustFall-Movement-at-South-African-universities.pdf
Loader, I. (2006). Policing, recognition, and belonging. *Annals of the American Academy of Political and Social Science*, 605(1): 202–221.
Loftus, B. (2009). *Police Culture in a Changing World*. Oxford: Oxford University Press
Makou, G., I. Skosana, and R. Hopkins (2017). Factsheet: The state of South Africa's prisons. *Africa Check*. Available at: https://africacheck.org/factsheets/factsheet-the-state-of-south-africas-prisons/

Mare, G. (2001). Race counts in contemporary South Africa: "An illusion of ordinariness". *Transformation*, 47: 75–93. Available at: http://citeseerx.ist.psu.edu/viewdoc/download;jsessionid=B8A86BA2FE657241CC2E6A3267924663?doi=10.1.1.569.5775&rep=rep1&type=pdf

Marks, M. (2005). *Transforming Robocops: Changing Police in South Africa*. Scottsville, South Africa: University of Scottsville, KwaZulu-Natal Press.

Masuku, T. (2006). Targeting foreigners: Xenophobia among Johannesburg's police. *South African Crime Quarterly*, 15: 19–24. Available at: https://issafrica.s3.amazonaws.com/site/uploads/CQ15MASUKU.PDF

McGreal, C. (2001). Policemen jailed for racist dog attack. *The Guardian*. Available at: www.theguardian.com/world/2001/nov/30/chrismcgreal

Mlachila, M. and T. Moeletsi (2019). Struggling to make the grade: A review of the causes and consequences of the weak outcomes of South Africa's education system. IMF Working Paper. Available at: www.imf.org/en/Publications/WP/Issues/2019/03/01/Struggling-to-Make-the-Grade-A-Review-of-the-Causes-and-Consequences-of-the-Weak-Outcomes-of-46644

Mlangeni, P. (2017). Police minister calls for white youth to join SAPS. *SABC News*, 22. Available at: www.sabcnews.com/sabcnews/police-minister-calls-white-youth-join-saps/

Myburgh, P. (2018). *Gangster State*. Johannesburg: Penguin Random House South Africa. Kindle Edition.

Netschitenzhe, J. (2015). Interrogating the concept and dynamics of race in public policy. In X. Mangcu (ed.), *The Colour of Our Future: Does Race Matter in Post-Apartheid South Africa?* Johannesburg: Wits University Press.

Newham, G., T. Masuku, and J. Dlamini (2006a). *Diversity and Transformation in the South African Police Service*. Braamfontein, South Africa: Centre for the Study of Violence and Reconciliation. Available at: www.csvr.org.za/docs/policing/diversity.pdf

Newham, G., T. Masuku, and J. Dlamini (2006b). *Diversity and Transformation in the South African Police Service: A Study of Police Perspectives on Race, Gender and the Community in the Johannesburg Policing Area*. Johannesburg: CSVR. Available at: www.csvr.org.za/docs/policing/diversity.pdf

O'Regan, J.C. and A.V. Pikoli (2014). *Towards a Safer Khayelithsa: Report of the Commission of Inquiry into Allegations of Police Inefficiency and a Breakdown in Relations Between SAPS and the Community of Khayelitsha*. Cape Town: Khayelitsha Commission of Inquiry.

Perkins, G. (2018). *Danger and Death: Organisational and Occupational Responses to the Murder of Police in South Africa: A Case Study*. Unpublished Doctoral Thesis. Department of Public Law, Faculty of Law, University of Cape Town.

Politics Web (2016) 'Hendrik Verwoerd: 10 quotes', https://www.politicsweb.co.za/documents/hendrik-verwoerd-10-quotes

Population Registration Act No. 30 of 1950.

Posel, D. (2001). Race as common sense: Racial classification in twentieth-century. *African Studies Review*, 44(2): 87–113.

Rauch, J. (2004). *Police transformation and the South African TRC*. Johannesburg: Centre for the Study of Violence and Reconciliation.

Reiner, R. (2010). *The politics of the police*, 4th ed. Oxford: Oxford University Press.

Republic of South Africa (1996). *The Constitution of the Republic of South Africa*. Available at: www.justice.gov.za/legislation/constitution/SAConstitution-web-eng.pdf

Republic of South Africa (1998). *Employment Equity Act, No. 55 of 1998*.

Republic of South Africa (2001). *Government Gazette: Regulation Gazette No. 7142*, Vol. 434. Pretoria, 24 August, No. 22583.
Republic of South Africa (2003). *Broad-Based Black Economic Empowerment Act, No. 53 of 2003*.
Roberts, D. (2011). *Reading the Riots: Investigating England's Summer of Disorder*. London: Guardian Books.
SAHRC (1999). *Racism in the SAPS Vryburg District*. Report by the South African Human Rights Commission. Available at: www.sahrc.org.za/home/21/files/Reports/INQUIRY%20INTO%20Racism%20in%20the%20SAPS%20Vryburg%20district.pdf1999.pdf (Accessed 15 May 2019).
SAPS (2018). *South African Police Service Annual Report 2017/18*.
Seekings, J. and N. Nattrass (2005). *Class, Race, and Inequality in South Africa*. New Haven, CT: Yale University Press.
Shaw, M. and S. Shearing (1998). Reshaping security: An examination of the governance of security in South Africa. *African Security Review*, 7(3): 3–12. DOI: 10.1080/10246029.1998.9627860
Shear, K.S. (1998). *Constituting a State in South Africa: The Dialectics of Policing, 1900–1939*. Unpublished Dissertation. Northwestern University.
Sklansky, D. (2007). Seeing blue: Police reform, occupational culture, and cognitive burn-in. In M. O'Neill, M. Marks, and A. Singh (ed.), *Police Occupational Culture (Sociology of Crime, Law and Deviance, Vol. 8)*. Bingley: Emerald Group Publishing Limited. 19–45.
Skolnick, J.H. (2008). Enduring issues of police culture and demographics. *Policing and Society: An International Journal of Research and Policy*, 18(1): 35–45.
Skolnick, J.H. (2011). *Justice without Trial: Law Enforcement in Democratic Society*, 4th ed. New Orleans: Quid Quo Pro Books.
South African Police Union (SAPU) (2017). *Police Promotions Policy an Insult to Officers*. Available at: www.politicsweb.co.za/news-and-analysis/police-promotions-policy-an-insult-to-officers--sa
Stanko, B., J. Jackson, B. Bradford, and K. Hohl (2012). A golden thread, a presence amongst uniforms, and a good deal of data: Studying public confidence in the London Metropolitan Police. *Policing and Society*, 22(3): 317–331.
Statistics South Africa (2012). *Statistical Release (Revised) Census 2011*. Pretoria. Available at: www.statssa.gov.za/publications/P03014/P030142011.pdf
Statistics South Africa (2016). *Mortality and Causes of Death in South Africa, 2016: Findings from Death Notification: Statistical Release P0309.3*. Available at: www.statssa.gov.za/publications/P03093/P030932016.pdf
Statistics South Africa (2017). *Poverty Trends in South Africa: An Examination of Absolute Poverty between 2006 and 2015*. Pretoria: Statistics South Africa.
Statistics South Africa (2018a). *Statistical Release P0318 General Household Survey 2017*. Available at: www.statssa.gov.za/publications/P0318/P03182017.pdf
Statistics South Africa (2018b). *Victims of Crime Survey 2017/18: Statistical Release P0341*. Pretoria: Stats SA. Available at: www.statssa.gov.za/?p=11632
Steinberg, J. (2008a). *South Africa's Xenophobic Eruption*. ISS Paper 169. Pretoria: Institute for Security Studies.
Steinberg, J. (2008b). *Thin Blue: The Unwritten Rules of Policing South Africa*. Cape Town: Jonathan Ball.
Steinberg, J. (2011). Security and disappointment: Policing, freedom and xenophobia in South Africa. *British Journal of Criminology*, 52(2): 345–360.

Steinberg, J. (2014). *A Man of Good Hope*. Johannesburg: Jonathon Ball.

Super, G.J. (2013). *Governing through Crime in South Africa: The Politics of Race and Class in Neo-Liberalizing Regimes*. Dorchester: Ashgate.

TRC (1998a). *Truth and Reconciliation Commission of South Africa Report*, Vol. 2. Available at: www.justice.gov.za/trc/report/finalreport/Volume%202.pdf

TRC (1998b). *Truth and Reconciliation Commission of South Africa Report*, Vol. 3. Available at: www.justice.gov.za/trc/report/finalreport/Volume%202.pdf

Tyler, T.R. (2004). Enhancing police legitimacy. *The Annals of the American Academy of Political and Social Science*, 593(1): 84–99.

Vigneswaran, D. and M. Duponchel (2009). *One Burden Too Many? A Cost-Benefit Analysis of Immigration Policing in Gauteng*. Johannesburg: Forced Migration Studies Programme, University of Witwatersrand.

Wilkinson, R. and R. Picket (2010). *The Spirit Level: Why Equality Is Better for Everyone*. London: Penguin Books.

Wolby, E., S. Hassim, and T. Kup (eds.) (2008). *Go Home or Die Here: Violence, Xenophobia and the Reinvention of Difference in South Africa*. Johannesburg: University of the Witwatersrand Press.

van der Bijl, A. (2017). Huge settlement for member in SAPS due to affirmative action: Solidarity. *Politics Web*. Available at: www.politicsweb.co.za/politics/huge-settlement-for-member-in-saps-due-to-affirmat

Van Doren, J. (1989). Critical legal studies and South Africa. *South African Law Journal*, 106: 648–665.

Zimring, F.E. (2015). *When Police Kill*. Cambridge, MA: Harvard University Press.

Zulu, P., V. Gounden, T. Oakley-Smith, and M. Seleoane (1999). *The Independent Committee of Inquiry into Racism in the South African Police Service Report*. Pretoria: South African Police Service.

Chapter 8

Gender (in)equality within the SAPS

Introduction

Although women made up 37% of all SAPS personnel in 2019 (SAPS, 2019 p. 34), they have not always been so well represented in South African policing. The same is true of policing around the globe. While women have been employed in police agencies in the United States and Europe since the early 1900s, their role was primarily that of social workers (Brown and Heidensohn, 2000). Until the 1970s, policing was predominantly the professional domain of males—usually white in Europe and its former colonies (Stroshine and Brandl, 2011), with women represented in only small numbers in some police agencies worldwide or not at all in others (Heidensohn, 1998). This was also true of the South African Police force (SAP). The SAP employed its first women in clerical positions in 1960 (Potgieter, 2012) and its first female officers in 1972 (Cawthra, 1993). While it is not uncommon to observe women in all functions of the police and at all levels of rank in the twenty-first century, these accomplishments continue to fall short of numerical equity, with the exception of a few organizations. Heidensohn (1992) argues that much of this shift can be traced to three sources: pragmatism (shortages of sufficient male citizens in the workforce), police public relations, and legal pressures, although these do not function independently of one another. As we will show, this also holds true in South Africa.

As women began entering the operational domain of policing in greater numbers during the 1970s, they faced significant opposition from their male peers and often from the public in South Africa and elsewhere. As Brown (1996, p. 1) has noted, "Countries vary in the dates of admitting women into the police, but the opposition to their admission was almost universal." For many male officers, the larger-scale entry of women into policing in the 1970s was viewed as an assault on the police culture (Brown, 1996) and as an act of "breaking and entering" (Martin, 1980). As a result, early research on gender differences in police work focused on whether women could "hold their own" in the male-dominated profession (Dejong, 2005). Prior to the 1990s, only a relatively small number of studies focused on the effect of gender on police attitudes. This decade would see increased interest in the study and publication of projects investigating the effect of gender on police attitudes (Worden, 1993) and the examination of whether female officers carry out and/or perceive the job differently than male officers (Dejong, 2005).

In 2009, Poteyeva and Sun published a systematic review of 33 research articles published between 1990 and 2006 that studied attitudinal differences and similarities (e.g., the police role; community policing; job satisfaction; stress; domestic violence) between male and female officers (Poteyeva and Sun, 2009). According to the authors, only two studies found significant main gender effects, while ten studies found mixed effects and nineteen found no such significant gender effects (Poteyeva and Sun, 2009). However, all of the articles included in the analyses were published in English, in U.S.-based academic journals, and have employed multivariate regression analysis using gender merely as a control variable. In conclusion, Poteyeva and Sun (2009) noted that, although there is "very limited support for the proposition that the increased hiring of women would transform the face of American policing," there is some evidence (the National Center for Women and Policing, 2001; Sun, 2007) that female officers "behave differently from their male counterparts, with policewomen less likely to use excessive force and engage in misconduct and more likely to provide support to victims of domestic violence" (Poteyeva and Sun, 2009, p. 520). That women are less likely to abuse their power and attract allegations of misconduct, and that increasing numbers of women police officers reduces the levels of police corruption is also an argument put forward by Prenzler et al. (2010) about gender equity in Australian and New Zealand policing.

This chapter explores the relation between police integrity and gender in South Africa. Although women have been represented in South African policing since the early 1970s, recent data indicate that women are yet to be completely incorporated within the modern police agency. The chapter discusses the specific experience of South Africa's integration of women into the Service. It also includes a discussion of prior studies' investigations into the relation between police culture and gender in general and gender and police integrity in particular. The chapter concludes with an empirical analysis of gender in relation to SAPS officers' views regarding misconduct seriousness, appropriate and expected discipline, and willingness to report.

Gender in South Africa

Policing shapes, but is also shaped by, the sociocultural norms of society. One cannot separate gender dynamics within the SAPS from those outside the organization, especially when it comes to integrity violations such as gender-based and sexual violence by and against police officers, sexual misconduct by police officers, expressions of homophobia by police officers, and abuse of force by officers. In this section, we provide an overview of key gender dynamics in South Africa's recent past and present with bearing on the construction and enactment of gender in the contemporary SAPS, with a focus on those aspects most relevant to police integrity.

As described in Chapter 7, South Africa is a racialized society, with race predicting many of life's outcomes. Similarly, it is highly gendered, with race and sex intersecting to generate compounded fragilities. Black women bear the brunt of the country's poverty, unemployment, and HIV epidemic (e.g., Hunter, 2010; StatsSA, 2019), with those in partnerships with violent men more likely to be infected with HIV (Dunkle et al., 2004). On the other hand, while violence against women is widespread—with President Ramaphosa calling it a national crisis in September 2019 (Francke, 2019)—most victims of violent crimes are men (Bruce, 2020). Both male and female experiences are thus the products of how gender is constructed in South Africa (Seedat et al., 2009).

The discrepancies in lived experiences between men and women in contemporary South Africa are in part the product of shifts in social relations set in motion during the colonial and apartheid periods and are closely linked with the emergence of the capitalist state (e.g., Baden et al., 1998). Whereas women in pre-colonial communities achieved some status by participating in food and craft production (Nattrass 2017, Loc 650), this shifted as colonial and apartheid authorities restructured social and familial norms to serve the white state and its industries. Following British governance of the Cape by 1814, the region was incorporated into global capitalism, with great bearing on social identity, including gender. In the subsequent two centuries, most inhabitants were reduced to a semi-proletariat (Landau, 2009, p. 392). For example, by forcing Xhosa communities to pay a land tax, British authorities coerced Xhosa men into the labor market (see Chapter 2). This included agricultural work, which the Xhosa considered a woman's lot (Legassick and Ross, 2009, p. 303). Many Xhosa men migrating to Cape Town were forced into indentured labor. Unable to afford *labola* (the "bride price"), some men could not marry, affecting their social standing as males. It is likely that in this and other related ways, the colonial and apartheid states could systematically emasculate black men.

With the discovery of diamonds near Kimberley in 1870, and gold in 1886, migration as well as efforts to control African masculinity increased. When the diamond price collapsed in 1880, mine owners erected closed compounds for black workers to prevent smuggling and desertion. Compound-bound men could not return home at will or cohabit with their families, and some began second families with women living close to the mines. When workers did return to their rural homes, it was with experience, wages, and new ideas about social norms (Etherington et al., 2009, p. 375). Exposure to urban life and cash allowed some to break free of the hierarchies and traditions of the countryside (Etherington et al., 2009, p. 377). Meanwhile, big industry invested in white South Africa, promoting and protecting white labor and masculinities over black.

White governments positioned black men below white women and children in the status hierarchy (Ashforth, 2005). The colonial and apartheid states co-opted black traditional leaders (chiefs), using a traditional system of "customary law" that positioned black women as inferior to black men (Robinson, 1996). By mid-century, many migrating men were leaving wives and children in the countryside for much of the year while they lived in all-male, often violent dormitories in the cities and on mines. Black men, infantilized as "boys" by the white state and employer, could reclaim status in the rural homestead by providing for their families, proving fertility, and through demonstrating bravery (at times through violence). Similarly, from the 1950s, black women increasingly migrated to cities to work as live-in domestic workers. Access to money freed some from the systems of patriarchy more commonly associated with the rural areas. By the 1960s, families in black urban communities were becoming women-led, displacing the role of the man as head of the familial unit (Mager and Mulaudzi, 2011, p. 380).

In parallel to these developments that saw family and community life profoundly altered and black men and women formally subjugated, the state took care to promote and protect opportunities for the white population. In 1930, white women were granted the right to vote, and with the economy expanding tenfold in real terms between 1930 and 1970 (Nattrass and Seekings, 2011, p. 518), white men were provided increasingly rewarding employment. As a result, the SAP struggled to fill its ranks with white men, for whom better remunerated work was often easily available (see Chapter 2).

Coupled with these shifts was the framing of black urban youth as violent and unruly, both by the white state (Super, 2013) and by the black communities (Glaser, 2000). Glaser shows how the narratives used to explain urban African juvenile criminality in Soweto from the 1930s-1970s suggested a breakdown in morals brought about by the move from rural to urban life. He suggests that similar gangs and discourses probably existed in other urban settings with concentrations of poor city-bred youth with limited employment possibilities, housing, schooling, and recreational facilities (Glaser, 1998, p. 720). Faull (2018) found the same narratives alive and well among the police he shadowed for eight months in 2012–2013. Whereas rural youth were deemed respectful and subservient, city youth were seen as corrupted by crime, prostitution, and the media.

Similar forces continue to shape masculinities in twenty-first-century South Africa. Several scholars (Seedat et al., 2009; Letsela and Ratele, 2009) have argued that the way masculinity is constructed in South Africa has resulted in young men experiencing a disproportionate amount of violence. For example, writing about contemporary Cape Town gangs, Jensen (2008) notes that they provide men with dignity and belonging, where a failing economy cannot. Similarly, Pinnock (2016) notes that Cape Town does not have a "gang problem," but, rather, a "youth problem," of which gangs are only one manifestation. It appears that, because it is unable to offer its young men rewarding lives of status within the law, South Africa continues to produce violent masculinities and gangs.

Against this background, Faull (2013, 2018), Hornberger (2011), and Perkins (2018) have suggested that some male SAPS officers turn to violence as a means of earning respect. Although they are well employed and have escaped gang life, many are recruited from poor, violent areas, and have extended families whom they must support, which can significantly reduce their material quality of life. As such, their stable employment can both boost and undermine their ability to "succeed" as men.

While white male officers generally do not face the same "black tax" (Ngwadla, 2018), white masculinity has also historically been wed to violence—ranging from the devastating conflict between the Boer and the British in the nineteenth century to the conscripting of all white men into the armed forces from 1967 to 1990 to defend the apartheid state (see Chapter 2). More recently, in 2014 the para-Olympian Oscar Pistorious shot and killed his girlfriend, Reeva Steenkamp, by firing bullets through his bathroom door. Much publicized theories and opinions abound around the facts of this tragic incident. One such viewpoint has Langa and colleagues arguing that Pistorious—who claimed he thought Steenkamp was an intruder—built his defense around the idea that black men are threatening and violent, and of white, gun-owning men as responsible for defending women (Langa et al., 2018). Pistorious was at first found guilty of manslaughter, suggesting that the judge (a black woman), accepted his version. However, the state appealed the case and Pistorius was later found guilty of murder.

The historical uplifting of some men and emasculation of others, the positioning of women as lesser than men, and the associated distribution of access to work and freedoms, has shaped South African society, including its police and the violence many face through their employment.

Gender in South African policing

In 1972, the SAP hired its first female officers and created a "women police force." Although the public service commission pronounced that female and male officers should be treated equally, women were not allowed to perform patrol duties until the late 1970s. Instead, they mostly provided victim and administrative support. They were also all white. In the early 1980s, the first colored, Indian, and black female officers began their police training (Potgieter, 2012). In the 1980s, women were increasingly deployed to historically male roles. However, they remained vastly outnumbered, junior in rank, and governed by different conditions of service than men until the early 1990s.

The basis for complete integration of women into policing in post-apartheid South Africa has followed the general pattern observed in other countries, with a unique difference in that its foundations are explicit in the South African Constitution, affecting the establishing legislation of the South African Police Service. Section 27(2) of the Police Service Act states "shall not preclude compliance with measures designed to achieve the objects contemplated in Sections

8(3)(a) and 212(2) of the Constitution" (South African Police Service [SAPS] Act 68 of 1995). These references are actually to the Interim Constitution (Act 200 of 1993), with Section 8(3)(a) reflecting the initial formulation of the constitutional right to substantive equality and Section 212(2) addressing the administration of public organizations.

Although the formulation of the SAPS Act follows initially from the Interim Constitution, subsequent implementation of the above provisions is now interpreted in light of Chapter 2, Section 9, Equality, and Chapter 10, Section 195(1)(i), Public Administration of the Constitution of the Republic of South Africa (1996). Under Section 9(1), "Equality includes the full and equal enjoyment of all rights and freedoms. To promote the achievement of equality, legislative and other measures designed to protect or advance persons, or categories of persons, disadvantaged by unfair discrimination may be taken." In addition, Section 9(3) states that, "The state may not unfairly discriminate directly or indirectly against anyone on one or more grounds, including race, gender, sex, pregnancy, marital status, ethnic or social origin, color, sexual orientation, age, disability, religion, conscience, belief, culture, language and birth."

The authorization to "promote the achievement of equality" found in Section 9(2) is also given expression in the Employment Equity Act, Act 55 (1998). Such countenance may be advanced through an affirmative action policy as justified through the interpretation of Chapter 10, Section 195(i), of the Constitution wherein, "[P]ublic administration must be broadly representative of the South African people, with employment and personnel management practices based on ability, objectivity, fairness, and the need to *redress the imbalances of the past to achieve broad representation* (emphasis added)." In fact, Chapter I, Section 2, of the Employment Equity Act (1998), identifies the purpose of the Act as to

> achieve equity in the workplace by (a) promoting equal opportunity and fair treatment in employment through the elimination of unfair discrimination; and (b) implementing affirmative action measures to redress the disadvantages in employment experienced by designated groups, in order to ensure their equitable representation in all occupational categories and levels in the workforce.

Under Chapter III of the Act, Section 15, affirmative action measures include preferential treatment and numerical goals but exclude quotas.

Based on the SAPS's annual reports from 2011–2012 and 2015–2016, the Service has been comparatively successful (internationally) in the hiring and promotion of female officers. Of the 152,000 Police Service Act employees in the 2011–2012 reporting period, 47,000 (25.4%) were women, and, among those holding the rank of a commissioned officer or higher, 16,500 of 54,000 (30.5%) were women, including nine of the 32 top management positions (SAPS, 2012). For the 2015–2016 Annual Report period, 41,000 of the

152,000 SAPS Act employees were women (27%) with 8,300 of 25,400 (33.3%) holding a rank of commissioned officer or higher, including 11 of 30 top management positions (SAPS, 2016). In the 2018–2019 report period, 43,412 (29%) of SAPS Act employees and 8,161 (34%) of commissioned officers were women (SAPS, 2019, p. 34).

The goal of achieving gender equality in the SAPS, based on gender representation, is consistent with such a goal for South Africa as a whole. A May 2014 media article indicated that South Africa was ranked 17th out of 136 countries in terms of gender equality based on data from the World Economic Forum. Furthermore, in terms of political engagement, South Africa was ranked 8th in gender equality. In addition, South Africa ranked 5th in the percentage of female members in parliament, and 11th for women in ministerial positions (eNCA, May 17, 2014). The article also noted that only 36% of South African women believed that the state was doing enough to empower women (eNCA, May 17, 2014).

Based on interviews with 26 female officers in the SAPS, Morison (2004) reported mixed results of the status of women in the SAPS regarding feelings, attitudes, outlooks, and views about their job situation. While several respondents indicated that women affected the culture of the police in regard to handling of domestic violence and reducing brutality, and that there was "very big" improvement both in the acceptance of women by their male colleagues and in work conditions, they also indicated that there was discrimination in promotion and that, because of gender imbalance in senior positions, transformation in the SAPS had not kept pace with the increased number of women in the SAPS (Morison, 2004). Morison (2005) also observed that the attitude of many male officers that women were not "physically or emotionally tough enough" to meet the demands of the police profession still persisted in the SAPS.

These observations are supported by Meyer and Reyneke-Cloete (2013), who compared responses from SAPS officers at two different times (non-identical samples), 2005 and 2013, regarding their attitudes about female officers and their acceptance of women in the police. Meyer and Reyneke-Cloete (2013) observed that for the 2005 group, 30% of respondents either strongly disagreed or disagreed with the statement: "A female can be just as good a police official as a male." Despite the further increase of women in the SAPS, in the 2013 survey, this number had declined to 23%. Respondents were also asked whether they would want a female police official as their partner ("I would not want a female police official to be my partner."). In the 2005 survey, 22% of the respondents agreed with the statement, while 25% of respondents agreed with the statement in 2013. When asked whether female police officers had a difficult time being accepted by their male peers, 31% of respondents said "yes" in 2005, as did 45% of the respondents in 2013. Both of these increases are likely tied to the remilitarization of the SAPS in the early 2010s (see Chapter 3).

Meyer and Reyneke-Cloete (2013) compared responses for male and female participants in the 2013 survey and observed several gender-based differences.

While 70% of female respondents agreed that a female could be just as good a police official as a male, only 52% of male respondents agreed with this statement. On the question of not wanting a female police official to be their partner, 14% of female respondents agreed, compared to 33% of male respondents.

In 2004, Newham and colleagues (2006) conducted a study of police officers, exploring their perspectives on race, gender, and the community in Johannesburg. Their sample for quantitative analyses consisted of 580 respondents, 17.4% of whom were women, drawn from 21 police stations (Newham et al., 2006). They also conducted in-depth qualitative interviews of 40 participants, 17 of whom were women. Similar to the findings of Meyer and Reyneke-Cloete (2013), Newham and colleagues found that, while 87% of female respondents agreed with the statement that women generally make good police officers, this was the case for only 59% of male respondents. In addition, 36.6% of male respondents expressed negative views of women as police officers (Newham et al., 2006). At that time, Newham and colleagues (2006) also noted that women constituted only 15% of sworn police officers in the SAPS, far from the SAPS achieving its objective of having women make up 40% of the organization's members.

In a review of the status of affirmative action policy at the time, Montesh (2010) concluded that women had not been properly integrated into mainstream policing in South Africa. He noted that, although the proportion of female representation had increased in the SAPS, the increase was concentrated in Career Services, Management, Legal Services, Training, and Personnel Services, while the Detective Service and Crime Intelligence were still male-dominated (Montesh, 2010, p. 75). Based on his personal interview with N. Seabi in 2007, then Director and Head of the SAPS Employment Equity Unit, Montesh (2010) noted that the isolation and machismo were the most common elements of police culture prevalent in the South African Police Service, and that these characteristics of the culture contributed directly to the resistance among male officers to greater integration of women into the SAPS.

Meyer and Steyn (2009) compared male and female police recruit perceptions and experiences regarding isolation. Their analysis identified three dimensions of isolation: isolation from the community, isolation from the criminal justice system, and isolation from police administration. Meyer and Steyn (2010) found that, although there were no significant differences overall across the entire period, female recruits did experience a sense of increased social isolation from the community during the period of police college training. This sense of increased isolation disappeared upon completion of formal college training and reengagement with former social networks, suggesting that the experience of basic police training was different for female recruits, as it shaped their opinions reflecting isolation. Based on these findings, Meyer and Steyn (2010) concluded that the SAPS basic training needed to take into account the different experiences of women upon their entry into the organization. A later study by Steyn (2017) that compared survey data on male and female SAPS officers' attitudes over a ten-year period regarding isolation, solidarity, and cynicism found that women's attitudes intensified compared to men's attitudes. Steyn concluded that this finding showed that the introduction of additional women to the SAPS had not altered these key traits of the organizational culture.

During 16 months of fieldwork conducted between 2000 and 2003, Ulicki (2011) conducted a field observation study of sexual harassment in the SAPS, relying on 114 semi-structured interviews. Her study was carried out to investigate the tension between the formal gender equity policies of the SAPS and the experiences of harassment of female officers in the field, noting that "the experiences of female police officers are shaped not only by the numerical predominance of male officers, but also . . . by the structural organization of the SAPS and its informal culture" (Ulicki, 2011, p. 115). Based on her analysis of field observations and interviews, Ulicki (2011, p. 115) concluded that, although significant strides have been made in the SAPS toward the achievement of numerical gender parity, the SAPS response to sexual harassment has been "superficial and ineffective." Further, she concluded that such a response is "emblematic of the organization's lack of awareness of and commitment to the more substantive aspects of gender equity" (Ulicki, 2011, p. 115).

In Faull's 2008 analysis of a SAPS diversity training course and its effect at a single police station, trainers claimed that, after race, gender was the most important element of the training (Faull, 2008, p. 64). However, when a trainer asked SAPS officers to raise their hand if they identified as women, the trainees burst out laughing. Rather than use their response as an opportunity to discuss the constructed nature of gender, Faull notes, the trainer joined the laughter and moved on. Faull also noted that, while women were represented across the rank structure at the study station, their roles reflected historical gender norms. For example, 100% of finance and human resource positions were occupied by women, and 92% of kitchen staff, 76% of administration and communication staff, and 68% of logistics staff were women. Only 26% of uniformed officers and 36% of detectives were women (Faull 2008, p. 65). Potgieter (2012, p. 95) carried out in-depth interviews with SAPS officers to study masculinity within the organization and reported that "both male and female officers tended to regard certain duties as female while other duties were perceived as reserved for men." Women's duties were generally inside the station, where it was "safe," while men's work was outside, confronting danger. Similar attitudes have been noted by others researching the SAPS (e.g., Viljoen, 2015; Faull, 2018; Perkins, 2018).

Faull (2018) reports that only two cases of sexual harassment had been lodged at his study station in the three-year period preceding his fieldwork. However, a female officer interviewed for the study told him she did not think female officers would know where to report sexual harassment (Faull, 2008, p. 66). In this and other studies that employed participant observation with the South African police (Faull, 2018; Hornberger, 2011), researchers report witnessing open flirtation/sexual harassment by male officers toward female officers, and by male officers toward members of the public.

Using semi-structured interviews with SAPS officers to explore masculinity in the organization, Viljoen (2015) found that masculinity excluded women from field duties, undermined traditionally feminine qualities in officers, and allowed the acceptance of inappropriate sexual flirtation. In a separate study composed of in-depth interviews with SAPS officers (Faull, 2010, p. 230), the respondents told Faull not only how they and their colleagues used the power of their uniforms to seduce women, but also how they accepted sex from female suspects to release them: "It's easy to get sex when you're a policeman in uniform. Most police proposition ladies when they are in uniform or in a marked police car." Furthermore (Faull, 2010, p. 231), a respondent who was an Inspector with the police said: "I saw many police propositioning suspects. When a person is arrested, they will do anything for their release because they are under duress. I know many cases where policemen were arrested for rape." Faull (2010, pp. 236–237) also asked female officers about their experiences as women in the organization:

> You have to prove yourself much more than men have to. When I went for my first hostage negotiator interview. . . . They gave me so much trouble! "But you're a woman; how can you do this if you're a woman?" It's much more difficult to prove yourself as a good police person. People just accept that the policeman is good at what he does. You have to prove yourself more when you're a woman, definitely.
>
> —Inspector

> I wouldn't say sexual harassment doesn't happen, but each individual must handle the situation as he or she sees fit. I'm sure that if I had wanted to I could have made a dozen sexual harassment cases, but that's not the type of person I am. If I don't like the way you look at me or I don't like the way you touch me or I feel it's inappropriate, I'm going to tell you it's inappropriate and we're going to have a discussion and that's the end of it.
>
> —Inspector

In his 2018 study, Faull notes that many of the female officers he encountered were single parents, while some male officers unabashedly discussed the children they had with different women. According to Ashforth (2005, p. 210), this diminution of paternal capacity and authority may be one of the most important changes in South African families in the twentieth century. With rural-urban migration—a movement of youth away from elders, and men away from women—has come a shifting of family responsibility from fathers to mothers. This has resulted in a burden to women and a challenge to men who are still often seen as the rightful leaders of women and children and those with whom final authority should rest. In a 2019 IPSOS poll, 64% of female respondents agreed that "a woman must obey her husband or partner," and 12% of all respondents agreed that "A man has the right to hit his partner or wife" (Harris, 2019).

As already mentioned, South Africa has very high levels of domestic and sexual violence and rape (e.g., Gqola, 2015; Tlhabi, 2017). While it could be expected of female SAPS officers to be more sympathetic to female victims than their male colleagues, in some instances the opposite may be true. For instance, in a 2018 study exploring police officers' responses to requests from female victims of domestic violence for assistance in locating a shelter for battered women, Stone and Lopes (2018) note that female officers were often as unhelpful as males were. Some told victims the police cannot assist them in locating a shelter for battered women, that this was not a police matter, or that they did not know what a shelter was; they also insinuated that they must have provoked their partner; or they simply hung up on them (Stone and Lopes, 2018, pp. 21–22). Such treatment may indicate the degree to which some female SAPS officers buy into patriarchal and misogynistic gender norms, including the notion that women "have it coming" when beaten by male partners.

In 2009, in response to a high number of female officers being murdered by their partners, the ICD conducted a study of femicide in the SAPS (ICD, 2009). The study showed that 27% of the victims had filed reports of domestic violence prior to their murder and that 90% of the victims were married to or cohabiting with the perpetrator (ICD, 2009, p. 17). It also showed that, between 2004–2005 and 2006–2007, 30 police officers murdered their female partners. In 2018–2019, 16 SAPS officers were killed in "domestic related" incidents (SAPS, 2019), and in 2017–2018, the IPID received 105 complaints of rape by police officers (IPID, 2018, p. 36). According to the IPID, 41% of the alleged rapes committed by the police take place while they are on duty, suggesting that the relatively large numbers of women in the service do not significantly temper attitudes of machismo and sexual entitlement among their male colleagues.

These studies, although limited in number, indicate that even with strong constitutional, legislative, and policy provisions in support of full integration of women into the SAPS and the organization being among the world's leaders in the proportion of women employed in policing as well as holding leadership positions, resistance to women and discrimination from within the formal organization persist. They also suggest that the violent and patriarchal gender norms found in South African society at large are present within the organization, with the potential to affect levels of integrity.

Gender and police integrity theory

According to Waugh et al. (1998, p. 288), it has been frequently asserted that female police officers are "less likely than their male counterparts to engage in misconduct or to approve of such behavior by fellow officers." In this section we address the literature exploring the validity of this assumption. Our search of the literature yielded 17 published studies addressing gender and one or more of the dimensions of the theory of police integrity. Although the body of research on police integrity is relatively large (e.g., Kutnjak Ivković and Haberfeld, 2019), the number of police integrity studies addressing gender is quite limited. One of the potential reasons is the fact that, to entice the respondents to participate and do so honestly, both the police corruption questionnaire and the police integrity questionnaire (see, e.g., Klockars et al., 2000, 2004, 2006) had a very limited number of demographic questions, which did not include the question about the respondents' gender.

Overall, the findings indicate that there is little, if any, influence of gender on the dimensions of police integrity, a finding consistent with that of Poteyeva and Sun's meta-analysis of gender and policing (2009, p. 516). Of the studies reviewed, several (e.g., Othman et al., 2014; Rothwell and Baldwin, 2007; Kutnjak Ivković et al., 2018) were found to have included gender of participants in the description of sample demographics but did not refer to it in the analysis, yielding no further insight into the influence of gender on police integrity.

Three studies (Kucukuysal, 2008; Westmarland and Rowe, 2018; Wu and Makin, 2019) reported only that gender was not a significant factor in their findings. Kucukuysal (2008) investigated the effect of organizational culture on Turkish police officers' perceptions of integrity, observing no significant differences between male and female respondents. Westmarland and Rowe (2018) analyzed the perspectives of 520 police officers serving in three U.K. police forces regarding their views on the seriousness of potential misconduct and factors that might influence their willingness to reports colleagues' infractions. Their sample was comprised of about 30% female participants, a slightly higher rate than the representation of females in the U.K. police nationally. Based on their analysis, they concluded that there were no differences between male and females officers in their estimates of misconduct seriousness and in their expressed willingness to report misconduct. Similarly, Wu and Makin (2019) examined individual, organizational, and ecological factors influencing Chinese police officers' adherence to the code of silence. They reported the results of their regression analyses for the series of scenarios used to measure "willingness to report" that included a gender variable. In none of the scenarios analyzed was gender found to be significant, and no discussion of that result was included in their conclusion.

In contrast to these publications, some studies have found generally small association between attitudes toward the seriousness of described misconduct and gender, as well as willingness to report the misconduct of fellow police officers and gender. Waugh et al. (1998) administered a survey of police attitudes and ethical issues to 59 recruits and 106 first-year constables in the Queensland, Australia, Police Service. Based on their analysis of seven scenarios depicting ethical violations, they concluded that there were no consistent gender differences in either perceptions of seriousness or in stated willingness to report an observed violation committed by other officers. Similarly, Pagon et al. (2004) explored differences between male and female participants on a sample of 95 police officers and 247 students in Slovenia. Participants were provided 11 scenarios describing a range of corrupt behaviors and asked to respond to seven questions regarding each scenario. The authors' analysis produced mixed results. In two scenarios, female respondents rated the seriousness of the violation as less serious than their male counterparts did. By contrast, female respondents rated three scenarios as more serious than their male counterparts did. Respondents differed in regard to their stated willingness to report on three of the 11 scenarios, with female respondents indicating they would be more willing to report other police officers for use of unnecessary or excessive force or for a preferential treatment of an off-duty police officer's traffic accident.

Bjerregaard and Lord (2004) conducted a study of 443 students enrolled in an American university to examine not only the ethical and value orientation primarily of criminal justice students but also differences in responses by gender orientation. Using a scenario-based approach employing ten scenarios, Bjerregaard and Lord (2004) divided the scenarios into three groups: taking gratuities, abuse of authority, and criminal behavior. They reported that "gender was consistently related to a lower perceived likelihood of engaging in unethical behavior and a perception of these behaviors (with the exception of taking gratuities) as more serious" (Bjerregaard and Lord, 2004, p. 273).

Based on hypothetical scenarios of police misconduct, Andreescu and colleagues (2012) examined gender differences in the attitudes of 293 students in their final year of study at the Romanian Police Academy. They observed several significant differences between male and female respondents with respect to the mean scores of the perceived seriousness of the misconduct, appropriate sanction, and willingness to report transgressions. Although male and female respondents were almost identical in their views, they differed on mean scoring of seriousness of misconduct on four of the ten scenarios evaluated. Two of the differences pertained to the use of excessive force and the cover-up of a police DUI and accident, identical to the findings of Pagon et al. (2004) in Slovenia. In regard to the views on the appropriate sanctions for transgressions, both male and female respondents were in agreement regarding the most appropriate sanctions. However, female respondents overall tended to support more severe sanctions. As to the stated willingness to report infractions, differences were observed for three scenarios, with female respondents indicating less reluctance to report on another officer.

Although Kutnjak Ivković et al. (2016) focuses primarily on an exploration of the relation between police integrity (measured by misconduct seriousness) and views on community policing among 1,315 Croatian police officers, their analysis included a control variable for the respondents' gender. Findings on the relation between the seriousness of misconduct described in 11 scenarios and gender showed significant differences only for the scenarios depicting unnecessary/excessive force and verbal abuse of a citizen, with female officers rating the behavior as a more serious transgression than their male counterparts did.

Again, these findings are consistent with the earlier findings of Pagon et al. (2004) in Slovenia and Andreescu and colleagues (2012) in Romania.

On the other hand, Lobnikar and colleagues (2016) observed that female Slovenian police officers were less willing to report police misconduct in nine out of the 14 case scenarios presented. This finding is in direct contrast to the findings of Pagon et al. (2004) 15 years earlier. Lobnikar and colleagues observed not only this change, but also a change in the circumstances influencing motivations to report. Whereas Pagon et al. (2004, p. 15) argued that the moral reasoning for male officers to report was as follows:

> If they deserve to be punished, and other officers would report it, and I believe that this a serious offense, and other officers believe that this is a serious offense, and I think it is a violation of official policy, then I'll report it.

For female officers, they reported that the moral reasoning process was much simpler, "If I believe this is a serious offense, and if they deserve to be punished, then I'll report it." Fifteen years later, Lobnikar and colleagues (2016) report that the perception of most officers' willingness to report had become the primary consideration for whether to report misconduct for their female sample and attribute this change to the practical experience of female police officers in integrating into the Slovenian police over the 15-year period between the two studies.

In 2017, Wagner and colleagues (2017) attempted to ascertain whether feminization of public service can improve service quality and lower corruption in the Ugandan police. To answer the question, they examined the responses from 600 Ugandan police officers to 12 scenarios depicting possible violations of official rules. A variation in their study compared to previous studies was that they randomly varied the gender of the police officer and the gender of the victim depicted in the scenarios. They also controlled for the gender of the 600 respondents to the survey. Wagner and colleagues (2017) found that neither the gender of the officer nor the gender of the victim depicted in the cases are related to respondent assessments of police malpractice or suggested discipline. On the other hand, Wagner and colleagues (2017) found that male respondents, overall, rated the 12 scenarios as more serious than their female counterparts did; they also argued that the male officers were more likely to indicate that they would report the misbehavior of a fellow colleague than the female officers were. Male respondents were also observed to support more severe sanctions for violations than female respondents were.

The extant research on the potential effect of gender on police integrity offers multiple conclusions. First, the evidence is mixed about the potential differences between male and female officers in their assessment of the seriousness of violations presented through the use of scenarios. There are also mixed findings regarding their response about the appropriate sanctions for perceived violations, as well as willingness to report violations by their colleagues. But the general absence of differences is not straightforward. Gender differences often appear to be influenced by the content of the scenarios; female officers tend to see depictions of unnecessary or excessive force as more serious than their male counterparts do.

Several studies point to the possible effect of the degree of integration of female officers into their respective organizations and to the degree of representation relative to the gender composition of the organization as a whole. Interactions between gender and rank or gender and length of service have not been studied, even though rank and length of service have both been found to be related to police integrity. Two U.S. studies (Stinson et al., 2015; Stinson et al., 2016) shed some light on the extent of criminal behavior among U.S. police officers. In an analysis of 6,724 police officers arrested nationwide between 2005 and 2011, Stinson and colleagues (2016) found that 367 (5.5%) were female police officers, while at the same time, female police officers constituted around 12% of the American police officers nationwide. In an analysis of 2,119 cases in which officers were arrested, Stinson and colleagues (2015) reported significant differences in the types of crimes with which they were charged across male and female officers; whereas 40% of female officer arrests were for profit-motivated offenses, the same was true for only 24% of male officer arrests. Similarly, while 23% of arrests of female officers were for violence-related offenses, the same was true for about 50% of male officer arrests. While limited in their scope, these studies do provide suggestive evidence for the argument that employing more female officers might reduce the incidence of police crime and/or change the nature of the police crime that does occur.

Survey findings

The empirical part of our study is based on a survey of 868 SAPS police officers from all nine provinces and territories. The police integrity questionnaire contained 14 hypothetical scenarios covering a range of forms of police misconduct. The questionnaire was designed by Klockars and colleagues (2006) and revised by Kutnjak Ivković (2009). For details about the methodology, please see Chapter 1.

After the respondents had read each of the hypothetical scenarios, they were asked to answer a series of seven questions. One of the first questions asked them to assess whether the behavior described in the questionnaire violated the SAPS official rules ("Would this behavior be regarded as a violation of official policy in your agency?"). Possible answers ranged on a 5-point scale from 1 = "definitely not" to 5 = "definitely yes." The next two questions asked the respondents to assess misconduct seriousness ("How serious do YOU consider this behavior to be?" "How serious do most police officers in your agency consider this behavior to be?"). The respondents could have selected an answer on a 5-point scale from 1 = "not at all serious" to 5 = "very serious." The questions that followed inquired about the appropriate discipline and expected discipline ("If an officer in your agency engaged in this behavior and was discovered doing so, what, if any, discipline do you think should follow?" "If an officer in your agency engaged in this behavior and was discovered doing so, what, if any, discipline do you think would follow?"). The respondents could have selected one of the six disciplinary options, ranging from 1 = "no discipline" to 6 = "dismissal." Finally, two questions inquired about the code of silence ("If an officer in your agency engaged in this behavior and was discovered doing so, what, if any, discipline do you think should follow?" "If an officer in your agency engaged in this behavior and was discovered doing so, what, if any, discipline do you think would follow?"). The respondents could have selected one of the six disciplinary options, ranging from 1 = "no discipline" to 6 = "dismissal."

To encourage honest answers while filling out the questionnaire, Klockars and Kutnjak Ivković (2004) purposely limited the number of demographic questions. In our version of the questionnaire, we added a question about the respondents' gender. About a quarter of our respondents identified as women (28%) and about three-quarters identified as men (71%), a pattern consistent with gender representation in the SAPS.

Familiarity with the official rules

In the majority of the scenarios (9 out of 14, or 64%; Table 8.1), there were no statistically significant differences between female and male police officers' recognition of the rule-violating nature of misconduct. In only five scenarios (Scenario 2: Failure to arrest, Scenario 7: Verbal abuse, Scenario 8: Cover-up of DUI, Scenario 12: Failure to report hate crime, Scenario 14: Failure to react to graffiti), mostly the least likely to be evaluated as rule-violating behaviors by both female and male police officers (Table 8.1), were there statistically significant differences in the mean evaluations of seriousness by gender.[1] In all but one of these scenarios, female officers were more likely to recognize these behaviors as rule-violating than male officers were (Table 8.1). However, these differences are neither large[2] nor substantive.[3]

The relative order of familiarity with official rules—how familiar respondents are with each scenario compared to the other scenarios in the questionnaire (Table 8.1)—was also very similar for male and female officers (Spearman's rho = .914, $p < .001$).

Perceptions of misconduct seriousness

The respondents' evaluations of misconduct seriousness did not differ by gender in most of the scenarios (8 out of 14 or 57%; Table 8.2). In the majority of the scenarios featuring statistically significant differences, women tended to evaluate misconduct as somewhat more serious than men did (Scenario 5: Supervisor corruption, Scenario 7: Verbal abuse, Scenario 12: Failure to report hate crime, Scenario 14: Failure to react to graffiti).[4] Although these differences are statistically significant, they are neither large[5] nor substantively important.[6]

The number of scenarios in which the respondents' estimates of how serious most police officers would evaluate the misconduct differed by their gender

Table 8.1 Violation of official rules by gender

Scenario	*Men*		*Women*		*t-test*	*Difference*
	Mean	*Rank*	*Mean*	*Rank*		
Scenario 1: Acceptance of gratuities	4.22	5	4.34	6	−1.44	−.12
Scenario 2: Failure to arrest	4.63	12	4.47	10	2.20*	.16
Scenario 3: Theft of knife	4.77	14	4.69	14	1.43	.08
Scenario 4: Abuse of deadly force	4.45	9	4.54	11.5	−1.34	−.09
Scenario 5: Supervisor corruption	4.31	7	4.45	9	−1.78a	−.14
Scenario 6: Striking arrestee	4.38	8	4.26	4.5	1.45	.12
Scenario 7: Verbal abuse	3.88	2	4.09	2	−2.44*	−.21
Scenario 8: Cover-up of DUI	3.91	3	4.22	3	−3.21**	−.31
Scenario 9: Kickback	4.30	6	4.35	7	−.58	−.05
Scenario 10: False report on drugs	4.53	11	4.39	8	1.91a	−.14
Scenario 11: Supervisor fails to stop excessive force	4.47	10	4.54	11.5	−.95	−.07
Scenario 12: Failure to report hate crime	4.10	4	4.26	4.5	−2.06*	−.16
Scenario 13: Bribe from reckless motorist	4.70	13	4.61	13	1.55	.09
Scenario 14: Failure to react to graffiti	3.66	1	3.94	1	−2.92**	−.28

Table 8.2 Own views about misconduct seriousness by gender

Scenario	*Men*		*Women*		*t-test*	*Difference*
	Mean	*Rank*	*Mean*	*Rank*		
Scenario 1: Acceptance of gratuities	3.91	4	3.91	2	.018	.00
Scenario 2: Failure to arrest	4.64	12	4.50	9	1.96a	.14
Scenario 3: Theft of knife	4.84	14	4.64	13	3.77***	.20
Scenario 4: Abuse of deadly force	4.56	10.5	4.58	11.5	−.34	.00
Scenario 5: Supervisor corruption	4.26	6	4.58	11.5	−4.33***	−.32
Scenario 6: Striking arrestee	4.29	8	4.29	5	−.09	.00
Scenario 7: Verbal abuse	3.80	2	4.05	4	−2.66**	−.25
Scenario 8: Cover-up of DUI	3.90	3	4.02	3	−1.13	−.12
Scenario 9: Kickback	4.28	7	4.42	8	−1.71a	−.14
Scenario 10: False report on drugs	4.56	10.5	4.41	7	2.02*	.15
Scenario 11: Supervisor fails to stop excessive force	4.43	9	4.52	10	−1.15	−.09
Scenario 12: Failure to report hate crime	4.11	5	4.32	6	−2.70**	−.21
Scenario 13: Bribe from reckless motorist	4.72	13	4.68	14	.69	.04
Scenario 14: Failure to react to graffiti	3.62	1	3.95	1	−3.46**	−.33

was even smaller than the number of cases in which their own evaluations of seriousness differed by their gender (Table 8.3). Put differently, in the overwhelming majority of the scenarios (11 out of 14 or 79%), female and male views of how serious most SAPS officers would evaluate misconduct were very similar. There were only three scenarios (Scenario 12: Failure to report hate crime, Scenario 13: Bribe from reckless motorist, Scenario 14: Failure to react to graffiti) with statistically significant differences,[7] which turned out to be on the small side[8] and substantively unimportant.[9]

We also examined the similarly of the relative order of seriousness evaluations across female and male respondents (Table 8.2). In a familiar pattern, how serious respondents evaluated each scenario compared to the other scenarios in the questionnaire was very similar for female and male respondents (Spearman's rho = .836, $p < .001$). As was the case for their own estimates of seriousness, the relative order of how serious most police officers would evaluate these same behaviors was very similar across female and male respondents (Spearman's rho = .846, $p < .001$).

Finally, we compared the respondents' own evaluations of seriousness with what they assumed would be most police officers' evaluations of seriousness (Table 8.4). Both female and male respondents assumed that most police officers would be significantly less likely to evaluate misconduct as serious than the respondents themselves would (14 out of 14 scenarios resulted in statistically

Table 8.3 Most police officers' views about misconduct seriousness by gender

Scenario	*Men*		*Women*		*t-test*	*Difference*
	Mean	*Rank*	*Mean*	*Rank*		
Scenario 1: Acceptance of gratuities	3.42	2	3.29	1	1.22	.13
Scenario 2: Failure to arrest	3.99	9	3.96	8	.33	.03
Scenario 3: Theft of knife	4.29	13	4.17	13	1.44	.12
Scenario 4: Abuse of deadly force	4.28	12	4.35	14	−.79	−.07
Scenario 5: Supervisor corruption	3.85	7	3.94	6.5	−.92	−.09
Scenario 6: Striking arrestee	3.79	6	3.80	5	−.11	−.01
Scenario 7: Verbal abuse	3.49	3	3.59	3	−.93	−.10
Scenario 8: Cover-up of DUI	3.56	4	3.64	4	−.74	−.08
Scenario 9: Kickback	3.94	8	3.94	6.5	.02	.00
Scenario 10: False report on drugs	4.16	11	4.09	11.5	.75	.07
Scenario 11: Supervisor fails to stop excessive force	4.01	10	3.99	9	.17	.02
Scenario 12: Failure to report hate crime	3.78	5	4.09	11.5	−3.34**	−.31
Scenario 13: Bribe from reckless motorist	4.41	14	4.08	10	3.85***	.33
Scenario 14: Failure to react to graffiti	3.31	1	3.55	2	−2.34*	−.24

Table 8.4 Views about misconduct seriousness by gender

Scenario	*Men Own*		*Men Most*		*t-test*	*Diff.*	*Women Own*		*Women Most*		*t-test*	*Diff.*
	Mean	*Rank*	*Mean*	*Rank*			*Mean*	*Rank*	*Mean*	*Rank*		
Scenario 1: Acceptance of gratuities	3.91	4	3.42	2	9.79***	.49	3.91	2	3.29	1	6.29***	.62
Scenario 2: Failure to arrest	4.64	12	3.99	9	14.14***	.65	4.50	9	3.96	8	7.24***	.54
Scenario 3: Theft of knife	4.84	14	4.29	13	13.04***	.55	4.64	13	4.17	13	7.18***	.47
Scenario 4: Abuse of deadly force	4.56	10.5	4.28	12	7.03***	.28	4.58	11.5	4.35	14	3.72***	.23
Scenario 5: Supervisor corruption	4.26	6	3.85	7	8.19***	.41	4.58	11.5	3.94	6.5	8.03***	.64
Scenario 6: Striking arrestee	4.29	8	3.79	6	11.80***	.50	4.29	5	3.80	5	6.69***	.49
Scenario 7: Verbal abuse	3.80	2	3.49	3	8.30***	.31	4.05	4	3.59	3	7.18***	.46
Scenario 8: Cover-up of DUI	3.90	3	3.56	4	8.62***	.34	4.02	3	3.64	4	6.30***	.38
Scenario 9: Kickback	4.28	7	3.94	8	8.12***	.34	4.42	8	3.94	6.5	7.71***	.48
Scenario 10: False report on drugs	4.56	10.5	4.16	11	10.43***	.40	4.41	7	4.09	11.5	5.38***	.32
Scenario 11: Supervisor fails to stop excessive force	4.43	9	4.01	10	9.87***	.42	4.52	10	3.99	9	8.32***	.53
Scenario 12: Failure to report hate crime	4.11	5	3.78	5	8.58***	.33	4.32	6	4.09	11.5	3.72***	.23
Scenario 13: Bribe from reckless motorist	4.72	13	4.41	14	8.71***	.31	4.68	14	4.08	10	8.41***	.60
Scenario 14: Failure to react to graffiti	3.62	1	3.31	1	8.64***	.31	3.95	1	3.55	2	6.74***	.40

significant differences; Table 8.4). However, while all these differences were statistically significant, they were substantively important in only three scenarios for both female and male respondents (Table 8.4).[10]

Views about discipline

We started our analyses with a comparison of the respondents' views about the appropriate discipline. In most scenarios (10 out of 14 or 71%; Table 8.5), there were no statistically significant differences between female and male respondents' views on the appropriate discipline. In the three scenarios with statistically significant differences (Scenario 3: Theft of knife, Scenario 4: Abuse of deadly force, Scenario 13: Bribe from reckless motorist; Table 8.5), all of which are typically evaluated as the most serious, male respondents were more likely to state that more severe discipline is appropriate than female respondents were.[11] On the other hand, in the fourth scenario with statistically significant differences (Scenario 1: Acceptance of gratuities), female respondents were more likely to state that more severe discipline is appropriate than male respondents were. However, none of these differences were substantively important.

Gender differences were even less pronounced for the respondents' views of the expected discipline (Table 8.6); in the overwhelming majority of the scenarios (11 out of 14 or 79%; Table 8.6), there were no statistically significant differences. In the three scenarios with statistically significant differences (Scenario 3: Theft of knife, Scenario 4: Abuse of deadly force, Scenario 13: Bribe from reckless motorist; Table 8.6), all of which are typically evaluated as the most serious, male respondents were more likely to say that they think that their police agency would issue more severe discipline than female respondents were.[12]

The relative order of both appropriate and expected discipline—how severe the discipline is for each scenario compared to the other scenarios in the questionnaire—indicates that male and female police officers provided very similar estimates (appropriate discipline: Spearman's rho $= .930$; $p < .001$; expected discipline: Spearman's rho $= .960$; $p < .01$).

Expected willingness to report

Respondents' own willingness to report did not vary considerably across our male and female respondents for most of the scenarios (9 out of 14, or 64%; Table 8.7). In the remaining five scenarios (Scenario 4: Abuse of deadly force, Scenario 5: Supervisor corruption, Scenario 7: Verbal abuse, Scenario 12: Failure to report hate crime, Scenario 14: Failure to react to graffiti), mostly evaluated as less serious, there were statistically significant differences between men and women in their own expressed willingness to report (Table 8.7).[13] In all but one of these scenarios, women were more likely to state that they would report than men were. However, these differences were neither large[14] nor substantively important.[15]

Table 8.5 Views about the appropriate discipline by gender

Scenario	*Men*		*Women*		*Chi-square/Phi*	*Difference*
	% serious	*Rank*	*% serious*	*Rank*		
Scenario 1: Acceptance of gratuities	23.7%	3	30.9%	7	4.76*/−.074	−7.2%
Scenario 2: Failure to arrest	34.3%	8	37.0%	8	.58 / −.026	−2.7%
Scenario 3: Theft of knife	66.6%	12	53.7%	12	12.53***/ .121	12.9%
Scenario 4: Abuse of deadly force	69.3%	13	55.7%	13	14.43***/ .129	13.6%
Scenario 5: Supervisor corruption	33.4%	7	30.5%	6	.70 / .028	2.9%
Scenario 6: Striking arrestee	28.9%	6	24.9%	3	1.40 / .040	4.0%
Scenario 7: Verbal abuse	13.8%	2	13.4%	1	.03 / .005	.04%
Scenario 8: Cover-up of DUI	28.4%	5	27.2%	4	.12 / .012	1.2%
Scenario 9: Kickback	43.0%	10	37.8%	9	1.97 / .048	5.2%
Scenario 10: False report on drugs	48.5%	11	48.0%	11	.02 / .005	0.5%
Scenario 11: Supervisor fails to stop excessive force	41.4%	9	42.7%	10	.120 / −.012	−1.3%
Scenario 12: Failure to report hate crime	27.1%	4	28.9%	5	.270 / −.018	−1.8%
Scenario 13: Bribe from reckless motorist	70.0%	14	63.0%	14	3.91*/.067	7.0%
Scenario 14: Failure to react to graffiti	13.5%	1	16.3%	2	1.12 / −.036	−2.8%

*Based on the results of bivariate analyses.

Table 8.6 Views about the expected discipline by gender

Scenario	*Men*		*Women*		*Chi-square/Phi*	*Difference*
	% serious	*Rank*	*% serious*	*Rank*		
Scenario 1: Acceptance of gratuities	23.2%	3	27.2%	4	1.54 / −.042	−4.0%
Scenario 2: Failure to arrest	34.9%	8	38.2%	9	.84 / −.031	−3.3%
Scenario 3: Theft of knife	65.6%	12	55.3%	13	7.97**/.096	10.3%
Scenario 4: Abuse of deadly force	71.6%	14	54.9%	12	22.15***/.160	16.7%
Scenario 5: Supervisor corruption	34.3%	7	32.9%	7	.14 / .013	1.4%
Scenario 6: Striking arrestee	30.0%	5	32.5%	6	.51 / −.024	−2.5%
Scenario 7: Verbal abuse	12.8%	2	15.4%	1	1.03 / −.035	−2.6%
Scenario 8: Cover-up of DUI	26.5%	4	22.4%	3	1.57 / .043	4.1%
Scenario 9: Kickback	40.6%	10	33.7%	8	3.47 / .063	6.9%
Scenario 10: False report on drugs	48.7%	11	48.8%	11	.00 / −.001	−0.1%
Scenario 11: Supervisor fails to stop excessive force	39.4%	9	42.7%	10	.76 / −.030	−3.3%
Scenario 12: Failure to report hate crime	31.2%	6	30.1%	5	.10 / .011	1.1%
Scenario 13: Bribe from reckless motorist	68.5%	13	56.9%	14	10.32**/.109	11.6%
Scenario 14: Failure to react to graffiti	12.2%	1	16.3%	2	2.54 / −.054	−4.1%

*Based on the results of bivariate analyses.

Table 8.7 Own willingness to report by gender

Scenario	*Men*		*Women*		*t-test*	*Difference*
	Mean	*Rank*	*Mean*	*Rank*		
Scenario 1: Acceptance of gratuities	3.72	6	3.61	4	1.06	.11
Scenario 2: Failure to arrest	4.25	11	4.11	11	1.51	.14
Scenario 3: Theft of knife	4.40	12	4.34	13	.86	.07
Scenario 4: Abuse of deadly force	4.42	13	4.17	12	2.88**	.25
Scenario 5: Supervisor corruption	3.79	7.5	4.08	8	−3.10**	−.29
Scenario 6: Striking arrestee	3.79	7.5	3.77	5	.20	.02
Scenario 7: Verbal abuse	3.18	1	3.60	3	−3.81***	−.42
Scenario 8: Cover-up of DUI	3.38	3	3.59	2	−1.86	−.21
Scenario 9: Kickback	3.71	5	3.83	6	−1.08	−.12
Scenario 10: False report on drugs	4.03	10	4.10	10	−.79	−.07
Scenario 11: Supervisor fails to stop excessive force	3.88	9	3.96	7	−.81	−.08
Scenario 12: Failure to report hate crime	3.69	4	4.09	9	−4.41***	−.40
Scenario 13: Bribe from reckless motorist	4.46	14	4.43	14	.46	.03
Scenario 14: Failure to react to graffiti	3.29	2	3.50	1	−2.05*	−.21

Similarly, the respondents' views of how strong the code of silence is among their colleagues, measured as the perceptions of most police officers' willingness to report (Table 8.8), did not differ significantly across gender in the majority of the scenarios (8 out of 14 or 57%; Table 8.8). On the other hand, there were six scenarios (Scenario 2: Failure to arrest, Scenario 5: Supervisor corruption, Scenario 7: Verbal abuse, Scenario 8: Cover-up of DUI, Scenario 12: Failure to report hate crime, Scenario 14: Failure to react to graffiti) with statistically significant differences, and in all but one scenario women expected that their colleagues would be more likely to report than men did (Table 8.8).[16] Although these differences were statistically significant, they were neither large[17] nor of substantive importance.[18]

The relative order of officers' own expressed willingness to report was very similar for male and female respondents (Spearman's rho = .893; $p < .001$), as was the relative order of their estimates of most police officers' willingness to report (Spearman's rho = .794; $p < .001$).

Finally, we also compared the respondents' own expressed willingness to report with how likely they anticipated most SAPS officers would be to report (Table 8.9). In all 14 scenarios, it seems that both male and female respondents' own expressed willingness to report was statistically significantly higher than how likely they assumed that their fellow colleagues would be to report. These differences were also large and substantively important in most of the scenarios for male respondents (10 out of 14 scenarios or 71%; Table 8.9),

Table 8.8 Most police officers' willingness to report by gender

Scenario	*Men*		*Women*		*t-test*	*Difference*
	Mean	*Rank*	*Mean*	*Rank*		
Scenario 1: Acceptance of gratuities	2.93	4	2.88	1	.43	.05
Scenario 2: Failure to arrest	3.54	11	3.24	4.5	2.93**	.30
Scenario 3: Theft of knife	3.61	12	3.59	10	.20	.02
Scenario 4: Abuse of deadly force	3.86	13	3.78	14	.73	.08
Scenario 5: Supervisor corruption	3.36	10	3.68	11.5	−3.07**	.32
Scenario 6: Striking arrestee	3.20	5	3.41	7.5	−1.89a	−.21
Scenario 7: Verbal abuse	2.82	2	3.24	4.5	−3.66***	−.42
Scenario 8: Cover-up of DUI	2.92	3	3.18	3	−2.28*	−.26
Scenario 9: Kickback	3.21	6	3.34	6	−1.19	−.13
Scenario 10: False report on drugs	3.39	9	3.48	9	−.82	−.09
Scenario 11: Supervisor fails to stop excessive force	3.31	8	3.41	7.5	−.92	−.10
Scenario 12: Failure to report hate crime	3.23	7	3.68	11.5	−4.59***	−.45
Scenario 13: Bribe from reckless motorist	3.87	14	3.76	13	1.10	.11
Scenario 14: Failure to react to graffiti	2.77	1	3.12	2	−3.20**	−.35

and in close to one-half of the scenarios for female respondents (6 out of 14 scenarios or 43%).[19]

Conclusion

Although women constitute over one-third of the SAPS's personnel in 2019 (SAPS, 2019, p. 34), studies indicate that the complete integration of women into policing has not occurred in South Africa. Since the 1970s when women started entering policing more systematically and in larger numbers, critics have challenged their ability to do police work (e.g., Brown, 1996), leading women to perpetually feel they had to prove they could "hold their own" (Dejong, 2005).

Despite the gender power and opportunity disparities that run through South African society, analysis of our data shows that, even where statistically significant differences between genders are observed, the differences are neither large nor substantive. We found very few differences across all measures of police integrity, including the respondents' familiarity with official rules, their evaluations of misconduct seriousness, their views about appropriate and expected discipline, and their willingness to report misconduct. These results are also consistent with empirical findings from the research studies discussed in the section on gender and police integrity theory (Waugh et al., 1998; Pagon et al., 2004; Andreescu et al., 2012; Lobnikar et al., 2016; Wagner et al., 2017),

Table 8.9 Willingness to report by gender

Scenario	*Men Own*		*Men Most*		*t-test*	*Diff.*	*Women Own*		*Women Most*		*t-test*	*Diff.*
	Mean	*Rank*	*Mean*	*Rank*			*Mean*	*Rank*	*Mean*	*Rank*		
Scenario 1: Acceptance of gratuities	3.72	6	2.92	3.5	13.57***	.80	3.61	4	2.88	1	8.27***	.73
Scenario 2: Failure to arrest	4.25	11	3.54	11	13.75***	.71	4.11	11	3.24	4.5	10.23***	.87
Scenario 3: Theft of knife	4.40	12	3.61	12	14.75***	.79	4.34	13	3.59	10	9.52***	.75
Scenario 4: Abuse of deadly force	4.42	13	3.86	13	12.52***	.56	4.17	12	3.78	14	6.05***	.39
Scenario 5: Supervisor corruption	3.79	7.5	3.36	9	10.12***	.43	4.08	8	3.68	11.5	6.58***	.40
Scenario 6: Striking arrestee	3.79	7.5	3.20	5	13.09***	.59	3.77	5	3.41	7.5	4.73***	.36
Scenario 7: Verbal abuse	3.18	1	2.82	2	8.93***	.36	3.60	3	3.24	4.5	5.93***	.36
Scenario 8: Cover-up of DUI	3.38	3	2.92	3.5	10.35***	.46	3.59	2	3.18	3	6.46***	.41
Scenario 9: Kickback	3.71	5	3.21	6	10.83***	.50	3.83	6	3.34	6	6.62***	.49
Scenario 10: False report on drugs	4.03	10	3.39	10	13.00***	.64	4.10	10	3.48	9	9.40***	.62
Scenario 11: Supervisor fails to stop excessive force	3.88	9	3.31	8	11.29***	.57	3.96	7	3.41	7.5	7.37***	.55
Scenario 12: Failure to report hate crime	3.69	4	3.23	7	9.87***	.46	4.09	9	3.68	11.5	6.23***	.41
Scenario 13: Bribe from reckless motorist	4.46	14	3.87	14	11.90***	.59	4.43	14	3.76	13	9.29***	.67
Scenario 14: Failure to react to graffiti	3.29	2	2.77	1	12.35***	.52	3.50	1	3.12	2	6.70***	.38

where scenario-specific gender differences might be observed, but not systematic differences across the scenarios. Like most extant police integrity studies, our study suggests that gender and attitudes of police integrity are not strongly and systematically related.

When we compared our findings about gender differences regarding the evaluation of the seriousness with the findings from previous research, we uncovered similarities among the studies related to the use of unnecessary/excessive force and covering up a police DUI and accident. In Romania (Andreescu et al., 2012), Uganda (Wagner et al., 2017), Croatia (Kutnjak Ivković et al., 2016), Slovenia (Lobnikar et al., 2016), and our own data, female officers see these specific events as more serious and deserving of more accountability than their male counterparts do. The differences between the evaluations provided by male and female officers are statistically significant, a tendency that stretches across different cultures and police organizations.

There also appear to have been some systematic differences, again neither large nor substantive, in how instances of a failure to protect (ignoring graffiti, domestic violence) are perceived and acted upon differently by male and female officers in their evaluation of offense seriousness and willingness to report. Overall, the evidence on these differences is not sufficient to discount a moderating effect of an organizational culture on gender differences. At the same time, however, small differences do exist, indicating that gender can, and does, produce a moderating effect on the organizational culture.

Notes

1 Analyzing gender differences in familiarity with the official rules while controlling for other demographic variables by means of linear regression preserved the statistically significant differences in familiarity with the official rules in 3 out of 5 scenarios previously detected in simpler, bivariate analyses (Scenario 7: Verbal abuse, Scenario 8: Cover-up of DUI, Scenario 14: Failure to react to graffiti).
2 Cohen's *d* indicated that the differences were small for all pairs, except for Scenario 13, in which the differences between 1 v. 2 and 1 v. 3 were of medium size.
3 We rely on the threshold of .50 to ascertain substantive differences, as established by Klockars and colleagues (2006).
4 Analyzing gender differences in perceptions of misconduct seriousness while controlling for other demographic variables by means of linear regression preserved all the statistically significant differences in perceptions of misconduct seriousness previously detected in simpler, bivariate analyses, with one exception (Scenario 10).
5 Cohen's *d* indicated that the differences were small.
6 We rely on the threshold of .50 to ascertain substantive differences, as established by Klockars and colleagues (2006).
7 Analyzing gender differences in police officers' estimates of most police officers' evaluations of seriousness while controlling for other demographic variables by means of linear regression preserved all the statistically significant differences in estimates of most police officers' evaluations of seriousness previously detected in simpler, bivariate analyses.
8 Cohen's *d* indicated that the differences were small.
9 We rely on the threshold of .50 to ascertain substantive differences, as established by Klockars and colleagues (2006).

10 We rely on the threshold of .50 to ascertain substantive differences, as established by Klockars and colleagues (2006).
11 Analyzing gender differences in police officers' views about appropriate discipline while controlling for other demographic variables by means of logistic regression preserved all but one statistically significant difference in appropriate discipline previously detected in simpler, bivariate analyses. The exception was Scenario 13, in which gender differences were only marginally significant in multivariate analyses. Also, differences were statistically significant in Scenario 1.
12 Analyzing gender differences in police officers' views about expected discipline while controlling for other demographic variables by means of logistic regression preserved all statistically significant differences in expected discipline previously detected in simpler, bivariate analyses.
13 Analyzing gender differences in own willingness to report while controlling for other demographic variables by means of linear regression preserved all the statistically significant differences in own willingness to report.
14 Cohen's *d* indicated larger differences in these scenarios.
15 They passed the threshold of substantive importance, as established by Klockars and colleagues (2006).
16 Analyzing gender differences in the estimates of most police officers' willingness to report while controlling for other demographic variables by means of linear regression preserved all the statistically significant differences in most police officers' willingness to report. The only exception was Scenario 8. In addition, the differences were statistically significant in Scenario 6.
17 Cohen's *d* indicated larger differences in these scenarios.
18 They passed the threshold of substantive importance, as established by Klockars and colleagues (2006).
19 We rely on the threshold of .50 to ascertain substantive differences, as established by Klockars and colleagues (2006).

References

Andreescu, V., D.G. Keeling, M.C. Voinic, and B.N. Tonea (2012). Future Romanian law enforcement: Gender differences in perceptions of police misconduct. *Journal of Social Research & Policy*, 3(1): 97–113.

Ashforth, A. (2005). *Witchcraft, Violence, and Democracy in South Africa*. Chicago, IL: University of Chicago Press.

Baden, S., S. Hasim, and S. Meintjes (1998). *Country Gender Profile: South Africa Report Prepared for the Swedish International Development Office (Sida) Pretoria, Republic of South Africa*. Available at: https://pdfs.semanticscholar.org/4b44/f2378f2441a7c8c9a8a553dfcfcd78a397ad.pdf?_ga=2.195153967.2017453163.1582346575-1885704529.1582346575

Bjerregaard, B. and V.B. Lord (2004). An examination of the ethical and value orientation of criminal justice students. *Police Quarterly*, 7(2): 262–284. https://doi.org/10.1177/1098611102250739

Brown, J. (1996). Integrating women into policing: A comparative European perspective. In *Policing in Central and Eastern Europe: Comparing Firsthand Knowledge with Experience from the West*. Slovenia: College of Police and Security Studies. Available at: www.ncjrs.gov/policing/int627.htm

Brown, J. and F. Heidensohn (2000). *Gender and Policing: Comparative Perspectives*. Basingstoke: Macmillan.

Bruce, D. (2020). *A Question of Gender? Violence against Women and Men in South Africa*. Southern Africa Report No. 24. Pretoria: ISS.

Cawthra, G. (1993). *Policing South Africa: The SAP and the transition from Apartheid.* London, U.K.: Zed Books Ltd.

Constitution of the Republic of South Africa (1996). Act No. 108 of 1996. Available at: www.gov.za/documents/constitution-republic-south-africa-1996

Dejong, C. (2005). Gender differences in officer attitude and behavior. *Women & Criminal Justice*, 15(3–4): 1–32. DOI: 10.1300/J012v15n03_01

Dunkle, K., R. Jewkes, H. Brown, G. Gray, J. McIntryre, and S. Harlow (2004). Gender-based violence, relationship power, and risk of HIV infection in women attending antenatal clinics in South Africa. *The Lancet*, 363: 1415–1421.

eNews Channel Africa (2014). *Is Gender Equality Still an Issue in SA*? Available at: www.enca.com/state-aid-and-impact-gender-equality

Etherington, N., P. Harries, and B. Mbenga (2009). From colonial hegemonies to imperial conquest, 1840–1880. In C. Hamilton, B. Mbenga, and R. Ross (eds.), *The Cambridge History of South Africa (Cambridge History of South Africa)*. Cambridge: Cambridge University Press. 319–391. doi:10.1017/CHOL9780521517942.008

Faull, A. (2008). *Policing Diversity: An Analysis of a Diversity Intervention and Its Effects on a South African Police Service Station.* Amsterdam: Rozenberg Publishers.

Faull, A. (2010). *Behind the Badge: The Untold Stories of South Africa's Police Service Members.* Cape Town: Penguin Random House South Africa. Kindle Edition.

Faull, A. (2013). Fighting for respect: Violence, masculinity and legitimacy in the SAPS. *South African Crime Quarterly*, 44: 5–14.

Faull, A. (2018). *Police Work and Identity: A South African Ethnography.* Abingdon, U.K.: Routledge.

Francke, R.L. (2019). South Africa in a crisis of violence against women, says president. *The Guardian.* Available at: www.theguardian.com/world/2019/sep/06/south-africa-faces-national-crisis-of-violence-against-women-says-president

Glaser, C. (1998). Swines, hazels and the dirty dozen: Masculinity, territoriality and the youth gangs of Soweto, 1960–1976. *Journal of Southern African Studies*, 24(4): 719–736.

Glaser, C. (2000). *Bo-Tsotsi: The Youth Gangs of Soweto, 1935–1976.* Social History of Africa Series. Oxford: James Currey Ltd.

Gqola, P.D. (2015). *Rape: A South African Nightmare.* Jacana: Auckland Park.

Harris, M. (2019). South Africans weigh in on attitudes towards Women. *Ipsos.* Available at: www.ipsos.com/en-za/south-africans-weigh-attitudes-towards-women

Heidensohn, F. (1992). *Women in Control? The Role of Women in Law Enforcement.* Oxford: Clarendon.

Heidensohn, F. (1998). Theoretical review papers: Comparative models of policing and the role of women officers. *International Journal of Police Science and Management*, 1(3): 215–226.

Hornberger, J. (2011). *Policing and Human Rights: The Meaning of Violence and Justice in the Everyday Policing of Johannesburg.* Abingdon: Routledge.

Hunter, M. (2010). *Love in a Time of HIV: Inequality, Rights and Gender in South Africa.* Bloomington, IN: Indiana University Press.

ICD (2009). *Femicide: A Case Study on Members of the South African Police Service.* Pretoria: Independent Complaints Directorate.

Independent Police Investigative Directorate (IPID) (2018). *Annual Report 2017/2018.*

Jensen, S. (2008). *Gangs, Politics and Dignity in Cape Town.* Chicago: Chicago University Press.

Klockars, C.B. and S. Kutnjak Ivković (2004). Measuring police integrity. In M.J. Hickman, A.R. Piquero, and J.R. Greene (eds.), *Police Integrity and Ethics.* Belmont, CA: Wadsworth Publishing.

Klockars, C.B., S. Kutnjak Ivković, and M.R. Haberfeld (eds.) (2004). *The Contours of Police Integrity.* Newbury Park, CA: Sage Publication.

Klockars, C.B., S. Kutnjak Ivković, and M.R. Haberfeld. (2006). *Enhancing Police Integrity.* Dordrecht, The Netherlands: Springer.

Klockars, C.B., S. Kutnjak Ivković, W.E. Harver, and M.R. Haberfield (2000). *The Measurement of Police Integrity.* U.S. Department of Justice, Office of Justice Programs. Available at: www.ncjrs.gov/pdffiles1/nij/181465.pdf

Kucukuysal, B. (2008). *Determinants of Turkish Police Officers' Perception of Integrity: Impact of Organizational Culture.* Electronic Theses and Dissertations. 3777. Available at: https://stars.library.ucf.edu/etd/3777

Kutnjak Ivković, S. (2003). To serve and collect: Measuring police corruption. *The Journal of Criminal Law and Criminology*, 93(2–3): 593–649.

Kutnjak Ivković, S. (2009). The Croatian police, police integrity, and transition toward democratic policing. *Policing: An International Journal of Police Strategies and Management,* 32(3): 459–488.

Kutnjak Ivković, S. and M.R. Haberfeld (2019). Exploring empirical research on police integrity. In S. Kutnjak Ivković and M. Haberfeld (eds.), *Exploring Police Integrity: Novel Approaches to Police Integrity Theory and Methodology.* New York: Springer.

Kutnjak Ivković, S., M. Haberfeld, and R. Peacock (2018). Decoding the code of silence. *Criminal Justice Policy Review*, 29(2): 172–189.

Kutnjak Ivković, S., I.C. Mraović, and K. Borovec (2016). Does community policing matter for police integrity? *Revija za Kriminalistiko in Kriminologijo/Journal of Criminal Investigation and Criminology*, 67(4): 313–325.

Landau, P.S. (2009). Transformations in consciousness. In C. Hamilton, B. Mbenga, and R. Ross (eds.), *The Cambridge History of South Africa*, Vol. 1. Cambridge: Cambridge University Press.

Langa, M., A. Kirsten, B. Bowman, G. Eagle, and P. Kiguwa (2018). Black masculinities on trial in absentia: The case of oscar pistorius in South Africa. *Men & Masculinities.* Available at: https://doi.org/10.1177/1097184X18762523

Legassick, M. and R. Ross (2009). From slave economy to settler capitalism: The Cape colony and its extensions, 1800–1854. In C. Hamilton, B. Mbenga, and R. Ross (eds.), *The Cambridge History of South Africa (Cambridge History of South Africa).* Cambridge: Cambridge University Press. 253–318. doi:10.1017/CHOL9780521517942.007

Letsela, L. and K. Ratele (2009). *Masculinity and Perceptions of Risk: Factors to Premature Male Mortality in South Africa.* Institute of Social and Health Sciences, UNISA. Available at: www.brothersforlife.org/sites/default/files/docs/Men_and_their_perceptions_of_Risks.pdf

Lobnikar, B., K. Prislan, B. Čuvan, and G. Meško (2016). The code of silence and female police officers in Slovenia: Gender differences in willingness to report police misconduct. *Policing: An International Journal,* 39(2): 387–400. https://doi.org/10.1108/PIJPSM-10-2015-0118

Mager, A. and M. Mulaudzi (2011). Popular responses to apartheid: 1948-1975. In R. Ross, A. Mager, and B. Nasson (eds.), *The Cambridge History of South Africa, Volume 2: 1885-1994.* Cambridge: Cambridge University Press. 369–408.

Martin S. (1980). *Breaking and Entering: Policewomen on Patrol.* Berkeley, CA: University of California Press.

Meyer, M.E. and S. Reyneke-Cloete (2013). Current perspectives toward women serving in the South African police. Unpublished paper presented at the 50th Annual Meeting of the Academy of Criminal Justice Sciences, Dallas, TX, 19–23 March.

Meyer, M.E. and J. Steyn (2008). Developing cynicism among male and female recruits in the South African Police Service. *Acta Criminologica,* 2: 1–20.

Meyer, M.E. and J. Steyn (2009). Nurturing isolation in the South African police service. *Policing: An International Journal of Police Strategies and Management*, 32(1): 108–127. DOI: 10.1108/13639510910937148

Montesh, M. (2010). Transformation in the South African Police Service: The implementation of affirmative action and employment equity. *SAPS South African Journal of Criminal Justice*, 23(1): 55–77.

Morison, C.J. (2004). Towards gender equality in South African policing. In G. Mesko, M. Pagon, and B. Dobovsek (eds.), *Policing in Central and Eastern Europe: Dilemmas of Contemporary Criminal Justice*, pp. 238–244. Available at: www.fvv.um.si/conf2004/papers/morrison.pdf

Morison, C.J. (2005). Gender discrimination versus equality in the police. *Acta Criminologica: Southern African Journal of Criminology*, 18(3): 20–28.

National Center for Women and Policing (2001). *Equality Denied: The Status of Women in Policing*. Available at: http://womenandpolicing.com/PDF/2002_Status_Report.pdf

Nattrass, G. (2017). A *Short History of South Africa*. London: Biteback Publishing. Kindle Edition.

Nattrass, N. and J. Seekings (2011). The economy and poverty in the twentieth century. In R. Ross, A.K. Mager, and B. Nasson (eds.), *The Cambridge History of South Africa*, Vol. 2. Cambridge: Cambridge University Press.

Newham, G., T. Masuku, and J. Dlamini (2006). *Diversity and Transformation in the South African Police Service: A Study of Police Perspectives on Race, Gender and the Community in the Johannesburg Policing Area*. Centre for the Study of Violence and Reconciliation: Braamfontein, South Africa. Available at: https://csvr.org.za/docs/policing/diversity.pdf

Ngwadla, N. (2018). The burden of black tax. *Fin 24*. Available at: www.fin24.com/Money/the-burden-of-black-tax-20180309

Othman, R., N. Omar, A. Azam, S. Ibrahim, W.A. Farouq, N. Rustam, and N. Aris (2014). Influence of job satisfaction and codes of ethics on integrity among police officers. *Procedia: Social and Behavioral Sciences*, 145: 266–276.

Pagon, M., B. Lobnikar, and D. Anelj (2004). Gender differences in leniency towards police misconduct. In G. Mesko, M. Pagon, and B. Dobovsek (eds.), *Policing in Central and Eastern Europe: Dilemmas of Contemporary Criminal Justice*. Ljubljana, Slovenia: Faculty of Criminal Justice, University of Maribor, pp. 1–16.

Pinnock, D. (2016). *Gang Town*. Cape Town: Tafelberg.

Perkins, G. (2018). *Danger and Death: Organisational and Occupational Responses to the Murder of Police in South Africa: A Case Study*. Unpublished PhD Thesis. University of Cape Town.

Poteyeva, M. and I.Y. Sun (2009). Gender differences in police officers' attitudes: Assessing current empirical evidence. *Journal of Criminal Justice*, 37: 512–522.

Potgieter, L. (2012). *Are You Man Enough? A Case Study of How Masculinity Is Represented and Experienced in the South African Police Service*. Unpublished PhD Dissertation. University of Pretoria.

Prenzler, T., J. Fleming, and A.L. King (2010). Gender equity in Australian and New Zealand policing: A five-year review. *International Journal of Police Science and Management*, 12(4): 584–595.

Robinson, J. (1996). The Power of Apartheid: State, Power, and Space in South African Cities. Oxford, United Kingdom: Butterworth-Heinemann.

Rothwell, G.R. and J.N. Baldwin (2007). Whistle-blowing and the code of silence in police agencies: Policy and structural predictors. *Crime & Delinquency*, 53(4): 605–632.

Seedat, M., A. van Niekerk, R. Jewkes, S. Suffla, and K. Ratele (2009). Violence and injuries in South Africa: Prioritising an agenda for prevention. *The Lancet*, 374(9694): 1011–1022.

South African Police Service (2012). *Annual Report 2011/2012*. Pretoria: South African Police Service. Available at: https://www.saps.gov.za/about/stratframework/annualreports_arch.php
South African Police Service (2016). *Annual Report 2015/2016*. Pretoria: South African Police Service. Available at: https://www.saps.gov.za/about/stratframework/annualreports_arch.php
South African Police Service (2018). *Annual Report 2017/2018*. Pretoria: South African Police Service. Available at: https://www.saps.gov.za/about/stratframework/annualreports_arch.php
South African Police Service (2019). *Annual Report 2018/2019*. Pretoria: South African Police Service. Available at: https://www.saps.gov.za/about/stratframework/annualreports_arch.php
South African Police Service Act (1995). *Act No 68 of 1995*. Pretoria: South African Police Service. Available at: https://www.saps.gov.za/about/stratframework/annualreports_arch.php
Stats SA (2019). General Household Survey 2018. Available at: http://www.statssa.gov.za/publications/P0318/P03182018.pdf
Steyn, J. (2017). Lilac? "Ten years of blue nurtured nature". *Police Practice and Research*, 18(3): 259–274. DOI: 10.1080/15614263.2017.1292745
Stinson, P.M., J. Liederbach, S.P. Lab, and S.L. Brewer, Jr. (2016). Police integrity lost: A study of law enforcement officers arrested. *Criminal Justice Faculty Publications*, 63. Available at: https://scholarworks.bgsu.edu/crim_just_pub/63
Stinson, P.M., N. Todak, and M. Dodge (2015). An exploration of crime by policewomen. *Police Practice and Research*, 16(1): 79–93. DOI: 10.1080/15614263.2013.846222
Stone, K. and C. Lopes (2018). Policing responses to domestic violence: Exploring reactions by the police to women in need of shelter. *Heinrich Boll Foundation & National Shelter Movement of South Africa*. Available at: https://za.boell.org/sites/default/files/hbf_saps_research_paper_web_1.pdf
Stroshine, M.S. and S.G. Brandl (2011). Race, gender, and tokenism in policing: An empirical elaboration. *Police Quarterly*, 14(4): 344–365.
Sun, I. (2007). Policing domestic violence: Does officer gender matter? *Journal of Criminal Justice*, 35: 581–595.
Super, G.J. (2013). *Governing through Crime in South Africa: The Politics of Race and Class in Neoliberalizing Regimes*. Oxford: Routledge.
Tlhabi, R. (2017). *Khwezi: The Remarkable Story of Fezekile Ntsukela Kuzwayo*. Johannesburg: Jonathan Ball Publishers. Kindle Edition.
Ulicki, T. (2011). "Just the way things are": Gender equity and sexual harassment in the South African Police Service. *Transformation: Critical Perspectives on Southern Africa*, 76: 95–119. DOI: 10.1353/trn.2011.0031
Viljoen, L. (2015). Are you man enough? A Case study of the represented and experienced masculinity in the South African Police Service. *South African Review of Sociology*, 46(3): 45–63. DOI: 10.1080/21528586.2015.1066709
Wagner, N., M. Rieger, A. Bedi, and W. Hout (2017). Gender and policing norms: Evidence from survey experiments among police officers in Uganda. *Journal of African Economies*, 26(4): 492–515.
Waugh, L., A. Ede, and A. Alley (1998). Police culture, women police and attitudes towards misconduct. *International Journal of Police Science and Management*, 1(3): 288–300.
Westmarland, L. and M. Rowe (2018). Police ethics and integrity: Can a new code overturn the blue code? *Policing and Society*, 28(7): 854–870. DOI: 10.1080/10439463.2016.1262365
Worden, A. (1993). The attitudes of women and men in policing: Testing conventional and contemporary wisdom. *Criminology*, 31: 203–235.
Wu, G. and D.A. Makin (2019). The quagmire that is an unwillingness to report: Situating the code of silence within the Chinese Police context. *Criminal Justice and Behavior*, 46(4): 608–627. https://doi.org/10.1177/0093854819831732

Chapter 9

Police integrity across rank and file

Introduction

Together with their uniforms, discretionary authority, legally sanctioned use of force, and 24-hour public presence, rank hierarchies are a defining feature of police organizations. Because of their semi-military nature, most police agencies are hierarchical bureaucracies with a wide base composed of numerous line officers, a narrower middle section with fewer supervisors, and a very narrow top with a handful of police managers. This organizational structure implies a clear division of labor between workers, first-line supervisors, and chief executives (Walker and Katz, 2011, p. 105).

Explorations of police culture suggest that cultural norms are more strongly adopted by line officers than by supervisors (e.g., Crank, 2014; Crank et al., 1993, 1995; Manning, 1989; Paoline and Terrill, 2014; Reuss-Ianni and Ianni, 1983). In their work on cops and managers, Reuss-Ianni and Ianni (1983) found that two separate subcultures developed as a result of somewhat different strains experienced by line officers and supervisors: The "street cop culture," which incorporates the values and attitudes of traditional police culture and is primarily shared by line officers, and the "management cop culture," which incorporates broader concerns about the police agency and the society at large and is primarily shared by supervisors (Reuss-Ianni and Ianni, 1983). Another U.S. study (Silver et al., 2017) supported the view that the supervisory status and the adherence to the norms of traditional police culture were negatively related. Waddington (1999) also found significant differences in professional values between British lower-ranking officers and their supervisors.

Like most police agencies, the SAPS uses a hierarchical structure that distinguishes managers from front-line officers (Batts et al., 2012). Its ten ranks range from Constable to General, with four subdivisions of top management (n = 31), senior management (n = 877), all commissioned officers (n= 23,993), and all non-commissioned officers (n = 125,890) in 2018. Those in training are called Student Constables, while civilian staff do not receive SAPS ranks (n = 42,470) (SAPS, 2018, p. 30).

Aside from the division of labor, rank within the SAPS serves three purposes. First, rank symbolizes aspiration, status, and remuneration, that is, progress toward the dignified life promised by the new South Africa. Promotion means having a better chance of escaping historical poverty and entering the middle class (see Chapters 2 and 3). Second, because officers often perceive rank promotion as arbitrary or unmeritocratic, rank also symbolizes a culture of organizational injustice, where many officers resent their supervisors and the organization. Finally, rank is a potent symbol of the kind of policing and culture expected by politicians and police leaders. This is clearest in the shift from military to civilian ranks at the end of apartheid, the return to military ranks in 2010, and the call for the "demilitarization" of the SAPS in the National Development Plan and White Paper on Policing (see Chapter 3). As such, rank shapes organizational values and integrity.

This chapter explores the contours of police integrity across the SAPS hierarchical organizational structure. We study the potential differences among non-commissioned officers, commissioned officers, and police managers. We begin by exploring potential similarities and differences in their familiarity with official rules and their evaluations of misconduct seriousness. We then proceed to analyze their views about the discipline they think appropriate for behaviors described in hypothetical scenarios and the discipline they think the SAPS would mete out for such behaviors. We conclude the analysis by focusing on the similarities and differences in the code of silence among non-commissioned officers, commissioned officers, and police managers.

Rank and the SAPS

Integrity and organizational justice

Rank in South African policing has rarely been earned solely through merit. Instead, the undemocratic nature of policing in the former colony and early state meant that seniority and promotions have been shaped by a combination of merit, politics, and nepotism (Shear, 2012; Cawthra, 1993). In recruitment and promotions, the early SAP first gave preference to white men of English descent, and later to those of Afrikaner ancestry (Goodhew, 1990; Nasson, 1992). Unable to attract a sufficient number of white officers, for much of the twentieth century black officers were intentionally under-trained to prevent rank progression (Seegers, 1991), and even when promoted, were often disrespected by white colleagues (Shear, 1998). The first black commissioned SAP officer was appointed in 1970, and in 1972 the first women were recruited into the organization.

Although key promotions within the SAP achieved a semblance of meritocracy by requiring the passing of written exams for the ranks of sergeant, warrant officer, and lieutenant, access to exams at times was supported for those already part of strong SAP networks (Faull, 2019). Generals, too, were politically appointed and white, while Afrikaans-speaking men held most authority and leadership roles in the SAP until the end of apartheid. Similarly, homeland

(Bantustan) police forces were ethnically homogenous and predominantly served their apartheid-affiliated bureaucratic elites and their patronage networks (e.g., Seegers, 1991; Hyslop, 2005; Francis, 2002; Cawthra, 1993).

As described in previous chapters (e.g., Chapters 7 and 8), since the mid-1990s, in an effort to redress the injustice of the organization's and the country's past, recruitment and promotion in the SAPS have emphasized the attainment of particular race and gender quotas. Although ostensibly supported by the country's Constitution (1996, pp. 5–6), such practices have resulted in numerous allegations of institutional racism and related court challenges (e.g., DOJ, 2017). Combined with a period of unusually rapid recruitment from 2002 to 2012, when 123,606 new officers were hired (Bruce, 2013), these policy choices, while deemed necessary by the organization, have certainly reduced morale among some of its members and possibly increased the risk of integrity violations by those demotivated. In his ethnography of policing in South Africa, Faull (2010, Loc 1466) reported these negative sentiments on promotion among the police as they were illustrated in a statement by a former white SAPS Captain, "You're promoted on color these days. The people in charge are incompetent. Not everybody, but most of them."

However, the challenge with affirmative action is that, even in the instances in which it has not been applied, some officers may perceive themselves as victims (Faull, 2010, Loc:1595):

> A lot of people—especially white people—said, "I'll never get promoted." I had a white colleague who said he would never get promoted but then he and his brother both got promoted. They looked at performance . . . [so] I think people are being promoted on performance, not race.
>
> —SAPS Colonel, colored, 2009

The perception that promotions are not based on merit—that there is no clear career pathing—is not limited to concerns about race (Faull, 2012):

> There are people with no qualification who have been promoted. Why me? Why am I not being considered? . . . I know I can work hard, I can do any job. . . . If I can just get a promotion I can continue being a policeman. I don't have a problem with it if they recognize my quality. The reason people are quitting the police, they need to consider the promotion policy. . . . These guys that were meant to be promoted a long time ago are not yet promoted. People should be promoted on performance and qualifications. If you work hard you deserve promotion.
>
> —SAPS Constable, black, 2012

Speaking in the parliament in 2010, then National Commissioner Bheki Cele reflected on the SAPS mass recruitment policy as making it an employer of last resort for job-seeking South Africans. He compared SAPS efforts to address the subsequent damage to organizational integrity with trying to fix an airplane in mid-flight (PMG, 2010). Perceived injustice in promotions and rank also affect the most senior managers. Patronage was central to the early South African state (e.g., Posel, 1999; van Vuuren, 2006) and has remained part of the business-politics nexus in the democratic era (Hyslop, 2005; Feinstein, 2007; Myburgh, 2018). As described in preceding chapters (e.g., Chapter 5), the Mbeki and Zuma administrations appointed their allies to leadership positions within the SAPS and, in the case of Zuma, within the broader criminal justice system. Zuma appointees in particular, in turn, irregularly promoted numerous allies, including over a hundred SAPS generals.

Not only are such appointments potential integrity violations, but they certainly promote an organizational culture in which officers do not know whether their efforts will be rewarded, or if their manager's rank is indicative of merit, luck, or nepotism. Similarly, at various times the SAPS has been accused of letting nepotism and corruption determine who is recruited into the organization. As recently as 2019, its head of Human Resources stated, "[R]ecruitment processes must be reviewed urgently to address alleged incidents of corruption and nepotism" (Mgwenya, 2019). Considering the importance of procedurally just policing for police effectiveness and legitimacy (e.g., Tyler, 2017; Donner et al., 2015; Stanko et al., 2012) and the recognition that one cannot expect such policing from officers when they do not feel fairly treated by their employer themselves (e.g., Roberts and Herrington, 2013; Quinton et al., 2015), such a state of affairs within the SAPS does not bode well for integrity within the organization.

Status and class

Post-apartheid South Africa is a highly aspirational society (see Chapter 3). Despite growing inequality and public protests (e.g., Duncan, 2016, Alexander et al., 2018), poor South Africans continue to expect that their lives will improve (Posel and Rogan, 2017). The same optimism is, however, not shared by all police officers (e.g., Steinberg, 2008; Hornberger, 2011; Faull, 2018; Perkins, 2018). This may be in part because, unlike the poor majority, police officers earn relatively good salaries but remain unable to fulfill their material ambitions (Faull, 2018), with many being the sole breadwinners in extended networks of poor and unemployed (e.g., Harper and Seekings, 2010). The resultant disaffection and perceptions of low-status partly drive officers' use of violence to earn respect (Hornberger, 2011; Faull, 2013a; Perkins, 2018) and may make them vulnerable to other abuses if their respect for the organization dwindles. In this context, rank promotion and the potential pay and quality of life improvements that it symbolizes are hallowed.

Most SAPS officers remain in a rank for at least seven years before they are promoted, though in exceptional circumstances promotions may be awarded in two years (PMG, 2017). Within each rank there exists a range of salary bands.

In 2017, the South African Police Union (SAPU) suggested that the SAPS was not promoting officers fast enough, adding that "the lack of promotions in the police may open room for bribes" (SAPU, 2017). In response, the SAPS announced in December 2018 that 69,000 officers would be promoted over a three-year period. In another indication of the politics of promotions, SAPU responded to the announcement, lamenting: "[A] member can spend 11 years in one rank. We [call] to speed up this strategy so that our members can be promoted" (South African Government News Agency, 2018). In addition to a lack of trust in such newly promoted supervisors, rapid promotions raise other integrity-related questions, including the effects of the lack of required experience on the new supervisors' ability to supervise and the consequences of the lack of availability of in-service training for all newly promoted supervisors.

While many in South Africa believe that policing would improve and corruption would decline if SAPS officers were paid better wages (e.g., Grobler, 2013), most do not realize that junior officers already earn more than 95% of the citizenry do. In 2018, the bulk of the SAPS's workforce—the non-commissioned officers—earned between R175,000 (US $11,660) and R407,000 (US $27,130) per year. In contrast, two-thirds of South Africans lived off approximately R21,600 (US $1,440) a year (SALDRU, 2019). South Africans' celebration of conspicuous consumption (e.g., Posel, 2010; Burger et al., 2015) likely generates disaffection among a majority that cannot consume. This includes most police who, despite earning relatively well, often begin life in poverty and, because of obligations to others, struggle to attain the kind of life with which status is associated. Commissioned officers, especially the approximately 900 senior and top-management ranks, are among the best paid workers in the country, earning up to R2 million per annum. Yet despite their generous salaries, top managers continue to be implicated in corruption scandals (see Chapters 3 and 5), once again illustrating that corruption is not easily curbed through pay interventions.

From commissioner to general

Finally, rank and related insignia have played an important role in attempts to reform the culture, integrity, and legitimacy of the SAPS. As described in earlier chapters (e.g., Chapters 2, 3, and 5), apartheid's SAP was an intentionally paramilitary force. It employed military ranks, engaged in military-like training, provided its special units with military-like uniforms, and routinely policed black communities as though it was an occupying army, often together with the country's actual military forces (Cawthra, 1993; Shaw and Shearing, 1998; Rauch, 2000). Police officers were also deployed with soldiers to fight wars in Angola and the former Rhodesia (see Chapter 2). As a result, during apartheid's final decade, police were widely feared and distrusted (Van der Spuy, 2005; Kynoch, 2003; Cawthra, 1993).

In an effort to break from this past, South Africa's 1996 Constitution intentionally provided for the creation of a police "service" rather than a police "force." Civilian ranks replaced military ones, and policies shifted from command and control to human rights and community relations (Hornberger, 2011, 2013). But, as fear of violent crime grew in the late 1990s and the early 2000s, the SAPS began referring to itself in 2010 as a "force" once again, promised to empower police to more readily use lethal force (Bruce, 2011), and reintroduced military ranks (Hornberger, 2013; Faull, 2013b).

The change in ranks and accompanying rhetoric was intended to communicate three key messages. First, we will no longer be soft on criminals. Second, our officers will be more disciplined henceforth. Third, the public should trust us. Unfortunately, little good came of the change. Instead, soon after the shift, a number of high-profile abuse of force incidents occurred, including the two 2011 killings by the police: Andries Tatane, during a service delivery protest, and 34 striking miners at Marikana (Burger, 2011; Farlam, 2015; Chapter 3).

At the same time, the government's flagship policy, the 2012 National Development Plan, called for the "demilitarizing" of the SAPS (The Planning Commission, 2012, p. 54):

> The decision to demilitarize the police force, moving away from its history of brutality, was a key goal of transformation after 1994. The remilitarization of the police in recent years has not garnered greater respect for the police or higher conviction rates. If anything, it has contributed to violence. The police should be demilitarized and managed towards a professional civilian service.

Four other NDP recommendations sought to correct the way rank mismanagement had eroded the legitimacy of the SAPS. First, the NDP recommended that that all officers be required to undergo competency assessments to determine whether they meet the standards of their rank. Second, the NDP recommended that the National Commissioner and his or her deputies be appointed on a competitive basis, with a panel of experts judging candidates against objective criteria. Third, the NDP recommended that recruitment be allowed directly into commissioned ranks in order to improve the caliber of applicants (The Planning Commission, 2012). Eight years later, however, the recommendations remain ignored. Similarly, the government's 2016 White Paper on Policing (Civilian Secretariat for Police, 2016, p. 14) implies that police abuses may be linked to the SAPS establishment of "para-military police units" and war-like rhetoric from leaders, both of which accompanied the reversion to military ranks. Like the NDP, the White Paper's call to "demilitarize" the police has had little influence.

Rank and the theory of police integrity

The definition of police integrity developed by Klockars and colleagues (2006, p. 1) asserts that police integrity is "the normative inclination among police to resist temptations to abuse the rights and privileges of their occupation." The police in this definition could be individual police officers, groups of police officers, or even whole police agencies. Such an approach creates opportunities to study police integrity at three different levels: the individual police officer level (i.e., micro-level), the level of subunits within each agency (i.e., meso-level), and the police agency level/country level (i.e., macro-level). The vast majority of prior research on police integrity almost exclusively targeted either the individual police officer level (for an overview, see Kutnjak Ivković, 2015; Kutnjak Ivković and Haberfeld, 2019) or the agency and/or country level (e.g., Klockars et al., 2000, 2004; Kutnjak Ivković and Haberfeld, 2015; Kutnjak Ivković et al., 2016; Micuci and Gomme, 2009). Extant literature on police integrity at the meso-level is limited (e.g., Hickman et al., 2016; Kutnjak Ivković et al., 2016, 2019; Lim and Sloan, 2016; Raines, 2010). The meso-level research need not only incorporate explorations of police integrity across different units of a police agency (i.e., horizontal meso-level research) but could instead focus on the exploration of police integrity across different levels of the police hierarchy (i.e., vertical meso-level research).

Extant research on police integrity has explored all three organizational dimensions of the theory of police integrity. The first dimension of the theory stipulates that the way in which the official rules are made and communicated to, and understood by, the police officers is critical for the level of police integrity (Klockars and Kutnjak Ivković, 2004; Klockars et al., 2000, 2004, 2006; Kutnjak Ivković, 2015). Studies that have examined the relation between perceptions of misconduct seriousness and rank in various countries (e.g., Bosnia and Herzegovina: Kutnjak Ivković, 2004; South Korea: Yun, 2003; Turkey: Bucak, 2012; Kucukuysal, 2008; United States: Marche, 2009; Micucci and Gomme, 2005; Raines, 2010; Lee et al., 2013) yielded quite heterogeneous conclusions about the effect of supervisory status on the evaluations of police integrity. While some studies found that supervisors and non-supervisors evaluate seriousness of police misconduct relatively similarly (e.g., Kucukuyal, 2008; Kutnjak Ivković, 2004; Lee et al., 2013; Yun, 2003), the results of other studies (Marche, 2009; Micucci and Gomme, 2005; Raines, 2010) suggested that supervisors evaluate police misconduct more seriously than non-supervisors do.

Raines's study (2010) was the only project that did not rely on the traditional dichotomy of supervisors/non-supervisors and, at the same time, found some statistically significant differences. Her results show that line officers evaluated misconduct as the least serious and the mid-level and senior managers provided the most serious evaluations (e.g., Raines, 2010). By contrast, Kutnjak Ivković and colleagues (2019) found that non-supervisors, first-line supervisors, and administrators had similar views about the measures of the first dimension of the police integrity theory.

The literature focusing on the second dimension of the police integrity theory—the relation between police integrity and the effectiveness of internal control mechanisms, and the differences across supervisors and line officers in their views about the appropriate and expected discipline—is not voluminous (e.g., Kutnjak Ivković et al., 2016, 2019; Kutnjak Ivković and Shelley, 2010; Micucci and Gomme, 2005; Raines, 2010). Nevertheless, the results of these studies are far from uniform. A Czech study (Kutnjak Ivković and Shelley, 2010) and a U.S. study (Kutnjak Ivković et al., 2016) found that supervisors and non-supervisors seem to express similar views of discipline and its fairness. By contrast, two U.S. studies analyzing the same national dataset (Micucci and Gomme, 2009; Raines, 2010) reported that, compared to non-supervisors, supervisors both expected and supported more severe discipline. A U.S. study by Kutnjak Ivković and colleagues (2019) pointed out that, while the non-supervisor, supervisor, and manager views about the appropriate discipline for misconduct differed, the respondents' views about discipline they assume that their police agency would mete out were very similar. Put differently, non-supervisors, supervisors, and managers agreed on what discipline their agencies would mete out but had quite different views about discipline that would be appropriate for cases of police misconduct.

The third dimension of the theory states that the level of police integrity is negatively related with the strength of the code of silence (Klockars and Kutnjak Ivković, 2004; Klockars et al., 2000, 2004, 2006; Kutnjak Ivković, 2015). Most studies involving rank focused on the third dimension of the police integrity theory, exploring the relation between police integrity and the code of silence (e.g., Australia: Porter and Prenzler, 2016; Bosnia and Herzegovina: Kutnjak Ivković and Shelley, 2008; Croatia: Kutnjak Ivković and Klockars, 2000; Czech Republic: Kutnjak Ivković and Shelley, 2008, 2010; Hungary: Kremer, 2000; USA: Kargin, 2009; Kutnjak Ivković et al., 2016, 2018; Long et al., 2013; Micucci and Gommer, 2008; Raines, 2010; Rothwell and Baldwin, 2006, 2007a, 2007b). Like the studies exploring the other two theoretical dimensions, studies analyzing the respondents' expressed willingness to report yielded mixed findings. Early studies conducted in Eastern European countries in the 2000s (e.g., Kremer, 2000; Kutnjak Ivković and Klockars, 2000; Pagon and Lobnikar, 2000) provided the first evidence that line officers were more likely to say that they would adhere to the code of silence than the supervisors were. More recent studies offered less homogeneous conclusions. Subsequent studies from Bosnia and Herzegovina and the Czech Republic (e.g., Kutnjak Ivković and Shelley, 2008, 2010) reported either no differences or only minimal differences in the code of silence across supervisors and non-supervisors. On the other hand, studies from the United States (e.g., Kutnjak Ivković et al., 2016, 2018, 2019; Long et al., 2013; Micucci and Gomme, 2009; Raines, 2010) and Australia (e.g., Porter and Prenzler, 2016) point to less uniformity in the code of silence between supervisors and non-supervisors. In the most recent U.S. study (Kutnjak Ivković et al., 2019), the differences among non-supervisors, supervisors, and managers were the largest for the measures of the third dimension of the police integrity theory.

In addition, a couple of studies also indicated that, while the supervisory status may appear relevant in bivariate analyses, the use of multivariate models with other demographic and organizational variables included in the model removed the statistical significance of this relation (e.g., Kargin, 2009; Porter and Prenzler, 2016).

Survey findings

Our empirical discussion is based on the survey of 868 SAPS officers from all provinces and territories. Following the questionnaire on police integrity developed by Klockars and colleagues (2006) and revised by Kutnjak Ivković (2009), participants in the study were asked to evaluate 14 hypothetical scenarios covering a range of forms of police misconduct. For details about the methodology, please see Chapter 1.

Upon reading the description of misconduct contained in each scenario, the respondents were asked to answer seven identical questions. To assess their familiarity with official rules and the extent of their support for legal rules, one of the questions instructed the respondents to evaluate whether the behavior described in the questionnaire violates the official rules ("Would this behavior be regarded as a violation of official policy in your agency?"). Possible answers ranged on a 5-point scale from 1 = "definitely not" to 5 = "definitely yes." In addition, the respondents were asked how serious misconduct is ("How serious do YOU consider this behavior to be?" "How serious do most police officers in your agency consider this behavior to be?"). The respondents could have selected an answer on a 5-point scale from 1 = "not at all serious" to 5 = "very serious." To ascertain their views about the disciplinary environment, the respondents were asked to select the discipline they think should be appropriate for misconduct described in each scenario and the discipline they think that the SAPS would mete out ("If an officer in your agency engaged in this behavior and was discovered doing so, what, if any, discipline do you think should follow?" "If an officer in your agency engaged in this behavior and was discovered doing so, what, if any, discipline do you think would follow?"). The respondents could have selected one of the six disciplinary options, ranging from 1 = "no discipline" to 6 = "dismissal." To assess their affinity for the code of silence, the respondents were asked to state how willing they and their fellow police officers, respectively, would be to report misconduct described in each of the hypothetical scenarios ("Do you think you would report a fellow police officer who engaged in this behavior?" "Do you think most police officers in your agency would report a fellow police officer who engaged in this behavior?"). The respondents could have selected one answer from a 5-point scale ranging from 1 = "definitely not" to 5 = "definitely yes."

Empirical studies of police integrity typically focus only on the dichotomy between non-supervisors and supervisors (e.g., Bucak, 2012; Kargin, 2009; Kremer, 2000; Kucukuysal, 2008; Kutnjak Ivković, 2004; Kutnjak Ivković and

Klockars, 2000; Kutnjak Ivković and Shelley, 2008a, 2010; Kutnjak Ivković et al., 2016; Kutnjak Ivković et al., 2018; Lee et al., 2013; Lim and Sloan, 2016; Long et al., 2013; Marche, 2009; Micucci and Gomme, 2005; Pagon and Lobnikar, 2000; Porter and Prenzler, 2016; Rothwell and Baldwin, 2006, 2007a, 2007b). That dichotomy may conceal any differences between first-line supervisors and police administrators. Very few studies have compared the contours of police integrity across three or more hierarchical levels[1] (e.g., Bucak, 2012; Kucukuysal, 2008; Kutnjak Ivković et al., 2019; Raines, 2010; Yun, 2003). The results of the most recent U.S. study (Kutnjak Ivković et al., 2019) clearly highlight the need to differentiate between supervisory groups as well, pointing out that the relation between the respondents' views of police integrity and rank are connected in complex ways.

To entice the respondents to participate in the study and answer truthfully, Klockars and colleagues (2006) included only a few demographic questions. One of them inquired about the respondents' rank. Possible answers included four options describing non-commissioned officers (in our data: student constable: 4.5%; constable: 33.6%; sergeant: 9.8%; and warrant officer: 19.0%), five options describing commissioned officers (lieutenant:[2] 2.5%; captain: 18.9%; major: 0.6%; lieutenant colonel: 6.8%; colonel: 2.8%); and three options describing commissioned officers in senior management (only brigadier was selected of the three options: 0.5%). For the purposes of our empirical analyses, we have developed three categories: non-commissioned officers, commissioned officers, and senior management. Because of the small number of respondents in senior management, we augment that group by including in it the two highest ranks of commissioned officers (i.e., lieutenant colonel and colonel).

Familiarity with the official rules

In about one-half of the scenarios (7 out of 14 scenarios or 50%), there were statistically significant differences by rank in the respondents' evaluations of whether the behavior described in the scenarios violates the SAPS's official rules (Table 9.1).[3] Although statistically significant, however, these differences were neither very large[4] nor substantive.[5]

We analyzed the scenarios with statistically significant differences to explore which pairs (e.g., non-commissioned respondents and commissioned respondents; commissioned respondents and respondents in management) had statistically significantly different views. To begin, there were no statistically significant differences between non-commissioned and commissioned officers in their likelihood of evaluating misconduct as rule-violating (Table 9.1), so the statistically significant differences must be in the other comparisons. In fact, all of the scenarios with statistically significant differences included respondents in management as one of the two groups in the comparison. Specifically, when their answers were statistically significantly different, respondents in management were more likely to evaluate misconduct as rule-violating than

Table 9.1 Violation of official rules by rank

Scenario	*Non-Commissioned (1)*		*Commissioned (2)*		*Management (3)*		*F-test (Tukey b's post-hoc)*	*1 v. 2*	*1 v. 3*	*2 v. 3*
	Mean	*Rank*	*Mean*	*Rank*	*Mean*	*Rank*				
Scenario 1: Acceptance of gratuities	4.22	5.5	4.27	5	4.55	7	3.04* (1 v. 3; 2 v. 3)	−.05	−.33	−.28
Scenario 2: Failure to arrest	4.54	12	4.63	14	4.77	11	2.88	−.09	−.23	−.14
Scenario 3: Theft of knife	4.76	14	4.62	13	4.94	14	7.36*** (1 v. 3; 2 v. 3)	.14	−.18	−.32
Scenario 4: Abuse of deadly force	4.49	11	4.50	9	4.41	4	.25	−.01	.08	.09
Scenario 5: Supervisor corruption	4.33	8	4.34	6	4.54	6	1.43	−.01	−.21	−.20
Scenario 6: Striking arrestee	4.30	7	4.39	7	4.70	9.5	5.45** (1 v. 3; 2 v. 3)	−.09	−.40	−.31
Scenario 7: Verbal abuse	3.91	2	4.05	3	4.02	2	1.23	−.14	−.11	.03
Scenario 8: Cover-up of DUI	3.95	3	4.10	4	4.21	3	1.87	−.15	−.26	−.11
Scenario 9: Kickback	4.22	5.5	4.46	8	4.70	9.5	10.32*** (1 v. 3)	−.24	−.48	−.24
Scenario 10: False report on drugs	4.45	10	4.58	11	4.60	8	2.25	−.13	−.15	−.02
Scenario 11: Supervisor fails to stop excessive force	4.44	9	4.53	10	4.84	12	7.22** (1 v. 3; 2 v. 3)	−.09	−.40	−.31
Scenario 12: Failure to report hate crime	4.15	4	4.02	2	4.43	5	4.59* (1 v. 3; 2 v. 3)	.13	−.28	−.41
Scenario 13: Bribe from reckless motorist	4.68	13	4.59	12	4.87	13	4.25* (1 v. 3; 2 v. 3)	.09	−.19	−.28
Scenario 14: Failure to react to graffiti	3.74	1	3.85	1	3.61	1	1.15	−.11	.13	.24

either the respondents from non-commissioned positions or the respondents from the commissioned positions were (seven scenarios for non-commissioned respondents v. respondents in management; six scenarios for commissioned respondents v. respondents in management; Table 9.1).

The relative order of familiarity with official rules—how likely the respondents are to evaluate each scenario as rule-violating compared to the other scenarios in the questionnaire—is shown in Table 9.1 (ranks). Non-commissioned and commissioned respondents aligned in their relative assessments of familiarity with official rules (Spearman's rho = .937; $p < .001$). The extent of alignment, while fairly strong, was not as strong for the respondents in management and non-commissioned respondents (Spearman's rho = .796; $p < .01$) and for the respondents in management and commissioned respondents (Spearman's rho = .840; $p < .001$).

Perceptions of misconduct seriousness

Another set of questions invited the respondents to ascertain the seriousness of police misconduct described in the scenarios. The results suggest that, although there are scenarios and pairs for which the respondents' evaluations of seriousness were related to their rank (Table 9.2), the evaluations of seriousness overall are more similar than different. There are seven out of 14 scenarios in bivariate analyses (Table 9.2) and six out of 14 scenarios in multivariate analyses in which there were statistical significant differences by rank, but only four are significant in both types of models.[6] However, while statistically significant, only one of these differences was large[7] and only two were substantively important.[8] Following the threshold for substantive differences, as established by Klockars and colleagues (2006), the mean differences were important in the scenario describing the acceptance of gratuities (Scenario 1) and in the scenario describing a kickback (Scenario 9).

With one exception (Scenario 7: Striking arrestee), evaluations of seriousness provided by non-commissioned and commissioned respondents were very similar, resulting in only one scenario with a statistically significant difference (Table 9.2). While estimates of misconduct seriousness provided by the respondents in management were similar to the respondents from the two other groups in most of the scenarios, there were also a few scenarios (four for the non-commissioned respondents and three for the commissioned respondents; Table 9.2) in which the respondents from management had somewhat different views, and in all of these scenarios the respondents from management evaluated misconduct as more serious than either non-commissioned or commissioned respondents did (Table 9.2).

The relative order of seriousness evaluations—how serious respondents evaluated each scenario compared to the other scenarios in the questionnaire—is shown in Table 9.2 (ranks). Non-commissioned respondents, commissioned respondents, and the respondents in management had very similar views about

Table 9.2 Own views about misconduct seriousness by rank

Scenario	*Non-Commissioned (1)*		*Commissioned (2)*		*Management (3)*		*F-test (Tukey b's post-hoc)*	*1 v. 2*	*1 v. 3*	*2 v. 3*
	Mean	*Rank*	*Mean*	*Rank*	*Mean*	*Rank*				
Scenario 1: Acceptance of gratuities	3.80	2	4.10	5	4.39	5	9.88*** (1 v. 3)	−.30	−.59	−.29
Scenario 2: Failure to arrest	4.55	11	4.64	12	4.85	12	4.77** (1 v. 3)	−.09	−.30	−.21
Scenario 3: Theft of knife	4.78	14	4.76	14	4.90	14	1.89	.02	−.12	−.14
Scenario 4: Abuse of deadly force	4.58	12	4.58	10	4.48	8	.48	.00	.10	.10
Scenario 5: Supervisor corruption	4.38	8	4.26	6	4.43	7	1.09	.12	−.05	−.17
Scenario 6: Striking arrestee	4.22	6.5	4.50	9	4.41	6	5.83** (1 v. 2)	−.28	−.19	.09
Scenario 7: Verbal abuse	3.89	3	3.91	2	3.82	2	.17	−.02	.07	.09
Scenario 8: Cover-up of DUI	3.91	4	4.03	3	4.08	3	.95	−.12	−.17	−.05
Scenario 9: Kickback	4.22	6.5	4.41	7	4.79	11	11.74*** (1 v. 3; 2 v. 3)	−.19	−.57	−.38
Scenario 10: False report on drugs	4.46	10	4.64	12	4.68	9	4.17* (no pairs)	−.18	−.22	−.04
Scenario 11: Supervisor fails to stop excessive force	4.43	9	4.43	8	4.72	10	3.30* (1 v. 3; 2 v. 3)	.00	−.29	−.29
Scenario 12: Failure to report hate crime	4.18	5	4.07	4	4.33	4	1.98	.11	−.15	−.26
Scenario 13: Bribe from reckless motorist	4.71	13	4.64	12	4.89	13	3.16* (2 v. 3)	.07	−.18	−.25
Scenario 14: Failure to react to graffiti	3.73	1	3.79	1	3.57	1	.86	−.06	.16	.22

the relative order of scenario seriousness (Spearman's rho = .926; $p < .001$ for non-commissioned and commissioned; Spearman's rho = .884; $p < .001$ for non-commissioned and management; Spearman's rho = .903; $p < .001$ for commissioned and management).

We also explored the respondents' estimates of how seriously most SAPS officers would evaluate these scenarios. Bivariate analyses suggest that all three groups of respondents had very similar views of how seriously their fellow officers would evaluate misconduct. In fact, the respondents' rank was not systematically related to their perceptions of how seriously their colleagues would evaluate police misconduct (Table 9.3); in only one scenario (Scenario 9: Kickback) in bivariate analysis and in five scenarios in multivariate analyses, the respondents in management were more likely to assume that most police officers would evaluate misconduct as more serious than non-commissioned respondents did (Table 9.3).[9] The difference is neither large[10] nor substantively important.[11]

The relative order of most officers' seriousness evaluations—how serious the respondents perceived that most officers would evaluate each scenario compared to the other scenarios in the questionnaire—is shown in Table 9.3 (ranks). Non-commissioned and commissioned respondents were very similar in how seriously they assume most officers would evaluate each scenario compared to the other scenarios in the questionnaire (Spearman's rho = .956; $p < .001$). Correlations were similarly high, though just a bit smaller, for respondents in management and non-commissioned respondents (Spearman's rho = .880; $p < .01$) and for respondents in management and commissioned respondents (Spearman's rho = .873; $p < .01$).

Finally, we compared the respondents' own evaluations of scenario seriousness with how seriously they assumed most SAPS officers would evaluate the scenarios (Table 9.4). Although we assumed that, because of the self-serving bias, all groups expected that they would evaluate misconduct as much more serious than most police officers would, the differences between their own estimates and their estimates of most police officers' evaluations of seriousness were particularly large for respondents from management. The differences were statistically significant in all 14 scenarios for all three groups of respondents. However, only a handful of such scenarios for non-commissioned officers (two scenarios; Table 9.4) and commissioned officers (three scenarios; Table 9.4) were substantively important according to the criterion established by Klockars et al. (2006). On the other hand, for the respondents from management, those differences were large and substantively important in nine of out 14 scenarios (64%). Such a result is not surprising because the respondents from management are on the top of the hierarchy and are the farthest removed from "most" police officers—line officers.

Views about discipline

The first question concerning discipline inquired about the respondents' views of the discipline they thought appropriate for the behavior described in each

Table 9.3 Most police officers' views about misconduct seriousness by rank

Scenario	*Non-Commissioned (1)*		*Commissioned (2)*		*Management (3)*		*F-test (Tukey b's post-hoc)*	*1 v. 2*	*1 v. 3*	*2 v. 3*
	Mean	*Rank*	*Mean*	*Rank*	*Mean*	*Rank*				
Scenario 1: Acceptance of gratuities	3.39	1	3.46	2	3.23	2	.83	−.07	.16	.23
Scenario 2: Failure to arrest	4.04	10	3.91	8	3.82	7	2.08	.13	.22	.09
Scenario 3: Theft of knife	4.26	12	4.27	13	4.21	11	.12	−.01	.05	.06
Scenario 4: Abuse of deadly force	4.35	14	4.21	11	4.25	14	1.39	.14	.10	−.04
Scenario 5: Supervisor corruption	3.92	7	3.88	7	3.63	5	1.80	.04	.29	.25
Scenario 6: Striking arrestee	3.80	5	3.81	5	3.77	6	.02	−.01	.03	.04
Scenario 7: Verbal abuse	3.55	3	3.54	3	3.37	3	.66	.01	.18	.17
Scenario 8: Cover-up of DUI	3.60	4	3.57	4	3.55	4	.08	.03	.05	.02
Scenario 9: Kickback	3.87	8	4.04	9	4.23	12	3.83* (1 v. 3)	−.17	−.36	−.19
Scenario 10: False report on drugs	4.12	11	4.26	12	4.07	8	1.23	−.14	.05	.19
Scenario 11: Supervisor fails to stop excessive force	3.96	9	4.10	10	4.11	9.5	1.14	−.14	−.15	−.01
Scenario 12: Failure to report hate crime	3.85	6	3.82	6	4.11	9.5	1.95	.03	−.26	−.29
Scenario 13: Bribe from reckless motorist	4.34	13	4.29	14	4.24	13	.38	.05	.10	.05
Scenario 14: Failure to react to graffiti	3.44	2	3.35	1	3.13	1	2.11	.09	.31	.22

Table 9.4 Views about misconduct seriousness by rank

Scenario	*Non-C. Own*		*Non-C. Most*		*t-test*	*Diff.*	*Comm. Own*		*Comm. Most*		*t-test*	*Diff.*
	Mean	*Rank*	*Mean*	*Rank*			*Mean*	*Rank*	*Mean*	*Rank*		
Scenario 1: Acceptance of gratuities	3.80	2	3.39	1	7.66***	.41	4.10	5	3.46	2	6.34***	.64
Scenario 2: Failure to arrest	4.55	11	4.04	10	11.14***	.51	4.64	11	3.91	8	8.30***	.73
Scenario 3: Theft of knife	4.78	14	4.26	12	11.97***	.52	4.76	14	4.27	13	6.60***	.49
Scenario 4: Abuse of deadly force	4.58	12	4.35	14	5.74***	.23	4.58	9	4.21	11	4.96***	.37
Scenario 5: Supervisor corruption	4.38	8	3.92	7	8.62***	.46	4.26	6	3.88	7	5.07***	.38
Scenario 6: Striking arrestee	4.22	6.5	3.80	5	9.91***	.42	4.50	8	3.81	5	8.22***	.69
Scenario 7: Verbal abuse	3.89	3	3.55	3	8.73***	.34	3.91	2	3.54	3	4.82***	.37
Scenario 8: Cover-up of DUI	3.91	4	3.60	4	7.86***	.31	4.03	3	3.57	4	6.00***	.46
Scenario 9: Kickback	4.22	6.5	3.87	8	8.58***	.35	4.41	7	4.04	9	4.95***	.37
Scenario 10: False report on drugs	4.46	10	4.12	11	9.37***	.34	4.64	11	4.26	12	5.44***	.38
Scenario 11: Supervisor fails to stop excessive force	4.43	9	3.96	9	11.23***	.47	4.43	8	4.10	10	4.20***	.33
Scenario 12: Failure to report hate crime	4.18	5	3.85	6	8.33***	.33	4.07	4	3.82	6	3.79***	.25
Scenario 13: Bribe from reckless motorist	4.71	13	4.34	13	10.13***	.37	4.64	11	4.29	14	4.55***	.35
Scenario 14: Failure to react to graffiti	3.73	1	3.44	2	8.37***	.29	3.79	1	3.35	1	5.41***	.44

Scenario	*Man. Own*		*Man. Most*		*t-test*	*Diff.*
	Mean	*Rank*	*Mean*	*Rank*		
Scenario 1: Acceptance of gratuities	4.39	5	3.23	2	7.18***	1.16
Scenario 2: Failure to arrest	4.85	12	3.82	7	8.50***	1.03
Scenario 3: Theft of knife	4.90	14	4.21	11	5.89***	.69
Scenario 4: Abuse of deadly force	4.48	8	4.25	14	2.83**	.23
Scenario 5: Supervisor corruption	4.43	7	3.63	5	5.23***	.80
Scenario 6: Striking arrestee	4.41	6	3.77	6	4.71***	.64
Scenario 7: Verbal abuse	3.82	2	3.37	3	5.03***	.45
Scenario 8: Cover-up of DUI	4.08	3	3.55	4	4.37***	.53
Scenario 9: Kickback	4.79	11	4.23	12	4.94***	.56
Scenario 10: False report on drugs	4.68	9	4.07	8	4.76***	.61
Scenario 11: Supervisor fails to stop excessive force	4.72	10	4.11	9.5	5.01***	.61
Scenario 12: Failure to report hate crime	4.33	4	4.11	9.5	2.05*	.22
Scenario 13: Bribe from reckless motorist	4.89	13	4.24	13	5.08***	.65
Scenario 14: Failure to react to graffiti	3.57	1	3.13	1	4.65***	.44

scenario. Statistically significant differences in the views about the appropriate discipline among the respondents of different ranks surfaced in about two-thirds of the scenarios (9 out of 14 or 64%; Table 9.5) in bivariate analyses and 11 scenarios in multivariate analyses.[12] However, with a few exceptions (Scenario non-commissioned v. commissioned in Scenario 2; non-commissioned v. respondents in management in Scenario 9), these differences were not of substantive importance.[13]

In all the scenarios in which we uncovered statistically significant differences between pairs of rank categories, a larger proportion of commissioned officers than non-commissioned officers advocated for more serious discipline. In a few scenarios in which there were significant differences between non-commissioned respondents and respondents in management,[14] a larger proportion of respondents in management thought that a more severe discipline is appropriate. Finally, in a handful of scenarios in which there were statistically significant differences in their views of appropriate discipline between commissioned respondents and respondents in management,[15] surprisingly, a larger proportion of commissioned officers advocated for more severe discipline.

The relative order of appropriate discipline—how severe respondents evaluated the appropriate discipline to be for each scenario compared to the other scenarios in the questionnaire—is shown in Table 9.4 (ranks). While the relative order is very similar for non-commissioned and commissioned respondents (Spearman's rho = .939; $p < .001$), as well as for commissioned respondents and respondents in management (Spearman's rho = .889; $p < .01$), the degree of similarity in relative views was somewhat less congruent for non-commissioned respondents and respondents in management (Spearman's rho = .840; $p < .01$).

The second question related to discipline inquired about the respondents' views about the severity of the discipline their police agency would mete out in response to learning about the behavior described in each scenario. Overall, the respondents' views about appropriate discipline were somewhat more heterogeneous than what their views about expected discipline were. Although there are statistically significant differences in the respondents' perceptions of the expected discipline by rank in about one-half of the scenarios in both bivariate and multivariate analyses (7 out of 14 scenarios of 50%; Table 9.6),[16] these differences were large and of substantive importance in only one scenario (Scenario 9: Kickback).[17] In a number of scenarios with statistically significant differences in the views of expected discipline between non-commissioned respondents and commissioned respondents (Scenarios 1, 2, 6, 9, and 10), a larger proportion of commissioned respondents expected more severe discipline. Similarly, in the instances in which there are statistically significant differences in the views between the respondents in management and non-commissioned respondents (Scenarios 1, 9, and 10), a larger proportion of the respondents in management expected harsher discipline. Finally, as was the case with appropriate discipline, the commissioned respondents were more likely to expect harsher discipline

Table 9.5 Views about the appropriate discipline by rank

Scenario	*Non-Commissioned (1)*		*Commissioned (2)*		*Management (3)*		*Chi-square/Phi (Post-hoc)*	*1 v. 2*	*1 v. 3*	*2 v. 3*
	% serious	*Rank*	*% serious*	*Rank*	*% serious*	*Rank*				
Scenario 1: Acceptance of gratuities	22.2%	3	31.4%	4	34.5%	7	10.54** /.110	−9.2%	−12.3%	−3.1%
Scenario 2: Failure to arrest	29.1%	7	50.3%	9	39.1%	9	29.18*** /.181	−21.2%	−10.0%	11.2%
Scenario 3: Theft of knife	59.9%	12	74.9%	13.5	57.5%	11.5	15.08**/.131	−15.0%	2.4%	17.4%
Scenario 4: Abuse of deadly force	64.2%	13	71.2%	12	57.5%	11.5	5.57 / .080	−7.0%	6.7%	13.7%
Scenario 5: Supervisor corruption	31.0%	8	39.3%	7	24.1%	3	7.36*/.092	−8.3%	6.9%	15.2%
Scenario 6: Striking arrestee	25.0%	4	36.1%	6	27.6%	4	8.87*/.101	−11.1%	−2.6%	8.5%
Scenario 7: Verbal abuse	12.9%	1	18.3%	1	9.2%	2	5.22 / .078	−5.4%	3.7%	9.1%
Scenario 8: Cover–up of DUI	26.0%	5	33.0%	5	29.9%	6	3.68 / .065	−7.0%	−3.9%	3.1%
Scenario 9: Kickback	34.6%	9	51.8%	10	60.9%	13	33.36***/.193	−17.2%	−26.3%	−9.1%
Scenario 10: False report on drugs	43.4%	11	59.7%	11	51.7%	10	15.91***/.135	−16.3%	−8.3%	8.0%
Scenario 11: Supervisor fails to stop excessive force	39.9%	10	48.2%	8	35.6%	8	5.32 / .078	−8.3%	4.3%	12.6%
Scenario 12: Failure to report hate crime	27.2%	6	27.7%	3	28.7%	5	.10 / .011	−0.5%	−1.5%	−1.0%
Scenario 13: Bribe from reckless motorist	66.1%	14	74.9%	13.5	63.2%	14	5.98*/.083	−8.8%	2.9%	11.7%
Scenario 14: Failure to react to graffiti	13.9%	2	18.8%	2	5.7%	1	8.52*/.099	−4.9%	8.2%	13.1%

*Based on the results of bivariate analyses.

Table 9.6 Views about the expected discipline by rank

Scenario	*Non-Commissioned (1)*		*Commissioned (2)*		*Management (3)*		*Chi-square/Phi (Post-hoc)*	*1 v. 2*	*1 v. 3*	*2 v. 3*
	% serious	*Rank*	*% serious*	*Rank*	*% serious*	*Rank*				
Scenario 1: Acceptance of gratuities	19.1%	3	33.0%	4	36.8%	8	23.89***/.164	−13.9%	−17.7%	−3.8%
Scenario 2: Failure to arrest	31.0%	7	47.1%	10	41.4%	9	17.73***/.142	−16.1%	−10.4%	5.7%
Scenario 3: Theft of knife	61.8%	12	70.7%	13	51.7%	11	9.89**/.107	−8.9%	10.1%	19.0%
Scenario 4: Abuse of deadly force	65.4%	13	72.8%	14	59.8%	14	5.46 / .079	−7.4%	5.6%	13.0%
Scenario 5: Supervisor corruption	33.7%	9	37.7%	6	23.0%	4	5.84 / .082	−4.0%	10.7%	14.7%
Scenario 6: Striking arrestee	27.0%	5	41.4%	8	29.9%	5.5	13.96**/.126	−14.4%	−2.9%	11.5%
Scenario 7: Verbal abuse	12.7%	1	17.3%	1	11.5%	2	2.89 / .058	−4.6%	1.2%	5.8%
Scenario 8: Cover−up of DUI	24.3%	4	29.3%	3	20.7%	3	2.93 / .058	−5.0%	3.6%	8.6%
Scenario 9: Kickback	33.0%	8	45.5%	9	55.2%	12	21.55***/.156	−12.5%	−22.2%	−9.7%
Scenario 10: False report on drugs	45.1%	11	56.0%	11	51.7%	10	7.36*/.092	−10.9%	−6.6%	4.3%
Scenario 11: Supervisor fails to stop excessive force	40.4%	10	40.8%	7	34.5%	7	1.21 / .037	−0.4%	5.9%	6.3%
Scenario 12: Failure to report hate crime	29.6%	6	34.6%	5	29.9%	5.5	1.69 / .044	−5.0%	−0.3%	4.7%
Scenario 13: Bribe from reckless motorist	66.2%	14	64.9%	12	56.3%	13	3.25 / .061	1.3%	9.9%	8.6%
Scenario 14: Failure to react to graffiti	12.9%	2	17.8%	2	5.7%	1	7.75*/.095	−4.9%	7.2%	12.1%

*Based on the results of bivariate analyses.

in the scenarios with statistically significant differences (Scenarios 3, 5, and 14) than the respondents in management were.

The relative order of expected discipline—how severe expected discipline is perceived to be for each scenario compared to the other scenarios in the questionnaire—is shown in Table 9.5 (ranks). The relative order for all three groups is similar (non-commissioned respondents and commissioned respondents: Spearman's rho = .899; $p < .001$; commissioned respondents and respondents in management: Spearman's rho = .902; $p < .01$; non-commissioned respondents and respondents in management: Spearman's rho = .810; $p < .01$).

Expected willingness to report

The questionnaire also contained two questions probing the respondents' willingness to report. In about one-half of the scenarios (8 out of 14 or 57%; Table 9.7; 6 out of 14 in multivariate models),[18] in bivariate analyses rank was related to the respondents' expressed willingness to report. In all but one scenario with statistically significant differences in the expressed willingness to report between the respondents in management and non-commissioned officers (Scenario 4: Abuse of Deadly Force), the respondents in management seemed more likely to say that they would report than non-commissioned respondents were (Table 9.7). Similarly, in all six scenarios with statistically significant differences, the respondents in management were more likely to say that they would report than commissioned respondents were (Table 9.7). While these differences were statistically significant, only three (one pair in Scenario 2; two pairs in Scenario 9) passed the threshold of substantive differences established by Klockars et al. (2006). Most of them were also small, and a few (Scenario 2; Scenario 9) were medium-sized.[19] By contrast, there were no statistically significant differences in the expressed willingness to report between non-commissioned and commissioned respondents (Table 9.7).

The relative order of officers' own expressed willingness to report—how willing the respondents are to report behavior described in one scenario compared to the behaviors described in other scenarios—is shown in Table 9.7 (ranks). Non-commissioned and commissioned respondents' relative order of willingness to report was very similar (Spearman's rho = .904; $p < .001$), but the relative orders of willingness to report between the respondents in management and non-commissioned respondents (Spearman's rho = .818; $p < .001$) and the respondents in management and commissioned respondents (Spearman's rho = .710; $p < .01$) were not as similar.

The respondents also had an opportunity to assess how willing they thought most police officers would be to report rule-violating behavior (Table 9.8). Rank was related to perceptions of most officers' willingness to report misconduct in only two scenarios in bivariate analyses (Scenario 4: Abuse of deadly force; Scenario 9: Kickback) and three scenarios in multivariate analyses (Table 9.8).[20] These differences were statistically significant and of substantive

Table 9.7 Own willingness to report by rank

Scenario	*Non-Commissioned (1)*		*Commissioned (2)*		*Management (3)*		*F-test (Tukey b's post-hoc)*	*1 v. 2*	*1 v. 3*	*2 v. 3*
	Mean	*Rank*	*Mean*	*Rank*	*Mean*	*Rank*				
Scenario 1: Acceptance of gratuities	3.59	4	3.92	8	3.95	4	5.66** (1 v. 3)	−.33	−.36	−.03
Scenario 2: Failure to arrest	4.14	11	4.20	11	4.75	14	11.63*** (1 v. 3; 2 v. 3)	−.09	−.61	−.55
Scenario 3: Theft of knife	4.38	12	4.31	12	4.57	12	1.93	.07	−.19	−.26
Scenario 4: Abuse of deadly force	4.39	13	4.38	13	4.07	7	3.35* (1 v. 3; 2 v. 3)	.01	.32	.31
Scenario 5: Supervisor corruption	3.85	8	3.87	6	4.11	8	1.61	−.02	−.26	−.24
Scenario 6: Striking arrestee	3.70	6	4.01	9.5	4.03	6	5.67** (no pairs)	−.31	−.33	−.02
Scenario 7: Verbal abuse	3.27	1	3.30	1	3.71	2	3.39* (1 v. 3; 2 v. 3)	−.03	−.44	−.41
Scenario 8: Cover-up of DUI	3.43	3	3.38	2	3.79	3	2.41	.05	−.36	−.41
Scenario 9: Kickback	3.69	5	3.62	4	4.41	10.5	10.83*** (1 v. 3; 2 v. 3)	.07	−.72	−.79
Scenario 10: False report on drugs	4.02	10	4.01	9.5	4.41	10.5	4.19* (1 v. 3; 2 v. 3)	.01	−.39	−.40
Scenario 11: Supervisor fails to stop excessive force	3.88	9	3.90	7	4.24	9	3.03* (1 v. 3; 2 v. 3)	−.02	−.36	−.34
Scenario 12: Failure to report hate crime	3.80	7	3.74	5	3.99	5	1.20	.06	−.19	−.25
Scenario 13: Bribe from reckless motorist	4.42	14	4.47	14	4.68	13	2.77	−.05	−.26	−.21
Scenario 14: Failure to react to graffiti	3.34	2	3.51	3	3.20	1	1.95	−.17	.31	.14

Table 9.8 Most police officers' willingness to report by rank

Scenario	*Non-Commissioned (1)*		*Commissioned (2)*		*Management (3)*		*F-test (Tukey b's post-hoc)*	*1 v. 2*	*1 v. 3*	*2 v. 3*
	Mean	*Rank*	*Mean*	*Rank*	*Mean*	*Rank*				
Scenario 1: Acceptance of gratuities	2.91	2	2.94	3	3.02	2	.26	−.03	−.09	−.08
Scenario 2: Failure to arrest	3.49	11	3.38	7	3.49	8	.47	.11	.00	−.11
Scenario 3: Theft of knife	3.66	12	3.42	9	3.68	12	2.47	.24	−.02	−.26
Scenario 4: Abuse of deadly force	3.90	14	3.84	13	3.47	7	3.87* (1 v. 3; 2 v. 3)	.06	.43	.37
Scenario 5: Supervisor corruption	3.47	10	3.40	8	3.59	10	.54	.07	−.12	−.19
Scenario 6: Striking arrestee	3.21	5.5	3.45	10	3.36	5	2.24	−.24	−.15	.09
Scenario 7: Verbal abuse	2.93	3	2.98	4	3.08	3	.41	−.05	−.15	−.10
Scenario 8: Cover-up of DUI	3.03	4	2.84	1	3.26	4	2.49	.19	−.23	−.42
Scenario 9: Kickback	3.21	5.5	3.13	5	3.80	13	7.03** (1 v. 3; 2 v. 3)	.08	−.59	−.67
Scenario 10: False report on drugs	3.37	9	3.48	11	3.66	11	1.70	−.11	−.29	−.18
Scenario 11: Supervisor fails to stop excessive force	3.28	7	3.51	12	3.44	6	2.11	−.23	−.16	.07
Scenario 12: Failure to report hate crime	3.34	8	3.36	6	3.52	9	.67	−.02	−.18	−.16
Scenario 13: Bribe from reckless motorist	3.81	13	3.95	14	3.87	14	.77	−.14	−.06	.08
Scenario 14: Failure to react to graffiti	2.90	1	2.86	2	2.83	1	.14	.04	.07	.03

importance[21] in one scenario (Scenario 9), but were not large.[22] While in one scenario (Scenario 4: Abuse of deadly force) the respondents in management were *less* likely to say that they would report than the other two groups were, in the other scenario with statistically significant differences (Scenario 9: Kickback), the respondents in management were *more* likely to say that they would report.

The relative order of most officers' expressed willingness to report—how willing the respondents assumed that most officers would be to report behavior described in one scenario compared to the behaviors described in other scenarios—is shown in Table 9.8 (ranks). The relative order of most officers' expressed willingness to report was more similar between non-commissioned respondents and commissioned respondents (Spearman's rho = .788; $p < .01$) than between non-commissioned respondents and the respondents in management (Spearman's rho = .733; $p < .01$) or between commissioned respondents and the respondents in management (Spearman's rho = .569; $p < .05$).

Finally, we compared the respondents' own expressed willingness to report with how likely they anticipated most police officers would be to report (Table 9.9). As we projected, because of the self-serving bias, all three categories of respondents were statistically significantly likely to say that they would report more than they assumed most police officers would (Table 9.9). The number of substantive differences between the respondents' own willingness to report and their estimates of most police officers' willingness to report was the smallest for non-commissioned officers (6 out of 14; Table 9.9), who are the closest to "most officers" depicted in the question, and the largest (12 out of 14; Table 9.9) for respondents in management, who are the farthest from "most officers" (i.e., line officers).

Conclusion

Police ranks are central to police culture, identity, and function. Although historically coveted, some police organizations have in recent decades moved to downplay traditional power hierarchies and the associated military symbolism (e.g., Batts et al., 2012). This has also been true of the SAPS, where rank remains subject to contestation, aspiration, and organizational injustice.

The SAPS was formed by a merger of 11 police forces and their former enemies, members of the liberation movements (Chapter 3). Over a very short period of time, former SAP officers found themselves reporting to people with whom they had previously been in conflict. While some new managers were natural and experienced leaders, others did not have the competencies required for the job. This eroded organizational morale among some police officers.

Urgently needing to reform the organization to better align it with national demographics (Bruce, 2013), the SAPS embarked on a rapid period of recruitment from 2002 to 2012, recruiting more than 120,000 new officers. This rapid intake saw quality control suffer and an increase in lax discipline. In an

Table 9.9 Willingness to report by rank

Scenario	*Non-C. Own*		*Non-C. Most*		*t-test*	*Diff.*	*Comm. Own*		*Comm. Most*		*t-test*	*Diff.*
	Mean	*Rank*	*Mean*	*Rank*			*Mean*	*Rank*	*Mean*	*Rank*		
Scenario 1: Acceptance of gratuities	3.59	4	2.91	2	12.09***	.68	3.92	8	2.94	3	8.93***	.98
Scenario 2: Failure to arrest	4.14	11	3.49	11	12.83***	.65	4.20	11	3.38	7	7.87***	.82
Scenario 3: Theft of knife	4.38	12	3.66	12	13.31***	.72	4.31	12	3.42	9	8.89***	.89
Scenario 4: Abuse of deadly force	4.39	13	3.90	14	10.89***	.49	4.38	13	3.84	13	6.93***	.54
Scenario 5: Supervisor corruption	3.85	8	3.47	10	9.27***	.38	3.87	6	3.40	8	6.55***	.47
Scenario 6: Striking arrestee	3.70	6	3.21	5.5	10.70***	.49	4.01	9.5	3.45	10	6.63***	.56
Scenario 7: Verbal abuse	3.27	1	2.93	3	9.06***	.34	3.30	1	2.98	4	3.90***	.32
Scenario 8: Cover-up of DUI	3.43	3	3.03	4	9.02***	.40	3.38	2	2.84	1	7.76***	.54
Scenario 9: Kickback	3.69	5	3.21	5.5	10.34***	.48	3.62	4	3.13	5	5.77***	.49
Scenario 10: False report on drugs	4.02	10	3.37	9	13.60***	.65	4.01	9.5	3.48	11	6.29***	.53
Scenario 11: Supervisor fails to stop excessive force	3.88	9	3.28	7	11.74***	.60	3.90	7	3.51	12	4.35***	.39
Scenario 12: Failure to report hate crime	3.80	7	3.34	8	9.83***	.46	3.74	5	3.36	6	4.54***	.38
Scenario 13: Bribe from reckless motorist	4.42	14	3.81	13	12.68***	.61	4.47	14	3.95	14	5.60***	.52
Scenario 14: Failure to react to graffiti	3.34	2	2.90	1	11.02***	.44	3.51	3	2.86	2	7.89**	.65

Scenario	*Man. Own*		*Man. Most*		*t-test*	*Diff.*
	Mean	*Rank*	*Mean*	*Rank*		
Scenario 1: Acceptance of gratuities	3.95	4	3.02	2	5.32***	.93
Scenario 2: Failure to arrest	4.75	14	3.49	8	9.21***	1.26
Scenario 3: Theft of knife	4.57	12	3.68	12	7.20***	.89
Scenario 4: Abuse of deadly force	4.07	7	3.47	7	5.15***	.60
Scenario 5: Supervisor corruption	4.11	8	3.59	10	3.90***	.52
Scenario 6: Striking arrestee	4.03	6	3.36	5	4.68***	.67
Scenario 7: Verbal abuse	3.71	2	3.08	3	5.03***	.63
Scenario 8: Cover-up of DUI	3.79	3	3.26	4	3.84***	.53
Scenario 9: Kickback	4.41	10.5	3.80	13	4.32***	.61
Scenario 10: False report on drugs	4.41	10.5	3.66	11	5.47***	.75
Scenario 11: Supervisor fails to stop excessive force	4.24	9	3.44	6	5.93***	.80
Scenario 12: Failure to report hate crime	3.99	5	3.52	9	4.82***	.47
Scenario 13: Bribe from reckless motorist	4.68	13	3.87	14	5.93***	.81
Scenario 14: Failure to react to graffiti	3.20	1	2.83	1	3.74***	.37

effort to address the latter, the SAPS reintroduced military ranks, seeking to instill discipline in its officers and fear in criminals. Neither appears to have been achieved. Instead, under President Zuma's administration, numerous top managers were irregularly appointed, with some aiding the former president's unbridled looting of state resources. Once again, these activities signaled to honest officers that their hard work may not pay off, that rank promotion could be a result of whom you know, not what you know. As a result, rank remains a contentious issue within the SAPS.

Our empirical results suggest there are very few differences between non-commissioned respondents and commissioned respondents in their integrity-related attitudes, thus providing further evidence that line officer and supervisor views may not be as different as we assumed (Kucukuyal, 2008; Kutnjak Ivković, 2004; Kutnjak Ivković and Shelley, 2010; Kutnjak Ivković et al., 2016; Lee et al., 2013; Yun, 2003). The only systematic difference between non-commissioned and commissioned respondents was in their opinions about the appropriate discipline, and to a smaller extent, about the expected discipline. In such cases, commissioned officers—who are higher up in the hierarchy than non-commissioned officers are—both approved and expected more severe discipline. Such findings are in accordance with the prior police integrity research suggesting that if views differ, supervisors express attitudes associated with higher integrity (e.g., Kremer, 2000; Kutnjak Ivković and Klockars, 2000; Kutnjak Ivković et al., 2019; Marche, 2009; Micucci and Gomme, 2005; Pagon and Lobnikar, 2000; Raines, 2010).

Although there are differences in the integrity-related views between non-commissioned respondents and the respondents in management, these differences appear in only a few scenarios and, as a rule, the respondents in management expressed attitudes associated with higher integrity than non-commissioned respondents did. Finally, commissioned respondents and the respondents in management—both groups of SAPS supervisors—had mostly similar views across all measures of police integrity.

Our results suggest that rank is not systematically and consistently associated with police integrity. We found very few differences between non-commissioned and commissioned officers. The largest differences we discovered were between the non-commissioned officers, the lowest-ranking of the three groups, and the respondents in management, the highest-ranking of the three groups. These two groups represent the opposite ends of the SAPS hierarchy. Even in such extremes, the differences are far from systematic and consistent, indicating that promotion to the higher ranks of the hierarchy does not carry with it the expected change in attitudes toward official rules and their enforcement.

Notes

1 We do not count police assignments as different hierarchical levels.
2 Lieutenant as a rank has been officially abolished by the SAPS.

3 Analyzing rank differences in familiarity with the official rules while controlling for other demographic variables by means of linear regression preserved all the scenarios with statistically significant differences in familiarity with the official rules previously detected in simpler, bivariate analyses. The only exception was Scenario 1.
4 Cohen's *d* indicates that the differences are small for all pairs, except the comparison of non-commissioned and management in scenarios 6, 9, and 11. In these scenarios, they are about small-to-medium sized.
5 We rely on the threshold of .50 to ascertain substantive differences, as established by Klockars and colleagues (2006).
6 Analyzing rank differences in perceptions of misconduct seriousness while controlling for other demographic variables by means of linear regression preserved statistically significant differences in perceptions of misconduct seriousness previously detected in simpler, bivariate analyses, in only four scenarios (Scenario 1, Scenario 4, Scenario 6, and Scenario 9). Differences in three scenarios (Scenario 2, Scenario 10, and Scenario 11) were not statistically significant in multivariate models, while differences in two scenarios (Scenario 4, Scenario 7) were significant in multivariate, but not bivariate analyses.
7 Cohen's *d* indicates that the differences were small for all pairs, except for Scenario 9 (1 v. 3), in which the difference was of a medium size.
8 We rely on the threshold of .50 to ascertain substantive differences, as established by Klockars and colleagues (2006).
9 Analyzing rank differences in police officers' estimates of most police officers' evaluations of seriousness while controlling for other demographic variables by means of linear regression resulted in a marginally significant difference for Scenario 6. In addition, it also indicated a few more scenarios in which the respondents in management were more likely to assume that most police officers would evaluate behavior as serious than non-commissioned police officers were (Scenario 2, Scenario 4, Scenario 5, and Scenario 14).
10 Cohen's *d* indicates that the difference is small.
11 We rely on the threshold of .50 to ascertain substantive differences, as established by Klockars and colleagues (2006).
12 Analyzing rank differences in police officers' views about appropriate discipline while controlling for other demographic variables by means of logistic regression preserved all but one scenario that had statistically significant differences in bivariate comparisons and actually revealed three more scenarios with statistically significant differences (Scenarios 7, 8, and 11) in the respondents' views of appropriate discipline. In total, multivariate models revealed that there are 11 scenarios remained with statistically significant differences after we controlled for other demographic variables.
13 We rely on the Klockars et al. (2006) rule of differences larger than 20% of being substantively important.
14 There were statistically significant differences in the views about appropriate discipline between non-commissioned respondents and respondents in management in Scenario 9 in bivariate analyses and Scenarios 1, 2, 8, 9, 10, and 14 in multivariate analyses.
15 There were statistically significant differences in the views about appropriate discipline between commissioned respondents and respondents in management in Scenario 3 in bivariate analyses and Scenarios 3, 4, and 14 in multivariate analyses.
16 Analyzing rank differences in police officers' views about expected discipline while controlling for other demographic variables by means of logistic regression preserved all scenarios that had statistically significant relations in bivariate comparisons in the respondents' views of expected discipline.
17 We rely on the Klockars et al. (2006) rule of differences larger than 20% of being substantively important.
18 Analyzing rank differences in own willingness to report while controlling for other demographic variables by means of linear regression preserved statically significant

differences in own willingness to report in 5 out of 8 scenarios. The three scenarios in which differences were no longer significant in the multivariate models are Scenario 6 (in which no significant pairs were identified in the post hoc tests in the bivariate models), Scenario 7 (in which the differences were marginally significant), and Scenario 11. There was also one scenario (Scenario 3) in which the differences were not significant in bivariate analyses but are significant in multivariate analyses.

19 Cohen's *d* indicates that the differences between respondents in management and other two groups were medium-sized in Scenario 2 and Scenario 9.

20 Analyzing rank differences in the estimates of most police officers' willingness to report while controlling for other demographic variables by means of linear regression preserved all the statistically significant differences in most police officers' willingness to report. In addition, multivariate models revealed a statistically significant difference in Scenario 3 as well.

21 They passed the threshold of substantive importance established by Klockars and colleagues (2006).

22 Cohen's *d* indicates that these differences were small.

References

Alexander, P., C. Runciman, T. Ngwane, B. Moloto, K. Mokgele, and N. van Staden (2018). Frequency and turmoil: South Africa's community protests 2005–2017. *South African Crime Quarterly*, 63: 27–42.

Batts, A., S. Smoot, and E. Scrivner (2012). Police leadership challenges in a changing world. *New Perspectives in Policing, Harvard Kennedy School and National Institute of Justice*. Available at: www.ncjrs.gov/pdffiles1/nij/238338.pdf

Bruce, D. (2011). Beyond section 49: Control of the use of lethal force. *South African Crime Quarterly*, 36: 3–12.

Bruce, D. (2013). New blood: Implications of en masse recruitment for the South African Police Service. *South African Crime Quarterly*, 43: 17–28.

Bucak, S. (2012). *An examination of police cynicism in turkey and its impacts on officers' perception of corruption*. (Order No. 3558466, University of Illinois at Chicago). ProQuest Dissertations and Theses, 214. Available at: http://ezproxy.msu.edu/login?url=http://search.proquest.com/docview/1343688917?accountid=12598. (1343688917)

Burger, J. (2011). To protect and serve: Restoring public confidence in the SAPS. *South African Crime Quarterly*, 36: 13–22.

Burger, R., M. Louw, B. Barbara, I. de Oliveira Pegado, and S. van der Berg (2015). Understanding consumption patterns of the established and emerging South African black middle class. *Development Southern Africa*, 32(1): 41–56. DOI: 10.1080/0376835X.2014.976855

Cawthra, G. (1993). *Policing South Africa: The SAP and the Transition from Apartheid*. London: Zed Books Ltd.

Civilian Secretariat for Police (2016). *White Paper on Policing*. Pretoria: Department of Police.

Constitution of the Republic of South Africa (1996). Act No. 108 of 1996. Available at: www.gov.za/documents/constitution-republic-south-africa-1996

Crank, J.P. (2014). *Understanding Police Culture*, 2nd ed. London: Routledge.

Crank, J.P., B. Regoli, J.D. Hewitt, and R.G. Culbertson (1993). An assessment of work stress among police executives. *Journal of Criminal Justice*, 21: 313–324.

Crank, J.P., R. Regoli, J.D. Hewitt, and R.G. Culbertson (1995). Institutional and organizational antecedents of role stress, work alienation, and anomie among police executives. *Criminal Justice and Behavior*, 22: 152–171.

Department of Justice (2016). *Solidarity vs the Minister of Safety and Security and Others*. Case no: J 879/12. Available at: www.justice.gov.za/labourcourt/jdgm-lbc/2016lab/2016-J%20879-12.pdf

Donner, C., J. Maskaly, L. Fridell, and W. Jennings (2015). Policing and procedural justice: A state-of-the-art review. *Policing: An International Journal*, 38(1): 153–172. https://doi.org/10.1108/PIJPSM-12-2014-0129

Duncan, J. (2016). *Protest Nation: The Right to Protest in South Africa*. Pietermaritzburg: University of KwaZulu-Natal Press.

Farlam, I.G. (2015). *Marikana Commission of Inquiry: Report on Matters of Public, National and International Concern Arising Out of the Tragic Incidents at the Lonmin Mine in Marikana, in the North West Province*. Available at: www.sahrc.org.za/home/21/files/marikana-report-1.pdf.

Faull, A. (2010). *Behind the Badge: The Untold Stories of South Africa's Police Service Members*. Cape Town: Penguin Random House South Africa. Kindle Edition.

Faull, A. (2012). *Excerpt from Interview in Field Diary during 8-Months of Ethnographic Fieldwork with the SAPS*. Unpublished notes.

Faull, A. (2013a). Fighting for respect: Violence, masculinity and legitimacy in the SAPS. *South African Crime Quarterly*, 44: 5–14.

Faull, A. (2013b). Towards a new professionalism for the South African Police Service. *South African Review of Sociology*, 44(2): 18–35.

Faull, A. (2018). *Police Work and Identity: A South African Ethnography*. Abingdon: Routledge.

Faull, A. (2019). Personal communication with former head of the SAPS Anti-Corruption Unit. *Stef Grobler*, 8 October.

Feinstein, A. (2007). *After the Party: Corruption and the ANC*. Cape Town: Jonathan Ball.

Francis, E. (2002). Rural livelihoods, institutions and vulnerability in the North-West Province, South Africa. *Journal of Southern African Studies*, 28(3): 531–550.

Grobler, L. (2013). *Crossing the Line: When Cops Become Criminals*. Auckland Park, South Africa: Jacana Media.

Goodhew, D. (1990). Between the devil and the deep blue sea: Crime, policing and the western areas of Johannesburg, c. 1930–1962. Seminar presentation, University of the Witwatersrand History Workshop, Johannesburg, February.

Harper, S. and J. Seekings (2010). *Claims on and Obligations to Kin in Cape Town, South Africa*. Centre for Social Science Research, University of Cape Town. Available at: https://open.uct.ac.za/bitstream/handle/11427/19289/Harper_Claims_onobligations_to_2010.pdf?sequence=1

Hickman, M.J., A.R. Piquero, Z.A. Powell, and J. Greene (2016a). Expanding the measurement of police integrity. *Policing: An International Journal of Police Strategies & Management*, 39(2): 246–267.

Hornberger, J. (2011). *Policing and Human Rights: The Meaning of Violence and Justice in the Everyday Policing of Johannesburg*. Abingdon, U.K.: Routledge.

Hornberger, J. (2013). From general to commissioner to general: On the popular state of policing in South Africa. *Law & Social Inquiry*, 38(3): 598–614.

Hyslop, J. (2005). Political corruption: Before and after apartheid. *Journal of Southern African Studies*, 31(4): 773–789.

Kargin, V. (2009). *An investigation of factors proposed to influence police officers' peer reporting intentions*. ProQuest Dissertations and Theses. Available at: http://ezproxy.msu.edu/login?url=http://search.proquest.com/docview/305066516?accountid=12598. (305066516).

Klockars, C.B., and S. Kutnjak Ivković (2004). Measuring police integrity. In M.J. Hickman, A.R. Piquero, and J.R. Greene (eds.), *Police Integrity and Ethics*. Belmont, CA: Wadsworth Publishing.

Klockars, C.B., S. Kutnjak Ivković, and M.R. Haberfeld (eds.) (2004). *The Contours of Police Integrity*. Newbury Park, CA: Sage Publication.

Klockars, C.B., S. Kutnjak Ivković, and M.R. Haberfeld (2006). *Enhancing Police Integrity*. Dordrecht, The Netherlands: Springer.

Klockars, C.B., S. Kutnjak Ivković, W.E. Harver, and M.R. Haberfeld (2000). *The Measurement of Police Integrity*. Research in Brief. U.S. Department of Justice, Office of Justice Programs, National Institute of Justice: Washington, DC: Government Printing Office.

Kremer, F. (2000). Comparing supervisor and line officer opinions about the code of silence: The case of Hungary. In M. Pagon (ed.), *Policing in Central and Eastern Europe: Ethics, Integrity, and Human Rights*. Ljubljana, Slovenia: College of Police and Security Studies. 211–219.

Kucukuysal, B. (2008). *Determinants of Turkish Police Officers' Perception of Integrity: Impact of Organizational Culture*, Doctoral dissertation, Florida: University of Central Florida Orlando.

Kutnjak Ivković, S. (2004). Sharing the view: Line officer and supervisor evaluations of police corruption seriousness in Bosnia and Herzegovina. In G. Mesko, M. Pagon, and G. Dobovsek (eds.), *Dilemmas of Contemporary Criminal Justice*. Ljubljana, Slovenia: Faculty of Criminal Justice, University of Maribor. 312–322.

Kutnjak Ivković, S. (2009). The Croatian police, police integrity, and transition toward democratic policing. *Policing: An International Journal of Police Strategies and Management*, 32(3): 459–488.

Kutnjak Ivković, S. (2015). Studying police integrity. In S. Kutnjak Ivković and M. Haberfeld (eds.), *Police Integrity across the World*. New York: Springer.

Kutnjak Ivković, S., I. Cajner Mraović, and K. Borovec (2016). An empirical test of the influence of society at large on police integrity in a centralized police system. *Policing: An International Journal of Police Strategies and Management*, 39(2): 302–318.

Kutnjak Ivković, S. and M.R. Haberfeld (eds.) (2015). *Police Integrity across the World*. New York: Springer.

Kutnjak Ivković, S. and M.R. Haberfeld (2019). Exploring empirical research on police integrity. In S. Kutnjak Ivković and M. Haberfeld (eds.), *Exploring Police Integrity*. New York: Springer.

Kutnjak Ivković, S., M. Haberfeld, and R. Peacock (2016). Does discipline fairness matter for the police code of silence? Answers from the U.S. supervisors and line officers. *Policing: An International Journal of Police Strategies and Management*, 39(2): 354–369.

Kutnjak Ivković, S., M. Haberfeld, and R. Peacock (2018). Decoding the code of silence. *Criminal Justice Policy Review*, 29(2): 172–189.

Kutnjak Ivković, S., M.R. Haberfeld, and R. Peacock (2019). Overlapping shades of blue: exploring police officer, supervisor, and administrator cultures of police integrity. In S. Kutnjak Ivković and M. Haberfeld (eds.), *Exploring Police Integrity: Novel Approaches to Police Integrity Theory and Methodology*. New York: Springer.

Kutnjak Ivković, S. and C.B. Klockars (2000). Comparing police supervisor and line officer opinions about the code of silence: The case of Croatia. In M. Pagon (ed.), *Policing in Central and Eastern Europe: Ethics, Integrity, and Human Rights*. Ljubljana, Slovenia: College of Police and Security Studies. 183–195.

Kutnjak Ivković, S. and T. O'Connor Shelley (2010). The Code of silence and disciplinary fairness: A comparison of Czech police supervisor and line officer views. *Policing: An International Journal of Police Strategies and Management*, 33(3): 548–574.
Kutnjak Ivković, S. and T.O. Shelley (2008). The police code of silence and different paths toward democratic policing. *Policing and Society*, 18(4): 445–473.
Kutnjak Ivković, S. and T.O. Shelley (2010). The code of silence and disciplinary fairness: A comparison of Czech police supervisor and line officer views. *Policing: An International Journal of Police Strategies and Management*, 33(3): 548–574.
Kynoch, G. (2003). Friend or foe? A world view of community-police relations in Gauteng townships, 1947–77. *Canadian Journal of African Studies*, 37(2–3): 298–327. DOI: 10.1080/00083968.2003.10751270
Lee, H., H. Lim, D.D. Moore, and J. Kim (2013). How police organizational structure correlates with frontline officers' attitudes toward corruption: A multilevel model. *Police Practice and Research*, 14(5): 386–401.
Long, M., J.E. Cross, T.O. Shelley, and S. Kutnjak Ivković (2013). The normative order of reporting police misconduct: Examining the roles of offense seriousness, legitimacy, and fairness. *Social Psychology Quarterly*, 76(3): 242–267. DOI 10.1177/0190272513493094
Manning, P. (1989). Occupational culture. In W.G. Bailey (ed.), *The Encyclopedia of Police Science*. New York: Garland.
Marche, G.E. (2009). Integrity, culture, and scale: An empirical test of the big bad police agency. *Crime, Law, and Social Change* 51: 463–486.
Mgwenya, B.C. (2019). *Speaker Notes: National Recruitment Indaba*, 25 July. Available at: www.saps.gov.za/newsroom/msspeechdetail.php?nid=21432
Micucci, A.J. and I.M. Gomme (2005). American police and subcultural support for the use of excessive force. *Journal of Criminal Justice*, 33: 487–500.
Myburgh, P. (2018). *Gangster State: Unravelling Ace Magashule's Web of Capture*. Cape Town: Penguin Books.
Nasson, B. (1992). "Messing with coloured people": The 1918 police strike in Cape Town, South Africa. *The Journal of African History*, 33(2): 301–319.
Pagon, M. and B. Lobnikar (2000). Comparing supervisor and line officer opinions about the code of silence: The case of Slovenia. In Pagon, M. (ed.), *Policing in Central and Eastern Europe: Ethics, Integrity, and Human Rights.* Ljubljana, Slovenia: College of Police and Security Studies. 197–209.
Paoline, E.A. and W. Terrill (2014). *Police Culture: Adapting to the Strains of the Job*. Durham, NC: Carolina Academic Press.
Perkins, G. (2018). *Danger and Death: Organisational and Occupational Responses to the Murder of Police in South Africa: A Case Study*. Unpublished Doctoral Thesis. Department of Public Law, Faculty of Law University of Cape Town. Available at: https://open.uct.ac.za/bitstream/handle/11427/29287/thesis_law_2018_perkins_grainne.pdf?sequence=1&isAllowed=y
The Planning Commission (2012). *National Development Plan* (November). Pretoria: The Planning Commission.
PMG (2010). *South African Police Service Training Programmes, Adoption of Committee Reports & Minutes*, 13 September. Available at: https://pmg.org.za/committee-meeting/12022/
PMG (2017). *Question NW2334 to the Minister of Police*. Parliamentary Monitoring Group. Available at: https://pmg.org.za/committee-question/6991/
Porter, L.E. and T. Prenzler (2016). The code of silence and ethical perceptions: Exploring police officer unwillingness to report misconduct. *Policing: An International Journal of Police Strategies & Management*, 39(2): 370–386.

Posel, D. (1999). Whiteness and power in the South African Civil Service: Paradoxes of the apartheid state. *Journal of Southern African Studies*, 25(1): 99–119.

Posel, D. (2010). Races to consume: Revisiting South Africa's history of race, consumption and the struggle for freedom. *Ethnic and Racial Studies*, 33(2): 157–175. DOI: 10.1080/01419870903428505

Posel, D. and D. Rogan (2017). Inequality, social comparisons and minimum income aspirations: Evidence from South Africa. Paper prepared for the Biennia: Conference of the Economic Society of South Africa, Rhodes University, Grahamstown, South Africa, 30 August–1 September.

Quinton, P., A. Myhill, B. Bradford, A. Fildes, and G. Porter (2015). *Fair Cop 2: Organisational Justice, Behaviour and Ethical Policing, an Interpretative Evidence Commentary*. London: College of Policing.

Raines, J. (2010). *Ethics in Policing: Misconduct and Integrity*. Sudbury, Massachusetts: Jones & Bartlett Learning.

Rauch, J. (2000). Police reform and South Africa's transition. Conference Paper: South African Institute for International Affairs Conference. Available at: www.csvr.org.za/docs/policing/policereformandsouth.pdf

Reuss-Ianni, E. and F.A. Ianni (1983). *Street Cops and Management Cops: The Two Cultures of Policing*. New York, NY: Transaction Publishers.

Roberts, K. and V. Herrington (2013). Organisational and procedural justice: A review of the literature and its implications for policing. *Journal of Policing, Intelligence and Counter Terrorism*, 8(2): 115–130. DOI: 10.1080/18335330.2013.821737

Rothwell, G.R. and J.N. Baldwin (2006). Ethical climates and contextual predictors of whistle-blowing. *Review of Public Personnel Administration*, 26(3): 216–244.

Rothwell, G.R. and J.N. Baldwin (2007a). Whistle-blowing and the code of silence in police agencies policy and structural predictors. *Crime & Delinquency*, 53(4): 605–632.

Rothwell, G.R. and J.N. Baldwin (2007b). Ethical climate theory, whistle-blowing, and the code of silence in police agencies in the state of Georgia. *Journal of Business Ethics*, 70(4), 341–361.

SALDRU (2019). *Income Check*. Available at: www.saldru.uct.ac.za/income-comparison-tool/

Seegers, A. (1991). One state, three faces: Policing in South Africa (1910–1990). *Social Dynamics: A Journal of African Studies*, 17(1): 36–48. DOI: 10.1080/02533959108458501

Shaw, M. and S. Shearing (1998). Reshaping security: An examination of the governance of security in South Africa. *African Security Review*, 7(3): 3–12. DOI: 10.1080/10246029.1998.9627860

Shear, K.S. (1998). *Constituting a State in South Africa: The Dialectics of Policing, 1900–1939*. PhD Dissertation. Northwestern University.

Shear, K. (2012). Chiefs or modern bureaucrats? Managing black police in early twentieth-century South Africa. *Comparative Studies in Society and History*, 54(2): 251–274.

Silver, J.R., S.P. Roche, T.J. Bilach, and S.B. Ryon (2017). Traditional police culture, use of force, and procedural justice: Investigating individual, organizational, and contextual factors. *Justice Quarterly*, 34(7): 1272–1309.

South African Government News Agency (2018). Thousands of police officers get promoted. *Posts Regraded*. Available at: www.sanews.gov.za/south-africa/thousands-police-officers-get-promoted-posts-regraded

South African Police Service (2018). *Annual Report 2017/2018*. Pretoria: South African Police Service. Available at: https://www.saps.gov.za/about/stratframework/annualreports_arch.php

South African Police Union (SAPU) (2017). Police promotions policy an insult to officers. Statement issued by Oscar Skommere, General Secretary, SAPU, 30 October. Available at: www.politicsweb.co.za/news-and-analysis/police-promotions-policy-an-insult-to-officers--sa

Stanko, B., J. Jackson, B. Bradford, and K. Hohl (2012). A golden thread, a presence amongst uniforms, and a good deal of data: Studying public confidence in the London Metropolitan Police. *Policing and Society: An International Journal of Research and Policy*, 22(3): 317–331.

Steinberg, J. (2008). *South Africa's xenophobic eruption*. ISS Paper 169, Pretoria: Institute for Security Studies.

Tyler, T. (2017). Procedural justice and policing: A rush to judgment? *Annual Review of Law and Social Science*, 13(1): 29–53.

Van der Spuy, E. (2005). *South African Police Reform in the 1990s: Internal Processes and External Influences*. Unpublished Doctoral Thesis. Cape Town: University of Cape Town.

Van Vuuren, H. (2006). *Apartheid Grand Corruption: Assessing the Scale of Crimes of Profit in South Africa from 1976 to 1994*. Cape Town: Institute for Security Studies. Available at: https://opensecrets.org.za/site/wp-content/uploads/Apartheid-Grand-Corruption-2006.pdf

Waddington, P.A.J. (1999). Police (canteen) sub-culture. An appreciation. The British Journal of Criminology, 39(2): 287–309. Available at: https://doi.org/10.1093/bjc/39.2.287

Walker, S. and C. Katz (2011). *The Police in America*. New York: McGraw Hill.

Yun, I. (2003). *A study of police officers' perceptions of police deviance: From an occupational socialization perspective. ProQuest Dissertations and Theses*, 89–89. Available at: http://ezproxy.msu.edu/login?url=http://search.proquest.com/docview/305329643?accountid=12598. (305329643)

Chapter 10

Police integrity in the South African context

Introduction

Contemporary South Africa is the product of a turbulent past marked by injustice, inequality, and violence. Police integrity does not grow naturally from such soil. A country in transition with a legacy of apartheid and a history of inuring to violence (Chapter 2), South Africa also has a large centralized police system—the South African Police Service (SAPS)—characterized by consistent leadership failures (Chapter 5), perceptions about widespread police corruption, and declining levels of police legitimacy (Chapter 1). As such, the country is inherently an illuminating location for an in-depth exploration of police integrity and the factors related to it.

The early South African state was built on diamonds, gold, and agriculture, mined and harvested by mostly black, colored, and Indian men, in service of the white purse (Chapter 2). For most of the twentieth century, police facilitated this inequality by enforcing a slew of oppressive laws, regulating by racial classification where people could work and live. Even sex was policed based on race categories as defined in legislation. At the same time, the police were largely absent from black townships unless to enforce curfews, pass laws, or hunt "enemies of the state." In the absence of a state crime prevention presence in many communities, criminal and interpersonal violence flourished.

Through the mid-1970s to the early 1990s, international and domestic resistance to apartheid increased significantly, raising both fiscal restrictions on the country and violence within and outside its borders. Western powers who supported the South African government because of the country's geostrategic significance in the Cold War, began to publicly mute their association. In the shadows, in defiance of sanctions, some still continued to collaborate with the apartheid state—laundering its money and selling it weapons (van Vuuren, 2006, 2017). Within the country, police were joined by soldiers and later, hastily trained auxiliary law enforcers, especially in black townships where they imposed draconian laws, killed and wounded scores of protestors, and arrested many more. Engaged in a "total defence strategy" against a perceived "total communist onslaught," rogue elements within the police force engaged in covert operations, torturing and assassinating state enemies. In the anti-apartheid community, the African National Congress (ANC) ratcheted up intimidation and violence in the mid-1980s, aiming to make the country "ungovernable" and force the government into submission. Initially ignored by the West, the organization had turned to the Soviet Union for weaponry and training, strategic and tactical guidance, and financial support, and, by the 1980s, the ANC was waging low-level guerilla warfare within and outside the borders of South Africa. With very little hope to defeat the country's security forces in conventional warfare, the ANC's insurgency tactics focused rather on causing a climate of terror, anxiety, and disorder within the country. Political subterfuge escalated, and while fear and paranoia beset South Africans, trust between the citizens and the authoritarian government was on a moribund decline (Chapter 2).

Against this background, the promise of a "new" democratic South Africa appeared both impossible and imperative. Although violence soared in the transition years between 1990 and 1994, democracy would at last arrive. In April, 1994, the ANC won a resounding victory in the first all-inclusive election, and Nelson Mandela became the country's first black president. The ANC's successful propaganda strategy during the 1980s and early 1990s had obliterated its political competitors and, combined with the wave of election popularity and the perceived illegitimacy of historically white political parties, the organization would be assured to stay in power for the foreseeable future, but possibly not "till Jesus comes" as predicted in 2014 by Jacob Zuma (Chapter 3; Jeffery, 2019, Loc 432). The hope that a fairly elected, democratic government would stem violence and bring legitimacy to the police, however, did not fully materialize.

In 1995 the SAPS was formed amidst promises of human rights and community policing. Although murder began declining, fear of crime spread, and by 2000 the SAPS and government had reverted to a "tough on crime" posture (Kinnes, 2017). Over the next two decades, street-level policing became increasingly militarized, while covert policing was used by government leaders to wage battles within the ANC (Steinberg, 2014). Under President Zuma (2009–2018), the ANC actively defended its leader, while the president and his allies pillaged the state purse and captured its criminal justice institutions (Pauw, 2018; Basson and Du Toit, 2018).

In 2018, roughly half of the heads of South Africa's households were satisfied with the performance of the police in their area (StatsSA, 2018), while just 34% of a representative sample of South Africans trusted the police (Afrobarometer, 2018). The fact that the SAPS National Commissioners have been routinely dismissed and suspended since 2008 for transgressions including corruption, maladministration, and dishonesty (Chapter 5) does not reflect well on the organization. The same is true of the stories about police abuse of force, corruption, and political bias that are regularly reported in the daily news columns and social media feeds (e.g., Grobler, 2013).

Research on trust in policing suggests that legitimacy is generated through policing that is consistently fair (Trinkner et al., 2016). Fair policing can also promote nation building, as people who perceive police as legitimate authorities are more likely to identify with the authority they represent (Bradford et al., 2014). But policing in contemporary South Africa is not perceived by many as even-handed. In actions starkly reminiscent of apartheid policing, SAPS officers in 2012 shot 112 striking mine workers protesting for better wages at Marikana, killing 34 (Farlam, 2015). While a Commission of Inquiry found the shootings unnecessary and likely unlawful, at the time of writing, no police officers involved had suffered any sanctions. In 2018, the equality court ruled that the SAPS discriminated against poor and black communities in its resource allocation (Western Cape High Court, 2018).

The SAPS's legitimacy is further weakened by perceptions and routine experiences of police misconduct. Of the 25% respondents on the 2018 Afrobarometer who requested help from the police in the year preceding the survey, 12% said that they paid a bribe, gave a gift, or did a favor to police in order to receive police assistance. Of the 27% of respondents who interacted with the police at a checkpoint, during a traffic stop, investigation, or related encounter, 26% said that they gave a bribe, gift, or favor to avoid a problem (Afrobarometer, 2018). At the same time, oversight bodies such as the Independent Police Investigative Directorate (IPID) are flooded with complaints of police abuse each year. In 2017–2018, these grievances included allegations against police officers of 3,661 assaults, 105 rapes, and 217 incidents of torture by police (IPID, 2018, p. 38).

While public perceptions and experience surveys related to police integrity are fairly common, equivalent data from within the SAPS are far less prevalent. This book seeks to remedy this deficiency.

Theoretical background and research findings

The underlying model we use to describe and analyze the contours of police integrity in South Africa is the theory of police integrity (e.g., Klockars and Kutnjak Ivković, 2004; Klockars et al., 2000, 2006; Kutnjak Ivković, 2015). Of its four dimensions, three emphasize the organizational aspects of police integrity, while the fourth dimension expands the horizon beyond the police organization itself by incorporating societal context into thinking about police integrity (Chapter 1). The empirical part of our work is based on a police officer survey (see Chapter 1) that utilizes the police integrity questionnaire (e.g., Klockars et al., 2006) to measure police integrity among the SAPS non-commissioned officers, commissioned officers, and police officers in management.

Supporting official rules

The first dimension of the police integrity theory links police integrity with the organizational rules, specifically how they are made by the police managers, the extent to which police officers are familiar with them, and how strongly police officers support the official rules (Klockars and Kutnjak Ivković, 2004, p. 1.4). Our analyses of existing legal rules prohibiting misconduct in South

Africa (Chapter 4) clearly demonstrate that the problem of the country's police is not the lack of existence of legal rules prohibiting misconduct. In fact, the Constitution of the Republic of South Africa (1996), statutory and common law, the SAPS Code of Conduct (1995), and the SAPS Disciplinary Regulations (2016) all contain norms directly addressing police misconduct. A comparison of the forms of misconduct described in the questionnaire with the legal rules prohibiting police misconduct leads to the conclusion that, in addition to being violations of the SAPS Code of Conduct (1995), almost all forms of misconduct described in the questionnaire violate statutory and common law (see Table 4.1). Because our data collection occurred before the new set of SAPS Disciplinary Regulations were developed in 2016, we use the 2006 SAPS Disciplinary Regulations as a reference.

As part of their police training, police officers should have been made familiar with the official rules, particularly the SAPS Code of Conduct (1995) and the SAPS Disciplinary Regulations (2006 and 2016), which explicitly and unequivocally state that certain forms of behavior are prohibited and constitute police misconduct, while they also describe the procedures and possible outcomes if such rule violations occur. The good news is that our empirical results indicate that the respondents as a group were able to recognize all behaviors described in the questionnaire as violations of official rules (Chapter 4). Because our questionnaire was designed in a way that contains a variety of forms of misconduct and a variety of levels of seriousness of each form, it should not be surprising that the respondents found some forms of misconduct easier to recognize as rule-violating, and to evaluate as more serious, than other forms. On the one hand, the respondents were more likely to recognize as rule-violations and evaluate as serious those forms of misconduct and behaviors that constitute violations of not only the SAPS rules but also of various statutory and common laws, such as theft from a crime scene, acceptance of a bribe, and failure to execute an arrest warrant. On the other hand, the respondents were much less likely to label as rule-violating and recognize as serious misconduct the verbal abuse of a citizen, an omission to react to juveniles writing graffiti, and a cover-up of a police DUI accident, all behaviors that might be only violations of the SAPS rules and not necessarily constitute violations of statutory and common law.

Our analyses also focused on the strength of SAPS officers' conviction that a particular activity or omission was a violation of official rules. In an agency of high integrity, respondents should not only be able to classify these behaviors as violations of official rules but should also be certain that these behaviors indeed violate official rules. Without any doubt, *all* respondents should be able to recognize, without any hesitation, extremely serious forms of police misconduct such as stealing from a crime scene, abusing deadly force, accepting bribes and kickbacks, and falsifying official forms, behavior that violates not only the SAPS rules but also various statutory and common-law rules. The bad news is that our results show that even for such severe violations there is a strong minority of the respondents, from 15% to over 30%, who did not appear to be certain whether these behaviors violate official rules.

The principle that police officers in a police agency of high integrity should have no trouble recognizing that such serious forms of police misconduct violate official rules and that such behavior should be perceived as serious misconduct that carries grave consequences may not be deeply rooted in South Africa. On the one hand, one may find it shocking that one out of four respondents could not definitely confirm that abusing deadly force is a violation of the rules. On the other hand, in a country (1) with a history of violence (see Chapters 1, 2, and 3), in which the ruling party had once, as a revolutionary organization, called on citizens to make the country ungovernable (Chapter 2), (2) where police leaders and ministers openly encourage the use of deadly force by the SAPS (i.e., "shoot to kill," Chapter 1), and (3) in which, after police officers abused deadly force and killed 34 striking workers at the Marikana mine, the administration tried to cover it up instead of disciplining the officers (Chapter 1), SAPS officers may be correct—the line prohibiting the abuse of deadly force appears nebulous at best. Similarly, one in three respondents could not confirm that accepting kickbacks contradicts the official rules. When the most senior police leader, National Commissioner Selebi (Chapters 1 and 5) is convicted of corruption, and another two former National Commissioners formally found unfit and recommended for dismissal (Cele and Phiyega), and with widespread disaffection with police salaries (e.g., Chapter 9), accepting a kickback—a form of police corruption—could certainly soften the respondents' views about whether such behavior actually violates official rules.

The certainty of recognizing misconduct as a rule violation gets even fuzzier for behaviors perceived to be less serious. In line with South Africa's violent history and violence-promoting social norms (CSVR, 2008; Collins, 2013; Kynoch, 2016), and constructions of masculinity (Ratele, 2008; Letsela and Ratele, 2009; Chapter 8), it comes as no surprise that one out of three respondents were not ready to state that striking a handcuffed man in the kidneys would definitely be a violation of the rules. Such attitudes also apply to a case of a supervisory failure to intervene when subordinate officers repeatedly strike and kick a man arrested for child abuse; one out of three respondents did not find this to be a definite example of rule-violating behavior. The fact that about one-third of respondents either did not recognize these behaviors as rule violations or did not want to state this on a survey is without any doubt an indication of cultural norms supportive of violence and a lack of concern by police managers.

As the severity of the use of force that was abused decreases in our scenarios, even fewer respondents recognize with certainty that such behaviors are rule violations and are able to evaluate them accordingly as serious. In a case that describes the use of profanities directed at a citizen, the *majority* of the respondents were not sure whether such verbal abuse is rule-violating and serious. This may be understood by framing the scenario as another expression of violence, and by considering it in light of South Africa's historical inequality. Verbal "dueling" by way of profanity or belittling could, for instance, be a means to defend honor or assert dominance as a precursor to physical violence. Language can also be used as a weapon by itself (e.g., Morris, 2017), indeed—it is generally considered a step in the police use-of-force continuum, though profanity is not officially condoned. In 2014, a SAPS constable was dismissed after assaulting and swearing at a suspect during interrogation (Sapa, 2014), but it remains unclear whether any officers have been disciplined for profanity alone.

The fact that a substantial proportion of our respondents may not recognize some behaviors as rule-violating and may be uncertain that even severe examples of police misconduct are indeed violations of official rules may be indicative of two underlying causes. First, police officers are simply not as familiar with the official rules as they should be. The solution to this problem is straightforward: Not only should they be taught the content of the official rules, but they should also learn why these rules are important and why they should support them. Extant research has demonstrated the importance of police officers' understanding of police misconduct as serious; police officers' evaluations of police misconduct as serious are directly and positively related to the police officers' expressed willingness to report misconduct (e.g., Klockars et al., 1997, 2004, 2006; Kutnjak Ivković et al., 2013, 2016; Kutnjak Ivković and Sauerman, 2013; Reyneke-Cloete and Meyer, 2016). Furthermore, the police officers' internalization of official rules as critical aspects of their police work, when coupled with other changes in the SAPS geared toward higher integrity, should cumulatively translate into behavior, including building both resistance to engage in rule-violating behavior and willingness to report it when they observe others engaging in it.

Second, police officers may actually be familiar with the official rules, but the lack of systematic enforcement by the SAPS managers and the blatant disrespect for the official rules by the leaders of the SAPS, coupled with the fact that hard work and rule adherence in South Africa do little to address the structurally unjust contexts into which many police officers are born (Chapter 3), may have created an informal set of rules that trump and discredit the official rules. To investigate whether the latter might serve as a plausible explanation for the disturbingly high rates of the inability to recognize even blatant behaviors as rule-violating, as well as the lack of certainty in providing that classification, we now turn our attention to the results of our study about the perceptions of rule enforcement.

Policing the police

The second dimension of the theory of police integrity (e.g., Klockars and Kutnjak Ivković, 2004; Klockars et al., 2000, 2004, 2006) emphasizes the influence of the police agency's internal control efforts on the level of police integrity within the agency. If the police agency has developed official rules prohibiting misconduct (i.e., the first dimension of the theory), these official rules should not remain a dead letter, left on the books just to appease the dissatisfied public. The second dimension of the theory goes further and shifts the focus on the actual enforcement of the rules and the consistency with which such enforcement occurs.

Establishing internal systems of police control has been a central endeavor of both the newly established South African democracy of 1994 and the organization of 25 years later. As described in Chapters 3 and 5, the post-apartheid history of police accountability in South Africa is characterized by the introduction of a range of accountability mechanisms, such as the Civilian Secretariat for Police Services (CSPS), community police forums (CPF), the Independent Complaints Directorate (ICD), and the SAPS's Anti-Corruption Unit (ACU), as well as the latter's sudden disappearance and low-key reappearance in 2015. Throughout it all, the SAPS should have continuously maintained an internal disciplinary system, or "policed the police" from the inside. SAPS-generated data about the operation and outcomes of the internal system suggest that very few police officers have been disciplined for, and convicted of, serious misconduct in recent decades (see Chapter 5), while citizen surveys continue to show low levels of trust in the police and perceptions of widespread occurrences of police misconduct (Afrobarometer, 2018; Chapter 1). To illustrate the gravity with which the South African public perceives these problems, we now turn to the respondents' views about this internal control system.

Our analyses of the SAPS internal rules (Table 5.1) show that the behavior described in eight out of 14 scenarios should result in a dismissal of a police officer who was officially investigated and found responsible for such behavior. When asked to predict the discipline that the SAPS would mete out for such behaviors, the majority of our respondents—barely one-half—expected dismissal in only one out of eight scenarios for the acceptance of a bribe from a motorist caught speeding, and, in addition, close to one-half of the respondents expected to be dismissed if they stole from a crime scene. Viewed differently, while the official documents provide the grounds for a police officer's dismissal if found to engage in the behaviors described in *any* of the eight scenarios, the *majority* of the respondents did not expect to be dismissed even for behaviors as severe as abusing deadly force, failing to execute a search warrant, accepting a kickback, and falsifying an official report.

One possible explanation for the perceptions that very few forms of misconduct would result in dismissal may be related to the fact that SAPS officers may not know what discipline has resulted from prior disciplinary hearings within the organization. This theory would hold that disciplinary proceedings and outcomes are protected by privacy laws. While this may be true for the disciplined officer's work unit/station, it is not accurate of the organization as a whole. For example, the 2018–2019 Annual Report includes a list of the accused officer's rank, station, offense, and disciplinary outcome from that year (SAPS, 2019a,

p. 270). Although these reports can be accessed by any motivated SAPS officer, it is unlikely that they are consulted by many. Quite likely, officers may not have an accurate count of what really might happen if they are caught engaging in police misconduct. As such, SAPS administrators need to find more effective means to convey the message to the rank and file about the consequences of such transgressions.

A more plausible explanation might be that SAPS officers are correctly interpreting the fact that misconduct is generally not taken seriously by the SAPS managers and that, indeed, few officers are held accountable for breaches of official rules. Although it is disturbing to learn that behaviors that not only constitute a violation of the SAPS rules but are also violations of various statutes and the common law are estimated by our respondents not to result in a dismissal from the SAPS, public perceptions about widespread misconduct (e.g., Institute for Justice and Reconciliation, 2017; Plus 94 Research, 2018) and the SAPS official disciplinary data about dismissals (Chapter 5) suggest that our respondents' perceptions may not be far from the truth. In fact, the 2012 National Development Plan (NDP) compelled the SAPS to improve its processes, and in its 2017–2018 Annual Report the SAPS noted that it continued to struggle with "consequence management" (SAPS, 2018), while citizens surveys showed constantly low levels of trust in the police (Afrobarometer, 2018).

To place our respondents' attitudes and beliefs in context, it is helpful to consider them together with developments in the organization and the country during recent decades. As shown in Figure 10.1, murder declined precipitously from 1995 to 2011, together with public perceptions of safety (Kriegler and Shaw, 2015; Pharoah, 2008). Trust in the police rose from a very low base in the mid-1990s, to a high of 51% in 2011 (Chingwete, 2016). With politicians calling on police to forget the Constitution and "shoot to kill" criminals, the SAPS, in an apparent effort to reign in ill-disciplined officers, rhetorically remilitarized in 2010, readopting military rank titles and referring to itself as a "force" rather than a "service." While the shift was presented as a means to reclaim respect for police and reinstill discipline within the ranks, disciplinary action actually increased far more in the year preceding this turnaround than in the years subsequent to it. And while the rise in disciplinary cases between 2007 and 2010 may appear productive, the organization expanded rapidly with 60,000 new officers during this period, meaning managers had 30% more employees to train, socialize, and potentially discipline, spreading the supervisor span of control even thinner.

In 2008, Bruce (p. 9) proposed that "not only has the SAPS actively undermined its corruption control mechanisms but it has done so while management systems, which constitute the basic mechanism of control, have been undermined." When looking at evidence from elsewhere on the continent and further afield, it is clear that other police organizations exist that are much more compromised than the SAPS. It is plausible, therefore, that some of the mechanisms the SAPS has put in place do at times function to enable accountability

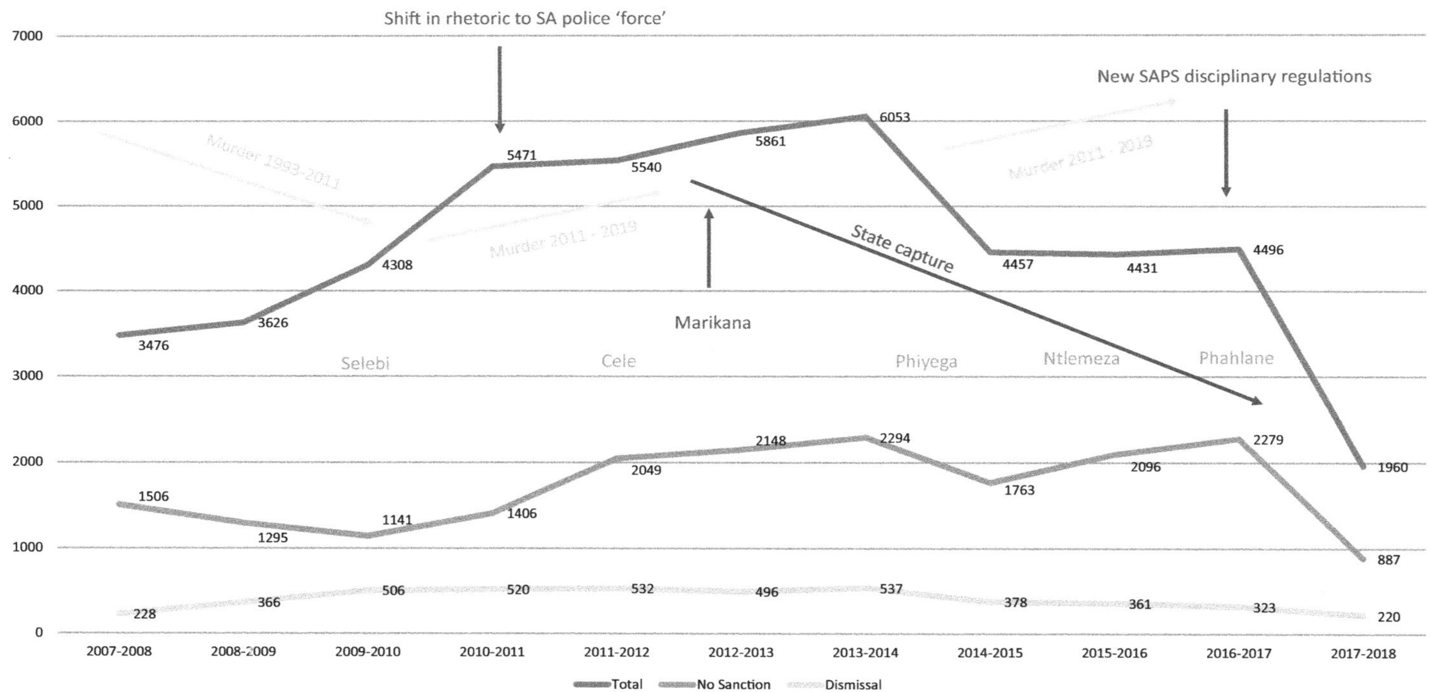

Figure 10.1 SAPS disciplinary data in context, 2007–2018

for a lack of professional conduct or integrity. Regardless, there can be no doubt that the SAPS is not the professional, legitimate police agency envisaged when it was formed in 1995, or longed for by the country's millions of anxious inhabitants.

The fundamental requirement for effective police reform and professional, legitimate policing is the willingness and ability of the top leadership. Indeed, the National Development Plan (NDP) recognized a key challenge facing the SAPS as the "serial crises of top management" (Chapter 5). This was evidenced by the fact that in the decade ending in November 2017, six different people had occupied the post of SAPS National Commissioner, the most senior officer in the organization. Many of these had embarked on restructuring exercises or made numerous high-level appointments and promotions without assessing the ability of the appointee to perform the functions of the posts into which they were appointed. For example, in 2017, 55 people were appointed into senior management positions without any assessments of their suitability for the posts, which can seriously undermine not only their credibility and effectiveness as supervisors but also the integrity and efficacy of the SAPS. The NDP recommended therefore that an independent, multi-sectoral National Police Board be established to assess the performance of those occupying the top echelons of the SAPS. Those who are found unable to fulfill the functions of the posts they hold should either be given dedicated assistance within a specified time frame (e.g., six months) to improve their performance or should be removed or redeployed elsewhere. These vacant posts should be filled using a transparent, merit-based, and competitive process. Until such time as the SAPS has a top management cohort consisting of women and men with appropriate skills, expertise, and integrity, it is highly unlikely that the agency will be able to reform itself into the professional and accountable organization envisaged by the Constitution. The simple fact is that until the top SAPS managers behave with high levels of integrity (see Kutnjak Ivković, 2006), they will either not want to demand integrity-driven behavior from their subordinates or, if they do indeed command such behavior without leading the charge, they would likely lose whatever credibility they have with those subordinates and be labeled as hypocritical. Sadly for the South African public and police officers who strive to achieve the professional standards of high integrity, the net effect of either outcome is the same.

Controlling the code of silence

The third dimension of the theory of police integrity (e.g., Klockars and Kutnjak Ivković, 2004; Klockars et al., 2000, 2004, 2006) directly links a strong code of silence with low levels of integrity. Extant research on police integrity found that the code is not a flat prohibition of reporting (e.g., Hickman et al., 2016a, 2016b; Khechumyan and Kutnjak Ivković, 2019; Kutnjak Ivković and Sauerman, 2016; Lim and Sloan, 2016; Marche, 2009; McDevitt et al., 2011; Micuccui and Gomme, 2005; Raines, 2010; Rothwell and Baldwin, 2007;

Schafer and Martinelli, 2008) and that it can vary across forms of misconduct, police agencies, and countries (e.g., Klockars et al., 2004; Kutnjak Ivković and Haberfeld, 2015). Hence, we expected that the code of silence would differ across different forms of police misconduct and across different levels of misconduct seriousness in South Africa.

The code of silence seems to be entrenched in South Africa. The apartheid SAP, the progenitor of the SAPS (Chapter 2), was an exceptionally secretive organization. Although breaking the code of silence was critical to fulfill the vision of professional policing envisaged by the ANC in the early 1990s (Haysom, 1993), many police officers appear unwilling to report or speak against their colleagues' abuses, even when they disagree with them (e.g., Grobler, 2013; Faull 2010, 2018). A South African police officer employed by both the SAP and SAPS from 1991 to 2006, succinctly described it: "What happens on the shift stays on the shift. What happens in the station stays in the station. It always stays inside with the police or in the station. It's got nothing to do with outside. Nothing, nothing, nothing'" (Faull, 2010, p. 179).

As part of the post-apartheid efforts to improve criminal justice accountability, the Independent Complaint Directorate (ICD; Chapter 3) was established in 1997. Progressive for the time, the ICD remained under-resourced and ultimately unable to significantly impact police abuses, which in part was also a result of the lack of cooperation by possible police witnesses (Bruce et al., 2000). In 2011, it was reorganized and renamed the Independent Police Investigative Directorate (IPID) in an effort to make it more effective. And while the ICD had failed to bring charges against Selebi, the IPID had charged former National Commissioner Phahlane. However, between April 2012 and March 2018, only 440 convictions were secured out of the 36,500 criminal complaints received (Viewfinder, 2019).

Extant research on police integrity in South Africa detected contours of a strong code of silence. In 2004, Newham's study (2004) of police stations in Johannesburg revealed that police officers would adhere to the code of silence for most forms of misconduct, except for a few very serious examples. Similarly, in a 2005 study of SAPS supervisors, Kutnjak Ivković and Sauerman (2011) detected the presence of a strong code of silence, covering even the most serious forms of police misconduct. Matlala and colleagues (2016) came to the same conclusion based on their survey of police officers in the Gauteng police stations. Meyer and colleagues' results (Meyer et al., 2013; Reyneke-Cloete and Meyer, 2016) of a survey of non-commissioned officers in the Gauteng Province provided further evidence of the code of silence among the SAPS officers, albeit at a somewhat weaker level than was previously reported.

In accordance with the previous research exploring police integrity in South Africa (e.g., Kutnjak Ivković and Sauerman, 2011; Matlala et al., 2016; Meyer et al., 2013; Reyneke-Cloete and Meyer, 2016; Newham, 2004), the results of our empirical study provide further evidence that the code of silence exists, that it is strong, and that it does not cover all behaviors equally (Chapter 6). The good news is that only about 10% of the respondents said they definitely would not report examples of the behavior evaluated as the most serious (i.e., failing to execute an arrest warrant, stealing from a crime scene, abusing deadly force, accepting a bribe from a reckless motorist). The bad news is that, for most of the other scenarios perceived to be less serious, at least 20% of the respondents or even more would protect such behavior in silence.

Worse still is that our respondents estimated that between 20 and 40% of their colleagues—a substantial minority of the police—would not be willing to report even the most serious forms of misconduct (Chapter 6). While we expected that because of the self-serving bias the respondents would tell us that they would be more likely to report than they anticipated their peers to be, the fact that the respondents assumed that one out of five colleagues would not report for *any* of the behaviors described in the questionnaire, regardless of how serious it is, is another piece of evidence of the code's strength.

The perception about fellow SAPS officers' willingness to report is critical for their own reporting; our empirical study clearly demonstrates that *the* key predictor of their own expressed willingness to report across all scenarios is the perception that most police officers would also report (Chapter 6). In terms of policy implications, our study provides clear evidence about the importance of teaching police officers about official rules and why misconduct should be viewed as serious, and supporting this with enforcement. As our empirical study shows, perceptions of misconduct seriousness are strongly and positively related to the respondents' own expressed willingness to report. The more seriously officers evaluate misconduct, the more likely they are to say that they would report it.

The strength of the code of silence is also closely associated with the severity of the expected discipline. Not only is the harshness of expected discipline negatively related to the respondents' willingness to adhere to the code of silence but their evaluation of discipline fairness was also directly and positively related to their expressed willingness to report. However, we are not advocating that the SAPS should start issuing draconian discipline and dismissing officers for the smallest infractions. Instead, we encourage the consistent application of discipline through a process that is considered fair by most SAPS officers (see e.g., Kutnjak Ivković et al., 2019; Sun et al., 2018, 2019; Van Craen, 2016a, 2016b; Van Craen and Skogan, 2017; Wu et al., 2017, 2019). The challenge here is not to degrade the severity of the expected discipline that is applied only occasionally to make it look fair(er) to the police, but to teach SAPS officers why a harsher discipline than what they expect may be more appropriate and, hence, change their opinions about the appropriate discipline. As Newham (2004, p. 247) concluded, "If most police officers do not believe that reporting the misconduct of colleagues will result in adequate disciplinary action being taken, then they are less likely to report misconduct themselves."

However, this is by no means an easy task, particularly in the South African context. Post-apartheid reforms may in some instances have inadvertently strengthened the code of silence, perhaps because some officers lacked the skills or resources required to successfully perform their duties and partly due to their often impossible mandate (Manning, 2005; Burger, 2006; Faull, 2017). The recent remilitarization of the police, a tendency that is at least in part politically driven, is in sharp contrast with the goal of narrowing the code of silence. Indeed, while the NDP and White Paper on Policing suggest an official government stance that rejects militarized policing, ambiguous "tough on crime" talk from the leaders remains fairly common. Indeed, the most recent SAPS Annual Performance Plan (2019/2020) lists "Stamping the authority of the state" as one of the first high-level policing objectives. This approach directly undermines a new strategic direction toward professionalism and accountability, strengthens the code of silence, and threatens to further damage trust between police and the public.

Furthermore, data performance deception (Faull, 2018, p. 80), describing the ways street-level officers focus on quantifying their activities, or faking the quantification thereof in the absence of actual activity, are effectively strengthening the code of silence, particularly bearing in mind that deception and the related code of silence function not only at the street level but also in the corridors of high policing, where crime intelligence officials plunder police funds and, in constant need of showing loyalty, spy on those deemed threats to the political elite. Senior police officers have clearly sought to protect one another during Selebi's corruption trial, Cele's board of inquiry into his fitness to hold office, and the Farlam Inquiry into the slaughter at Marikana (see Chapters 5 and 6). Judges have condemned these police managers, noting, for example, that Selebi had "complete contempt for the truth" (Joffee, 2010, p. 211) while characterizing the Hawks' head, Berning Ntlemeza, as "biased and dishonest . . . [lacking] integrity and honour" (SAPA, 2015). Faull found that the police he shadowed were so embedded in a culture of deceit that they distrusted both their colleagues and the public, and preferred to simply put their heads down and stay silent (Faull, 2018, p. 88).

Police integrity across the racial divide

In a country still recovering from 350 years of formalized race-based oppression, data reveal that the color of one's skin still strongly predicts many life outcomes, including health (Statistics South Africa, 2016, pp. 137–138), wealth (Statistics South Africa, 2012; Gumede, 2015), experiences of crime (Statistics South Africa, 2018), and police officers' living conditions (Steinberg, 2008; Faull, 2018), including their likelihood of being murdered (Perkins, 2018). The demographic composition of the SAPS changed rapidly after 1995. Although it is presently almost representative of the national demographics, it is not a race-neutral organization. Allegations of race- and sex-biased promotions and favoritism are not uncommon.

The analyses of our empirical data provide further evidence that race (still) matters in policing. Among the three racial groups in our sample (black, white, and others), the differences were by far the largest and most systematic between black and white respondents. In the majority of scenarios, black respondents were more likely than white respondents to evaluate misconduct as serious and expect fellow police officers to do the same (Chapter 7). Similarly, black respondents were more likely than white respondents to say that they would report in about one-half of the scenarios, and in the majority of them expect that fellow police officers would also report (Chapter 7). On the other hand, white respondents were more likely to support and expect harsher discipline than black respondents (Chapter 7).

While there are some differences in integrity-related attitudes between white and other respondents, particularly in familiarity with the official rules and seriousness evaluations, they are much smaller in magnitude than the differences between white and black respondents. Although the size of the differences between white and other (non-white) respondents is smaller than the size of the differences between white and black respondents, variances do exist.

Among all three racial groups, the differences in integrity-related attitudes are most similar between black respondents and other respondents (respondents who belong to other racial groups). Such findings are not surprising, keeping in mind that *all* "black" officers, be they black (African), colored, or Indian/Asian, were exposed to racism under the white minority government in the preceding two centuries or more (Brogden and Shearing, 1993). The apartheid ideology covered all aspects of police officers' lives, from black officers' lack of authority over whites, the issuing of different uniforms by race, and race-stratified pay scales and career opportunities (Newham et al., 2006).

Race has been central to the South African state since its formation, and to the politics, functions, and composition of its various police agencies. Not surprisingly, race remains strongly related to the police officers' views on police integrity, but the relationship is far from straightforward. Although there is some evidence that the racial divide in the SAPS is narrowing and that the wounds of apartheid are healing (e.g., Newham et al., 2006), the results of our study demonstrate that the task to achieve full equality is not accomplished. The country remains stratified by race, with opportunity, health, and wealth differing vastly between (but increasingly also within) historical race categories. The categories commonly referred to—and accepted by many South Africans (Mare, 2001)—as "races" have profoundly shaped and continue to shape South African society, including its police and criminal justice institutions.

Police integrity and gender

Integration of women into policing in South Africa, started in the 1970s, has followed a general pattern observed in other countries, with a distinguishable feature that its foundations are explicitly stated in the South African Constitution (1996) and further developed in the related legislation of the South African Police Service. Women in the SAPS constitute between one-quarter and one-third of all police officers across all ranks, from non-commissioned officers to top managers (Chapter 8). Although the jury is still out about the success of full integration of women into the SAPS (e.g., Faull, 2008; Meyer and Reyneke-Cloete, 2013; Meyer and Steyn, 2009; Montesh, 2010; Morrison, 2004, 2005; Newham et al., 2006; Steyn, 2017; Ulicki, 2011), the fact remains that women are no longer mere token appointees in the SAPS and are able to also climb through the SAPS hierarchy.

Despite the dramatic changes that occurred in policing since the 1970s, empirical studies measuring the effect of gender on police attitudes and behavior (e.g., Poteyeva and Sun, 2009) found a relatively modest gender effect. A few police integrity studies conducted in other countries that included gender into their analyses (Kucukuysal, 2008; Westmarland and Rowe, 2018; Wu and Makin, 2019) reported flatly that gender was not a significant factor in their findings. In a few studies that did find some gender effects (Andreescu et al., 2012; Bjerregaard and Lord, 2004; Kutnjak Ivković et al., 2016; Pagon et al., 2004; Waugh et al., 1998), the effect was limited.

Women constituted over one-quarter of respondents in our sample, enabling a seamless comparison of their views along all measures of police integrity with the views held by male respondents. Our results indicate that, broadly speaking, gender and views on police integrity are not strongly related; being a female police officer is not a powerful predictor of having attitudes of high police integrity. Specifically, we found very few differences between female and male respondents in their knowledge of official rules, perceptions of misconduct seriousness, expected discipline, and expressed willingness to report misconduct. In the rare instances where we did find some differences, female officers were somewhat more familiar with official rules, perceived misconduct as more serious, and were more likely to say that they would report misconduct than male officers were. It seems that these differences are contextual, because they are typically clustered around two scenarios: Abuse of deadly force and verbal abuse of citizens. On the other hand, male respondents tended to expect more severe discipline than female respondents. If our results are used as a measure of female integration in the SAPS, they would indicate that female police officers are "blue" first and female second.

Police integrity across police hierarchy

Previous explorations of police culture (e.g., Crank, 2014; Crank et al., 1993, 1995; Manning, 1989; Paoline and Terrill, 2014; Reuss-Ianni, 1983) suggested that the norms of the traditional police culture seemed less likely to be adopted by supervisors than by line officers. Because rank in South Africa symbolizes aspiration, status, and progress toward the dignified life promised by the new South Africa (Chapter 9), we expected that higher ranks would be associated with a greater likelihood that misconduct would be recognized, labeled, and addressed accordingly. However, because senior status and promotions in South Africa have been shaped by a combination of merit, politics, and nepotism

(Shear, 2012; Cawthra, 1993), a key issue to ponder is whether promotion to a higher rank could be reflected in attitudes associated with police integrity.

The results of our empirical study go both ways. On the one hand, non-commissioned and commissioned SAPS officers' familiarity with official rules and their evaluations of misconduct seriousness are very similar. In that sense, our results are fully aligned with extant research on police integrity finding no systematic and widespread differences between supervisors and non-supervisors (e.g., Kucukuyal, 2008; Kutnjak Ivković, 2004; Lee et al., 2013; Yun, 2003). On the other hand, officers in management positions tend to be somewhat more likely to recognize behaviors as rule-violating and assess them to be serious than either non-commissioned officers or commissioned officers, a finding in accordance with the results of other police integrity studies (Marche, 2009; Micucci and Gomme, 2005; Raines, 2010) that reported that rank and perceptions of official rules are related and that supervisors evaluate police misconduct more seriously than non-supervisors do.

Respondents from all three groups yielded a relatively uniform picture about the discipline they expect the SAPS to mete out (Chapter 9). Such a finding resonates well with the only other study that explored non-supervisor, supervisor, and manager opinions about expected discipline (Kutnjak Ivković et al., 2019) and in which the authors reported that the expectations of discipline were quite similar across the three rank categories. In a few scenarios in which we found differences, commissioned officers and officers in management expected somewhat harsher discipline than non-commissioned officers, in keep with our hypothesis. Unexpectedly, in three scenarios with differences between commissioned officers and officers in management, the commissioned officers were expecting the harsher discipline. Regardless, the fact remains that the views about the discipline the SAPS is expected to mete out as expressed by the respondents from different parts of the SAPS hierarchy are mostly similar, suggesting that they share common views about the discipline that would occur in response to misconduct officially investigated by the SAPS.

The adherence to the code of silence also does not appear strongly related to the respondents' rank (Chapter 9). Although we expected that non-commissioned officers would be more likely to support the code of silence than commissioned officers, our analyses revealed that rank does not matter for the respondents' expressed willingness to report. Most differences we found included respondents in management who seemed somewhat more likely to report than either non-commissioned officers or commissioned officers (Chapter 9). Although these differences are not strong and systematic, they are pointing in the right direction—employees at the top of the hierarchy should have views more supportive of the tendency to resist the temptations of their occupation than employees at the bottom of the hierarchy.

Indeed, our results suggest that rank is not systematically associated with police integrity. Even when we found differences, they were are far from consistent, indicating that promotion to the higher ranks of the hierarchy does not carry with it the required change in attitudes toward higher levels of police integrity. This could be a consequence of the rapid recruitment and promotion between 2002 and 2012 (Bruce, 2013) that might have resulted in a new breed of supervisors who are struggling to rise above the requirements of their previous rank as line officers to assume their new roles of police supervisors. Furthermore, promotion in the SAPS has been a political and contentious issue in the agency, signaling to honest officers that their hard work may not pay off, that rank promotion is sometimes a result of whom you know, and not what you know.

Conclusion

The 2018 election of President Cyril Ramaphosa ushered in a brief period of public "Ramaphoria"—a hope that the comprehensive corruption of the Zuma administration was damage of the past and that a new dawn may soon brighten South Africa. Although Ramaphosa acted swiftly on some matters, such as establishing judicial commissions of inquiry into high-level corruption and strengthening efforts to bolster the prosecuting authority and bring the corrupt elite to justice, more than 18 months into his presidency not a single person had been charged. And while steps have been taken to reform the damaged Hawks and Crime Intelligence sections of the SAPS, there is little indication that large-scale reform will take place in the near term. In February 2019, only 29% of South Africans believed the country was "going in the right direction" (Ipsos, 2019), and, by June that year, news sites proclaimed that Ramaphoria was "truly dead" (Staff Writer, 2019).

Trust is key to the function and productivity of the democratic state. When citizens trust in state institutions, such as the police, they can interact and trade with strangers, restrain violent impulses, and obey the law without fear of being cheated. South Africa has yet to attain this state. The results of the 2018 Afrobarometer (Plus 94 Research, 2018), the most recent one available, indicate that the *majority* of the South African respondents—66 percent—had either no trust in police at all or stated that they had "just a little" trust. This lack of faith in the police did of course not occur overnight—it was a gradual process in which the confidence in the SAPS has withered over time, from the mid-1990s to the mid-2010s. The percentage of respondents with confidence in the police started to decrease from about three-quarters in the immediate post-apartheid period (74% in 1994–1998) to about two-thirds in the early 2000s (60% in 1999–2004) to barely one-half in the late 2000s–early 2010s (52%; 2005–2014; Boateng and Lee, 2018). At the time of writing, it had hovered around the 50% mark for the best part of a decade.

Research on trust in policing suggests that legitimacy is generated through policing that is consistently fair (Oberwittler and Roche, 2018; Trinkner et al., 2016). Fair policing can also promote nation building, as people who perceive police as legitimate authorities are more likely to identify with the authority they represent (Bradford et al., 2014). The fact that SAPS National Commissioners have routinely been dismissed and suspended since 2008 for transgressions including corruption, maladministration, and dishonesty reflects poorly on the organization. The same is true of the stories about police abuse of force, corruption, and political bias that are regularly reported in the daily news columns and social media feeds.

In the short term, South Africa's government can increase public trust by demonstrably tackling police misconduct. As with the successful application of justice, the public must be allowed to witness this confrontation. The various commissions of inquiry underway, along with President Ramaphosa's efforts to reform the criminal justice system, must be supported. This includes the appointments of Lt-Gen. Godfrey Lebeya and Advocate Shamila Batohi to head the Hawks and National Prosecuting Authority (NPA), respectively; the establishment of a special directorate in the NPA to investigate allegations of state capture; and the creation of a special tribunal to fast-track civil claims linked to corrupt state contracts. With proper funding, support, and leadership, these entities could generate short-term wins that can help grow public trust. In the long run, however, major reforms are required within the SAPS to instill in its officers integrity and the respect for that which they are sworn to protect—a hard-fought-for rule of law.

In the end, the story of policing and police integrity in South Africa is a tale of a police agency that strives to adhere to the principles of democratic policing, yet remains at the mercy of numerous challenges. In a police agency observing the principles of democratic policing, including those of accountability and police integrity, police officers should have no difficulty in recognizing examples of police misconduct as rule violations, supporting the official rules, expecting consistent discipline for violations of official rules, and expressing willingness to report at least the serious violations of official rules. While the details of the story—be it the long-established systematic racism, political leaders with dubious agendas, a social culture supportive of the use of violence, or the consistent integrity failures of top police managers—may be more contextual and country-specific, the findings about the contours of police integrity and their correlates have a more universal appeal and should resonate well beyond the borders of South Africa with any scholar interested in learning about police integrity and any police manager intent on enhancing it.

References

Afrobarometer (2018). *Summary of Results: Afrobarometer Round 7 Survey in South Africa, 2018*. Available at: http://afrobarometer.org/sites/default/files/publications/Summary%20of%20results/saf_r7_sor_13112018.pdf

Andreescu, V., D.G. Keeling, M.C. Voinic, and B.N. Tonea (2012). Future Romanian law enforcement: Gender differences in perceptions of police misconduct. *Journal of Social Research & Policy*, 3(1): 97–113.

Basson, A. and P. Du Toit (2018). *Enemy of the People: How Jacob Zuma Stole South Africa and How the People Fought Back*. Johannesburg: Jonathan Ball Publishers. Kindle Edition.

Bjerregaard, B. and V.B. Lord (2004). An examination of the ethical and value orientation of criminal justice students. *Police Quarterly*, 7(2): 262–284. https://doi.org/10.1177/1098611102250739

Boateng, F.D. and J.Y. Lee (2018). Apartheid and post-apartheid analysis of public confidence in the police: A longitudinal analysis across time. *Policing: An International Journal of Police Strategies and Management*, 41(6): 766–781.

Bradford, B., K. Murphy, and J. Jackson (2014). Officers as mirrors: Policing, procedural justice and the (re)production of social identity. *The British Journal of Criminology*, 54(4): 527–550. https://doi.org/10.1093/bjc/azu021

Brogden, M. and C. Shearing (1993). *Policing for a New South Africa*. London: Routledge.

Bruce, D. (2013) The police are at war once again. South African Civil Society Information Service, 04 March. Available at: http://sacsis.org.za/site/article/1590 (Accessed 15 April 2014).

Bruce, D., K. Savage, and J. de Waal (2000). A duty to answer questions? The police, the independent complaints directorate and the right to remain silent. *South African Journal on Human Rights*, 16(1): 71–96.

Burger, J. (2006). Crime combating in perspective: A strategic approach to policing and the prevention of crime in South Africa. *Acta Criminologica*, 19(2): 105–118.

Cawthra, G. (1993). *Policing South Africa: The SAP and the Transition from Apartheid*. London, U.K.: Zed Books Ltd.

Chingwete, A. (2016). *In South Africa, Citizens' Trust in President, Political Institutions Drops Sharply*. Afrobarometer Dispatch No. 90. Afrobarometer and Institute for Justice and Reconciliation. Available at: https://afrobarometer.org/sites/default/files/publications/Dispatches/ab_r6_dispatchno90_south_africa_trust_in_officials.pdf

Collins, A. (2013). Violence is not a crime: A broader view of interventions for social safety. *South African Crime Quarterly*, 43: 29–37.

Constitution of the Republic of South Africa (1996). Act No. 108 of 1996. Available at: www.gov.za/documents/constitution-republic-south-africa-1996

Crank, J.P. (2014). *Understanding Police Culture*, 2nd ed. London: Routledge.

Crank, J.P., B. Regoli, J.D. Hewitt, and R.G. Culbertson (1993). An assessment of work stress among police executives. *Journal of Criminal Justice*, 21: 313–324.

Crank, J.P., R. Regoli, J.D. Hewitt, and R.G. Culbertson (1995). Institutional and organizational antecedents of role stress, work alienation, and anomie among police executives. *Criminal Justice and Behavior*, 22: 152–171.

CSVR (Centre for the Study of Violence and Reconciliation) (2010). *Why South Africa Is So Violent and What Should Be done about It: Statement by the Centre for the Study of Violence and Reconciliation*. Available at: www.csvr.org.za/old/index.php/media-articles/news-and-events/56-latest-news/2447-press-statement-the-study-on-the-violent-nature-of-crime-91110-.html

Farlam, I.G. (2015). *Marikana Commission of Inquiry: Report on Matters of Public, National and International Concern Arising Out of the Tragic Incidents at the Lonmin Mine in Marikana, in the North West Province*. Available at: www.sahrc.org.za/home/21/files/marikana-report-1.pdf

Faull, A. (2008). *Policing Diversity: An Analysis of a Diversity Intervention and Its Effects on a South African Police Service Station*. Amsterdam: Rozenberg Publishers.

Faull, A. (2010). *Behind the Badge: The Untold Stories of South Africa's Police Service Members*. Cape Town: Penguin Random House South Africa. Kindle Edition.

Faull, A. (2017). *Police Work and Identity: A South African Ethnography*. Abingdon: Routledge.

Faull, A. (2018). *Police Work and Identity: A South African Ethnography*. Abingdon: Routledge.

Grobler, L. (2013). *Crossing the Line: When Cops Become Criminals*. Auckland Park, South Africa: Jacana Media.

Gumede, V. (2015). Inequality in democratic South Africa. In X. Mangcu (ed.). *The Colour of Our Future: Does Race Matter in Post-Apartheid South Africa?* Johannesburg: Wits University Press.

Haysom, F. (1993). *Policing the Transition: Transforming the Police, African National Congress*. Available at: www.anc.org.za/content/policing-transition-transforming-police

Hickman, M.J., A.R. Piquero, Z.A. Powell, and J. Greene (2016a). Expanding the measurement of police integrity. *Policing: An International Journal of Police Strategies & Management*, 39(2): 246–267.

Hickman, M.J., Z.A. Powell, A.R. Piquero, and J. Greene (2016b). Exploring the viability of an attitudes toward ethical behavior scale in understanding police integrity outcomes. *Policing: An International Journal of Police Strategies & Management*, 39(2): 319–337.

Independent Police Investigative Directorate (IPID) (2018). Annual Report 2017/18. Pretoria: Independent Complaints Directorate.

Institute for Justice and Reconciliation (2017). *Afrobarometer*. Available at: www.ijr.org.za/home/wp-content/uploads/2012/07/SAF_R6_South-Africa-Round-Results-and-Analysis_20Mar17.pdf

Ipsos (2019). *Views on the South African Economy: Hope or Despair?*, 19 February. Available at: www.ipsos.com/en-za/views-south-african-economy-hope-or-despair

Jeffery, A. (2019). *People's War: New Light on the Struggle for South Africa*. Jeppestown, South Africa: Jonathan Ball Publishers. Kindle Edition.

Joffee, J. (2010). *Judgement: State vs Jackie Selebie, Case No 25/09*. Available at: www.saflii.org/za/cases/ZAGPJHC/2010/53.pdf

Khechumyan, A. and S. Kutnjak Ivković (2019). Exploring differences in police integrity within a centralized police system. In S. Kutnjak Ivković and M. Haberfeld (eds.), *Exploring Police Integrity: Novel Approaches to Police Integrity Theory and Methodology*. New York: Springer.

Kinnes, I. (2017). *Contested Governance: Police and Gang Interactions*. Unpublished PhD Thesis. Cape Town: University of Cape Town.

Klockars, C.B. and S. Kutnjak Ivković (2004). Measuring police integrity. In A.R. Piquero, J.R. Greene, and M.J. Hickman (eds.), *Police Integrity and Ethics*. Belmont, CA: Wadsworth Publishing, pp. 1.3–1.20.

Klockars, C.B., S. Kutnjak Ivković, & M.R. Haberfeld (eds.) (2004). *The Contours of Police Integrity*. Newbury Park, CA: Sage Publication.

Klockars, C.B., S. Kutnjak Ivković, & M.R. Haberfeld (2006). *Enhancing Police Integrity*. New York: Springer.

Klockars, C.B., S. Kutnjak Ivković, W.E. Harver, and M.R. Haberfeld (1997). *The Measurement of Police Integrity*. Final Report Submitted to the U.S. Department of Justice, Office of Justice Programs, National Institute of Justice, Washington, D.C.

Klockars, C.B., S. Kutnjak Ivković, W.E. Harver, and M.R. Haberfeld (2000). *The Measurement of Police Integrity*. Research in Brief. U.S. Department of Justice, Office of Justice Programs, National Institute of Justice. Washington, DC: Government Printing Office.

Kriegler, A. and M. Shaw (2016). *A Citizen's Guide to Crime Statistics in South Africa*. Johannesburg and Cape Town: Jonathan Ball.

Kucukuysal, B. (2008). *Determinants of Turkish Police Officers' Perception of Integrity: Impact of Organizational Culture*. Electronic Theses and Dissertations. 3777. Available at: https://stars.library.ucf.edu/etd/3777

Kutnjak Ivković, S. (2005). *Fallen Blue Knights: Controlling Police Corruption*. New York: Oxford University Press.

Kutnjak Ivković, S. (2015). Studying police integrity. In S. Kutnjak Ivković and M. Haberfeld (eds.), *Police Integrity across the World*. New York: Springer.

Kutnjak Ivković, S., and M.R. Haberfeld (2015). A comparative perspective on police integrity. In S. Kutnjak Ivković and M. Haberfeld (eds.), *Police Integrity across the World*. New York: Springer.

Kutnjak Ivković, S., M. Haberfeld, and R. Peacock (2013). Rainless West: The integrity survey's role in agency accountability. *Police Quarterly*, 16(2): 148–176.

Kutnjak Ivković, S., M. Haberfeld, and R. Peacock (2016). Does discipline fairness matter for the police code of silence? Answers from the U.S. supervisors and line officers. *Policing: An International Journal of Police Strategies and Management*, 39(2): 354–369.

Kutnjak Ivković, S., M.R. Haberfeld, and R. Peacock (2019). Overlapping shades of blue: Exploring police officer, supervisor, and administrator cultures of police integrity. In S. Kutnjak Ivković and M. Haberfeld (eds.), *Exploring Police Integrity: Novel Approaches to Police Integrity Theory and Methodology*. New York: Springer.

Kutnjak Ivković, S., I.C. Mraović, and K. Borovec (2016). Does community policing matter for police integrity? *Revija za Kriminalistiko in Kriminologijo/Journal of Criminal Investigation and Criminology*, 67(4): 313–325.

Kutnjak Ivković, S. and A. Sauerman (2011). Measuring the code of silence among the South African police: Findings from a SAPS supervisor survey. In C. Gould and G. Newham (eds.), *Toward a Coherent Strategy for Crime Reduction in South Africa beyond 2010*. Pretoria, South Africa: Institute for Security Studies, pp. 74–87.

Kutnjak Ivković, S. and A. Sauerman (2013). Curtailing the code of silence among the South African police. *Policing: An International Journal of Police Strategies and Management*, 36(1): 175–198.

Kutnjak Ivković, S. and A. Sauerman (2016). Police integrity in South Africa: A tale of three police agency types. *Policing: An International Journal of Police Strategies and Management*, 39(2): 268–283.

Kynoch, G. (2016). Apartheid's afterlives: Violence, policing and the South African state. *Journal of Southern African Studies*, 42(1): 65–78. DOI: 10.1080/03057070.2016.1087167

Lee, H., H. Lim, D.D. Moore, and J. Kim (2013). How police organizational structure correlates with frontline officers' attitudes toward corruption: A multilevel model. *Police Practice and Research*, 14(5): 386–401.

Letsela, L. and K. Ratele (2009). *Masculinity and Perceptions of Risk: Factors to Premature Male Mortality in South Africa*. Available at: www.brothersforlife.org/sites/default/files/docs/Men_and_their_perceptions_of_Risks.pdf

Lim, H. and J.J. Sloan (2016). Police officer integrity: A partial replication and extension. *Policing: An International Journal of Police Strategies & Management*, 39(2): 284–301.

Manning, P.K. (1989). Occupational culture. In W.G. Bailey (ed.), *The Encyclopedia of Police Science*. New York: Garland.

Manning, P.K. (2005). The police: Mandate, strategies and appearances. Reprinted in T. Newburn (ed.), *Policing: Key Readings*. Cullompton. Willan Publishing, pp. 191–214.

Marche, G.E. (2009). Integrity, culture, and scale: An empirical test of the big bad police agency. *Crime, Law, and Social Change*, 51: 463–486.

Mare, G. (2001). Race counts in contemporary South Africa: 'An illusion of ordinariness' *Transformation*, 47: 75–93.

Matlala, R.L.G., D. Mistry, and A. Phala (2016). Measuring the integrity of law enforcement officers in Gauteng Province. *The International Journal of Social Sciences and Humanities Invention*, 3(11): 2969–2980.

McDevitt, J., C. Posick, R. Zschoche, D.P. Rosenbaum, M. Buslik, and L. Fridell (2011). *Police Integrity, Responsibility, and Discipline*. National Institute of Justice. Available at: www.

nationalpoliceresearch.org/storage/updated-papers/Police%20Integrity%20Responsibility%20and%20Discipline.pdf
Meyer, M. and S. Reneyke-Cloete (2013). Current perspectives toward women serving in the South African police. Unpublished paper presented at the 5th Annual Meeting of the Academy of Criminal Justice Sciences, Dallas, TX, March 19–23, 2013.
Meyer, M.E. and J. Steyn (2009). Nurturing isolation in the South African Police Service. *Policing: An International Journal of Police Strategies and Management*, 32(1): 108–127. DOI: 10.1108/13639510910937148
Meyer, M.E., J. Steyn, and N. Gopal (2013). Exploring the public parameter of police integrity. *Policing: An International Journal of Police Strategies & Management*, 36(1): 140–156.
Micucci, A.J. and I.M. Gomme (2005). American police and subcultural support for the use of excessive force. *Journal of Criminal Justice*, 33: 487–500.
Montesh, M. (2010). Transformation in the South African Police Service: The implementation of affirmative action and employment equity. *SAPS South African Journal of Criminal Justice*, 23(1): 55–77.
Morison, C.J. (2004). Towards gender equality in South African policing. In G. Mesko, M. Pagon, and B. Dobovsek (eds.), *Policing in Central and Eastern Europe: Dilemmas of Contemporary Criminal Justice*, pp. 238–244. Available at: www.fvv.um.si/conf2004/papers/morrison.pdf
Morison, C.J. (2005). Gender discrimination versus equality in the police. *Acta Criminologica: Southern African Journal of Criminology*, 18(3): 20–28.
Morris, R.C. (2017). Mediation, the political task between language and violence in contemporary South Africa. *Current Anthropology*, 58(15): 123–134.
Newham, G. (2004). Out of step: Integrity and the South Africa Police Service. In C.B. Klockars, S. Kutnjak Ivković, and M.R. Haberfeld (eds.), *The Contours of Police Integrity*. Thousand Oaks, CA: Sage Publications, pp. 232–247.
Newham, G., T. Masuku, and J. Dlamini (2006). Diversity and transformation in the South African Police Service: A study of police perspectives on race, gender and the community in the Johannesburg policing area. *Centre for the Study of Violence and Reconciliation: Braamfontein, South Africa*. Available at: https://csvr.org.za/docs/policing/diversity.pdf
Oberwittler, D. and S. Roche (2018). *Police-Citizen Relations across the World*. New York: Routledge.
Pagon, M., B. Lobnikar, and D. Anelj (2004). Gender differences in leniency towards police misconduct. In G. Mesko, M. Pagon, and B. Dobovsek (eds.), *Policing in Central and Eastern Europe: Dilemmas of Contemporary Criminal Justice*. Ljubljana, Slovenia: Faculty of Criminal Justice, University of Maribor, pp. 1–16.
Paoline, E.A. and W. Terrill (2014). *Police Culture: Adapting to the Strains of the Job*. Durham, NC: Carolina Academic Press.
Pauw, J. (2018). *The President's Keepers: Those Keeping Zuma in Power and Out of Prison*. Johannesburg: Tafelberg. Kindle Edition.
Perkins, G. (2018). *Danger and Death: Organisational and Occupational Responses to the Murder of Police in South Africa: A Case Study*. PhD Thesis. Cape Town: University of Cape Town.
Pharoah, R. (2008). National victims of crime survey: Overview of key findings. Institute for Security Studies Papers 2008 (175). Available at: https://issafrica.org/south-african-crime-statistics/01-nov-2008-iss-paper-175-national-victims-of-crime-survey-robyn-pharoah
Plus 94 Research (2018). *Afrobarometer: South Africa Results*. Available at: www.afrobarometer.org/publications/south-africa-summary-results-2018

Poteyeva, M. and I.Y. Sun (2009). Gender differences in police officers' attitudes: Assessing current empirical evidence. *Journal of Criminal Justice*, 37: 512–522.

Raines, J. (2010). *Ethics in Policing: Misconduct and Integrity*. Sudburry, MA: Jones & Bartlett Learning.

Ratele, K. (2008). Masculinity and male mortality in South Africa. *African Safety Promotion: A Journal of Inquiry*, 6(22): 19–41.

Reuss-Ianni, E. and F.A. Ianni (1983). *Street Cops and Management Cops: The Two Cultures of Policing*. New York, NY: Transaction Publishers.

Reyneke-Cloete, S. and M.E. Meyer (2016). Viewing integrity in the South African Police Service among non-commissioned officers in Gauteng Province. *Acta Criminologica: Southern African Journal of Criminology*, 29(2): 87–106.

Rothwell, G.R., & J.N. Baldwin (2007). Whistle-blowing and the code of silence in police agencies policy and structural predictors. *Crime & Delinquency*, 53(4): 605–632.

SAPA (2014). Pretoria cop dismissed for assaulting, swearing at man during questioning. *Times Live*, 23 July. Available at: www.timeslive.co.za/news/south-africa/2014-07-23-pretoria-cop-dismissed-for-assaulting-swearing-at-man-during-questioning/

SAPA (2015). Sibiya: Judge Calls Acting Head of Hawks "Dishonest". *Mail & Guardian*, 23 March. Available at: https://mg.co.za/article/2015-03-23-judge-harsh-on-acting-head-of-hawks (Accessed 25 February 2018).

SAPS (2019a). *Annual Report 2018/2019*. Pretoria: Statistics South Africa.

SAPS (2019b). *Crime Situation in Republic of South Africa: Twelve (12) Months (April to March 2018_19*. Available at: www.saps.gov.za/services/april_to_march2018_19_presentation.pdf

Schafer, J.A. and T.J. Martinelli (2008). First-line supervisor's perceptions of police integrity. *Policing: An International Journal of Police Strategies and Management*, 31(2): 306–323.

Shear, K. (2012). Chiefs or modern bureaucrats? Managing black police in early twentieth-century South Africa. *Comparative Studies in Society and History*, 54(2): 251–274.

South African Police Service (2018). *Annual Report 2017/18*. Pretoria: South African Police Service. Available at: https://www.saps.gov.za/about/stratframework/annualreports_arch.php

Staff Writer (2019). Here's why Ramaphoria is now well and truly dead. *Business Tech*, 12 June. Available at: https://businesstech.co.za/news/finance/323015/heres-why-ramaphoria-is-now-well-and-truly-dead/

Statistics South Africa (2016). *Mortality and Causes of Death in South Africa, 2016: Findings from Death Notification: Statistical Release P0309.3*. Available at: www.statssa.gov.za/publi cations/P03093/P030932016.pdf

StatsSA (2012). *Statistical Release (Revised) Census 2011*. Pretoria. Available at: www.statssa.gov.za/Publications/P03014/P030142011.pdf

StatsSA (2018). *Victims of Crime Survey 2017/18*. Pretoria: Statistics South Africa.

Steinberg, J. (2008). *Thin Blue: The Unwritten Rules of Policing South Africa*. Cape Town: Jonathan Ball.

Steinberg, J. (2014). Policing, state power, and the transition from apartheid to democracy: A new perspective. *African Affairs*, 113(451): 173–191.

Steyn, J. (2017). Lilac? "Ten years of blue nurtured nature". *Police Practice and Research*, 18(3): 259–274. DOI: 10.1080/15614263.2017.1292745

Sun, I.Y., Y. Wu, J. Liu, and M. Van Craen. (2019). Institutional procedural justice and street procedural justice in Chinese policing: The mediating role of moral alignment. *Australian and New Zealand Journal of Criminology*, 52(2): 272–290.

Sun, I.Y., Y. Wu, M. Van Craen, and K.K.L. Hsu (2018). Internal procedural justice, moral alignment, and external procedural justice in democratic policing. *Police Quarterly*, 21(3): 387–412.

Trinkner, R., T.R. Tyler, and P.A. Goff (2016). Justice from within: The relations between a procedurally just organizational climate and police organizational efficiency, endorsement of democratic policing, and officer well-being. *Psychology, Public Policy, and Law*, 22: 158–172.

Ulicki, T. (2011). "Just the way things are": Gender equity and sexual harassment in the South African Police Service. *Transformation: Critical Perspectives on Southern Africa*, 76: 95–119. DOI: 10.1353/trn.2011.0031

Van Craen, M. (2016a). Understanding police officers' trust and trustworthy behavior: A work relations framework. *European Journal of Criminology*, 13(2): 274–294.

Van Craen, M. (2016b). Fair policing from the inside out. In M. Deflem (ed.), *The Politics of Policing: Between Force and Legitimacy*. Bingley, England: Emerald.

Van Craen, M. and W.G. Skogan (2017). Achieving fairness in policing: The link between internal and external procedural justice. *Police Quarterly*, 20(1): 3–23.

van Vuuren, H. (2006). *Apartheid Grand Corruption Assessing the Scale of Crimes of Profit in South Africa from 1976 to 1994*. Pretoria: Institute for Security Studies.

van Vuuren, H. (2017). *Apartheid Guns and Money*. Jacana Media: Auckland Park.

Viewfinder (2019). Viewfinder: IPID's cover-up of police brutality in SA. *Daily Maverick*, 7 October. Available at: www.dailymaverick.co.za/article/2019-10-07-viewfinder-ipids-cover-up-of-police-brutality-in-sa/

Waugh, L., A. Ede, and A. Alley (1998). Police culture, women police and attitudes towards misconduct. *International Journal of Police Science and Management*, 1(3): 288–300.

Western Cape High Court (2018). *Social Justice Coalition and Others v Minister of Police and Others* (EC03/2016) [2018] ZAWCHC 181; 2019 (4) SA 82 (WCC), 14 December. Available at: www.saflii.org/za/cases/ZAWCHC/2018/181.html

Westmarland, L. and M. Rowe (2018). Police ethics and integrity: Can a new code overturn the blue code? *Policing and Society*, 28(7): 854–870. DOI: 10.1080/10439463.2016.1262365

Wu, G. and D.A. Makin (2019). The quagmire that is an unwillingness to report: Situating the code of silence within the Chinese Police context. *Criminal Justice and Behavior*, 46(4): 608–627. https://doi.org/10.1177/0093854819831732

Wu, Y., I.Y. Sun, C.K.M. Chang, and K.K.L. Hsu (2017). Procedural justice received and given: Supervisory treatment, emotional states, and behavioral compliance among Taiwanese police officers. *Criminal Justice and Behavior*, 44(7): 963–982.

Wu, Y., I.Y. Sun, C.K.M. Chang, M. Van Craen, and J. Liu (2019). Linking supervisory procedural accountability to officer procedural accountability in Chinese policing. *Policing and Society*, 29(7): 749–764.

Yun, I. (2003). *A Study of Police Officers' Perceptions of Police Deviance: From an Occupational Socialization Perspective*. ProQuest Dissertations and Theses, pp. 89–89. Available at: http://ezproxy.msu.edu/login?url=http://search.proquest.com/docview/305329643?accountid=12598 (305329643).

Index

For Product Safety Concerns and Information please contact our EU
representative GPSR@taylorandfrancis.com
Taylor & Francis Verlag GmbH, Kaufingerstraße 24, 80331 München, Germany

www.ingramcontent.com/pod-product-compliance
Ingram Content Group UK Ltd.
Pitfield, Milton Keynes, MK11 3LW, UK
UKHW020154200926
475702UK00016B/2709

* 9 7 8 0 3 6 7 4 9 3 8 3 7 *